Women in

Microbiology

Women in Microbiology

Edited by

Rachel J. Whitaker
Department of Microbiology
University of Illinois Urbana-Champaign
Urbana, Illinois

and

Hazel A. Barton
Department of Biology
The University of Akron
Akron, Ohio

**ASM
PRESS**
Washington, DC

Library of Congress Cataloging-in-Publication Data

Names: Whitaker, Rachel J., editor. | Barton, Hazel A., editor.
Title: Women in microbiology / edited by Rachel J. Whitaker,
Department of Microbiology, University of Illinois Urbana, and
Hazel A. Barton, Department of Biology, The University of Akron.
Description: Washington, DC : ASM Press, [2018]
Identifiers: LCCN 2018010852 (print) | LCCN 2018012578 (ebook) | ISBN
9781555819545 (ebook) | ISBN 9781555819538 (pbk.)
Subjects: LCSH: Women microbiologists–Biography.
Classification: LCC QR30 (ebook) | LCC QR30 .W655 2018 (print) | DDC 579.092–dc23
LC record available at https://lccn.loc.gov/2018010852

Address editorial correspondence to
ASM Press, 1752 N St., N.W.,
Washington, DC 20036-2904, USA

Send orders to ASM Press, P.O. Box 605, Herndon, VA 20172, USA
Phone: 800-546-2416; 703-661-1593
Fax: 703-661-1501
E-mail: books@asmusa.org
Online: http://www.asmscience.org

Dedicated to the memory of
Congresswoman Louise M. Slaughter
(August 14, 1929–March 16, 2018)

Congresswoman Slaughter, a microbiologist,
was a powerful voice in the world whose career,
like many of the women profiled in this book,
took her in many directions.
She served in Congress from 1987–2018,
where she was a champion of women's rights
(coauthoring the Violence Against Women Act)
and a strong advocate for science, both in terms of funding
and the inclusion of women and minorities in scientific studies.

We are so grateful that she provided the
foreword for this book and hope that, in addition to
honoring the women microbiologists featured herein,
this book also serves as a memorial to Louise's legacy.

Contents

Contributors

Sonja-Verena Albers
Molecular Biology of Archaea
University of Freiburg
Institute of Biology II
Freiburg, Germany

Hazel A. Barton
Department of Biology
University of Akron
Akron, Ohio

Bonnie K. Baxter
Great Salt Lake Institute
Westminster College
Salt Lake City, Utah

Joan W. Bennett
Department of Plant Biology and
    Pathology
School of Environmental and Biological
    Sciences
Rutgers–The State University of
    New Jersey
New Brunswick, New Jersey

Graciela Brelles-Mariño
Center for Research and Development on
    Industrial Fermentations
Universidad Nacional de La Plata
Facultad de Ciencias Exactas
La Plata, Buenos Aires
Argentina

May C. Chu
Colorado School of Public Health
Anschutz Medical Center
Aurora, Colorado

Douglas E. Eveleigh
Department of Biochemistry and
    Microbiology
School of Environmental and Biological
    Sciences
Rutgers–The State University of
    New Jersey
New Brunswick, New Jersey

Rebecca V. Ferrell
Department of Biology
Metropolitan State University of Denver
Denver, Colorado

Lorraine A. Findlay
Department of Allied Health Sciences
Nassau County College
Garden City, New York

Bishara J. Freij
Division of Infectious Diseases
Beaumont Children's Hospital
Royal Oak, Michigan
Oakland University
William Beaumont School of
    Medicine
Rochester, Michigan

Wayne State University School of
Medicine
Detroit, Michigan

Joudeh B. Freij
Department of Microbiology and
Immunology
Johns Hopkins Bloomberg School of
Public Health
Baltimore, Maryland
Wayne State University School of Medicine
Detroit, Michigan

Millicent E. Goldschmidt
[Professor Emerita] Department of
Diagnostic and Biomedical Sciences
University of Texas Health Science
Center at Houston
Houston, Texas

Brian K. Hammer
School of Biological Sciences
Georgia Institute of Technology
Atlanta, Georgia

Caroline S. Harwood
Department of Microbiology
University of Washington School of
Medicine
Seattle, Washington

Jamie E. Henzy
Department of Biology
Boston College
Chestnut Hill, Massachusetts

Nina Molin Høyland-Kroghsbo
Department of Molecular Biology
Princeton University
Princeton, New Jersey

Crystal N. Johnson
Department of Environmental Sciences
Louisiana State University
Baton Rouge, Louisiana

Vanja Klepac-Ceraj
Department of Biological Sciences
Wellesley College
Wellesley, Massachusetts

Tamara Lewis Johnson
Office for Research on Disparities
and Global Mental Health
NIH/NIMH
Rockville, Maryland

Shirley Lowe
[Retired] Department of
Microbiology and Immunology
University of California,
San Francisco
San Francisco, California

Jeffrey Marlow
Department of Organismic and
Evolutionary Biology
Harvard University
Cambridge, Massachusetts

Norma C. Martinez-Gomez
Department of Microbiology and
Molecular Genetics
Michigan State University
East Lansing, Michigan

John P. McCutcheon
Division of Biological Sciences
University of Montana
Missoula, Montana

Margaret McFall-Ngai
Pacific Biosciences Research
Center
University of Hawaii at Manoa
Honolulu, Hawaii

Katherine McMahon
Department of Bacteriology
University of Wisconsin-Madison
Madison, Wisconsin

Rebecca E. Parales
Department of Microbiology and
  Molecular Genetics
University of California, Davis
Davis, California

Jennifer Pett-Ridge
Physical and Life Sciences
  Directorate
Lawrence Livermore National Lab
Livermore, California

Hannah T. Reynolds
Department of Biological &
  Environmental Sciences
Western Connecticut State University
Danbury, Connecticut

Candace N. Rouchon
Department of Microbiology and
  Immunology
Uniformed Services University of the
  Health Sciences
Bethesda, Maryland

Sophie Rowland
Department of Biological Sciences
Wellesley College
Wellesley, Massachusetts

Christa Schleper
Archaea Biology and Ecogenomics
  Division
University of Vienna
Vienna, Austria

Patrick D. Schloss
Department of Microbiology &
  Immunology
University of Michigan
Ann Arbor, Michigan

John R. Spear
Department of Civil and Environmental
  Engineering
Colorado School of Mines
Golden, Colorado

Kendall Tate-Wright
Great Salt Lake Institute
Westminster College
Salt Lake City, Utah

Carin K. Vanderpool
Department of Microbiology
University of Illinois Urbana-Champaign
Urbana, Illinois

Rachel J. Whitaker
Department of Microbiology
University of Illinois Urbana-Champaign
Urbana, Illinois

Natalie N. Whitfield
Clinical Services Laboratory
OpGen, Inc.
Gaithersburg, Maryland

Elizabeth G. Wilbanks
Department of Ecology, Evolution,
  and Marine Biology
University of California, Santa Barbara
Santa Barbara, California

Wendy J. Wilson
MSEPS
Las Positas College
San Leandro, California

Stephen H. Zinder
Department of Microbiology
Cornell University
Ithaca, New York

Foreword

The institutionalized problem of the disparity in the number of women versus men working in positions of power is no less cogent to the field of microbiology as it is to engineering, entertainment, and even the political sphere. As a policy maker and the only microbiologist in Congress, I have highlighted the importance of issues that touch or are touched by microbiology, from the critical growth of antibiotic resistance, to development of greener energy technologies, the importance of public health efforts, and advocacy for women's rights and equality in the workplace and elsewhere. *Women in Microbiology* is a beautifully written contemporary work covering both women throughout history who have advanced and impacted the field of microbiology, as well as women who are current leaders in the field. Some of the names are familiar and some are lesser known, but all of the stories serve to arouse a sense of excitement, driven by tales of new, important scientific insights, stories of overcoming adversity and breaking boundaries, and the inclusion of personal tips and advice from successful careers. Written by scientists, many chapters feature personal vignettes exemplifying how these women have taught, mentored, and inspired the authors through their careers. My hope is that this book will be broad-reaching, extending its inspiring stories to both men and women in the field of microbiology, and beyond. Microbiology is an area that touches all life as we know it, and I hope these stories touch your life in a very special way.

Congresswoman Louise Slaughter
United States Representative
25th Congressional District
New York
January 2018

Prefaces

In November of 2014, a small group of microbiologists gathered at a special symposium to celebrate Dr. Abigail Salyers, pioneer of human microbiome research, who had passed away the year before. Exciting ongoing science was discussed, but what made a lasting impression on me were the wonderful stories about Dr. Salyers as a person—the particular and often surprisingly strong and hilarious descriptions of her and her impact on our profession.

The next day, I met with Megan Angelini and Greg Payne from ASM Press to discuss ideas for potential new books. Moved by Abigail's story, I suggested a book profiling women in microbiology. We worked to develop the idea into a book focusing on women in microbiology who have had a positive impact on the field through their research accomplishments, their unique approaches to microbiology, or their actions as role models and mentors. Together, these chapters form a book that highlights the varied paths women microbiologists have taken to achieve professional recognition, explores their motivations as scientists, and considers the ways their lives have inspired others. We included both historical figures and active microbiologists, with names both familiar and less well known. Our hope was to write a book that would encourage current female scientists, inspire more women to enter the field of microbiology, and more widely share the diverse backgrounds and perspectives of these women. These were lofty goals, but as you will read, aspiring to lofty goals has often been key to the success of the women detailed here.

Against the advice of my mentors, colleagues, friends, and my husband, who feared it would take too much of the time I needed for other things, I agreed. That was three years ago and in a way they were right; creating this book was not an easy process. The hardest part was understanding that there was no way to make this compilation fully representative. There are many important, impactful, and inspirational women who are not included; along with many outstanding colleagues who could have contributed additional chapters. We barely scratch the surface of the stories that could and should be told. Over time I grew to accept this, and I hope others will too, with the understanding that it is not for lack of dedication or commitment, but simply because this is such a big topic. There is inevitable regret and disappointment at what has been left out, but I hope readers will see this as just a starting point toward achieving the goals outlined above.

With these caveats, I think you will find this first book on women in microbiology to be an exciting and eclectic mosaic of stories. Each chapter is told from the perspective of its author, profiling a woman in microbiology from near or far, in time and/or space, and each has a different focus, motivation, and style. As I read these stories, I at times laughed out loud. I find myself motivated by the many different pursuits of these women. I am surprised at the barriers and misunderstandings that women scientists continue to face but I am also inspired by the diversity of ways strong women respond to them with humor and persistence. I reflect on which characteristics resonate with me and wonder what will make other readers laugh and whose stories they will find most compelling. Above all, what I take away from this compilation is the unique combination of intellect, insight, integrity, passion, and persistence that each woman profiled has brought to the field of microbiology. I wonder if there is something about scientific endeavor or about being a woman in science that somehow selects for, or generates, this combination of traits. For women in microbiology and in science in general, the sample size is still small. There is no one path, one style, or one contribution that defines women in science, and yet, from my read of these amazing stories it is the combination of these traits that unites them.

I am extremely grateful for all of the authors willing and able to take the time to reflect on and write about a woman in microbiology who inspired them. I am also deeply appreciative of the many people who wanted to contribute but were unable to at this time. I am grateful for the wide-ranging and uniformly positive responses to this idea which encouraged us to continue. In particular I credit Megan Angelini who generated the spark for this book and who kept the fire burning through tough times with positive encouragement and the integrity, passion, and persistence necessary for its completion.

—*Rachel J. Whitaker*

When I've told people I was editing a book on the contributions of women in science, the question I often got from male colleagues was [to paraphrase] *"Why only about women? That seems a bit unfair. You should do it about both."* Yet no one mutters a word of objection when students are assigned textbooks that overwhelmingly highlight the careers and contributions of male scientists. This subtle bias in textbooks enforces the stereotype that the science of microbiology has been propelled forward by male scientists, with women relegated to supporting roles. It was time for a book that celebrated the role of women in microbiology.

Women obtain approximately 50% of all doctorates in biology, yet only 36% of Assistant Professors are women, a number that drops to 18% for Full Professors (1). The only place where women outnumber men in academia is in non-tenure track positions, where they hold 58% of the positions (2). There are many societal issues responsible for this disparity, and the potential solutions are certainly complex (1). One way forward is to encourage young women to see themselves as scientists; and once they can see themselves in that role, implement institutional changes to support and retain these women throughout their careers, with the opportunities necessary to attain high-level positions in academia, industry, and elsewhere. The purpose of this book is to profile women who have had a positive impact on the field of microbiology and highlight the many, diverse paths these women have taken to achieve professional recognition for their work.

Writing biographies that capture the motivations of women scientists has, historically, had varied results. On Twitter, Adela Torres (username @Daurmith) hilariously wrote the biographies of famous male scientists in the stereotypical way that many female scientists are written about: *"His dour personality made everyone think he'd never marry. Even so, Schrödinger got a wife and a Nobel Prize."* This dichotomy highlights the way in which social norms are perceived to play a greater role in the ability of women scientists to work, as if men were not equally subject to such societal pressure. In her critical review, Patricia Fara highlights the treatment of women in scientific biographies as *weird sisters*, the idea that any woman choosing a career in science did so as a misfit, or through choosing to sacrifice love and family, or (gasp!) at the cost of personal grooming (3). As Fara points out, given the disparity between men and women in science, such comparisons set scientific women apart and do little to encourage young women to see that it is entirely possible to have a rewarding scientific career, without the need to live as a multi-cat-owning spinster with frizzy hair and sensible shoes.

My coeditor Rachel Whitaker and I aimed to create a book exploring the interesting and inspiring lives of women in microbiology, helping to dispel common stereotypes and misconceptions for good. The biographies of women scientists in this book span almost 150 years, from the earliest studies of fungi under the microscope, through the microbiology of the Apollo moon missions, to the most cutting-edge environmental and computational science of today. The inspirational stories of these women are as variable as their science: there are those who inspired through a meticulous approach to science, others who demonstrated that it is possible to juggle innovative science with a family, others who were hell raisers, and those who nourished the creative talents of their students and helped them reach their full potential. There is as much breadth in our authors as the women they profile in terms of race and gender. After all, it is not solely the purview of women to be inspired by a female scientist, nor is such inspiration restricted in race.

We attempted to highlight the stories of women who found rewarding careers as microbiologists through a series of biographies chosen by the authors themselves. We asked, "Who inspired you, and how?" While it would be expected that a book such as this should feature the usual suspects of very famous women scientists, inspiration is a personal thing; some of our authors chose women who are indeed world famous, prize-winning scientists, whereas others chose women who drove change in scientific culture. Many chose women you might not have heard of but you may be thankful that you have now been introduced.

If we had highlighted every scientist we wished, this book would be far too long to publish. All of our authors are working professionals, and sitting down to compile a biography can be time consuming; for every three people we asked, two turned us down due to other commitments. Many of our authors spent months writing, interviewing the person they profiled at length and often interviewing their subject's family, collaborators, and students.

Some went to original archives or even visited the neighborhood where the scientist whom they profiled had herself received inspiration. It truly has been a labor of love and respect. As editors, we have hounded authors, with persistent (and probably annoying) emails. Despite the irritation of repeated, nudging requests, almost universally we received similar feedback: *"This was really a lot of work, but I'm so glad I did it. I have learned so much about X and what they accomplished, I am truly inspired. Thank you so much for the opportunity."*

We are honored and flattered that our authors would think to thank us; however, we owe a greater debt to them. I am truly grateful for the efforts and dedication of all the contributors, who have brought the stories of these truly inspirational women to life. I would also like to thank the staff at ASM Press, especially Megan Angelini and Christine Charlip, who made this book a reality.

We hope that the reader will be as inspired as we have been and see the hope, joy, and beauty in the stories of these women fascinated by the microbial world.

—*Hazel A. Barton*

1. **Shen H.** Inequality quantified: mind the gender gap. *Nature* **495:**22–24.
2. **Mason MA.** 9 March 2011. The pyramid problem. *Chronicle of Higher Education*.
3. **Fara P.** 2013. Weird sisters? *Nature* **495:**43–44.

About the Editors

 **Rachel J. Whitaker** is an Associate Professor of Microbiology in the School of Molecular and Cellular Biology at the University of Illinois at Urbana-Champaign and current leader of the Infection Genomics for One Health research theme at the Carl R. Woese Institute for Genomic Biology. Her research focuses on the molecular evolution of cellular microorganisms and their viruses in natural contexts ranging from hot springs to infectious diseases. Dr. Whitaker and her lab seek to integrate microbial biology with ecological and evolutionary principles using genomics as a primary tool. Dr. Whitaker attended Wesleyan University in Connecticut where she majored in biology and the interdisciplinary Science in Society Program (SiSP). After college she worked as a VISTA volunteer and as a research technician before graduate school. Dr. Whitaker received her Ph.D. at the University of California, Berkeley working in the lab of John Taylor. She extended her work to environmental genomics during postdoctoral research with Jillian Banfield from whom she learned a great deal about working as a woman in science. Dr. Whitaker's research has been funded primarily by NASA and the NSF. She was recently the recipient of the Paul G. Allen Distinguished Investigator Award. She is excited to serve as co-Director of the Microbial Diversity course at the Marine Biological Labs in Woods Hole, MA from summer 2018–2023. Outside of science, Dr. Whitaker enjoys being a mom to two boys whose boundless energy and curiosity keep her engaged. She is most grateful to her husband who has supported her science, career, and life of adventures, to her parents who taught her to care passionately about equality, justice and education, and to the inspiring men and women she has worked with as students and postdocs.

 **Hazel A. Barton** is a Professor and Director of the Integrated Bioscience Program at the University of Akron. Her research, geared toward understanding microbial interactions and adaptations to starvation in cave environments, has been funded by the U.S. National Science Foundation (NSF), U.S. National Institutes of Health, and the U.S. National Park Service. Dr. Barton also investigates the role the fungus *Pseudogymnoascus destructans* plays in the white-nose syndrome epidemic in bats, and has been funded in this venture by the U.S. Fish and Wildlife Service. Her research has been featured in *Sports Illustrated, Forbes, National Geographic Explorer, Outside, Science News, The Scientist, Popular Mechanics, Wired, Geo*, and *The Smithsonian* magazines; in the book *Extreme Scientists: Exploring Natures Mysteries from Perilous Places*; on NPR and BBC Radio; on Animal Planet, the History Channel, National Geographic, the CBS Early Show, BBC TV; and in the IMAX movie *Journey into Amazing Caves.*

Dr. Barton is also an avid caver, having explored caves on six continents. She is a past director of the National Speleological Society, the Quintana Roo Speleological Survey, and an award-winning cave cartographer. Dr. Barton is currently a Fellow of the National Speleological Society, a Kavli Fellow of the National Academy of Sciences, Chair of the Committee on the Status of Women in Microbiology for the American Society for Microbiology, and the recipient of an NSF CAREER Award.

Women in Microbiology
Edited by Rachel J. Whitaker and Hazel A. Barton
© 2018 American Society for Microbiology. All rights reserved.
doi:10.1128/9781555819545.ch1

Bonnie L. Bassler: The Group Accomplishes More than the Individual

1

Nina Molin Høyland-Kroghsbo[1]

THE BACTERIA WHISPERER

I was sitting, freezing cold, in an exceptionally large conference room in New Orleans at the 2011 General Meeting of the American Society for Microbiology. That year, Dr. Bonnie Bassler served as its president. I was especially excited for her talk. Finally, Dr. Bassler, also known as the "bacteria whisperer," confidently walked on stage with her characteristic fast pace. She smiled warmly. She stood in front of a large screen depicting her favorite microbe, *Vibrio harveyi*, a bioluminescent marine bacterium. With energetic gestures and contagious passion, she pointed to a slide showing a lonely bacterium swimming all by itself, releasing cell-cell signaling molecules called autoinducers. Dr. Bassler explained that at low cell density, the autoinducers float away and the bacterium acts as an individual. Next, she pointed to a group of bacteria, all producing the autoinducers. "The bacteria grow in number, and since they all participate in making autoinducers, the extracellular concentration of the signal molecules increase in step with increasing cell number. When the molecules hit a certain threshold level, it tells the bacteria that they have neighbors, and, in synchrony, all of the bacteria turn on light production." She could barely contain her excitement for the little critters. Her eyes sparkled as she described how, via a process called quorum sensing, many species of bacteria use different chemical languages to work together like multicellular organisms to accomplish amazing or terrifying things, depending on the bacterium. As the information soaked in, I was certain that I was finally in the right place in science.

---

[1]Department of Molecular Biology, Princeton University, Princeton, NJ 08544

For over 25 years, Bonnie Bassler (Fig. 1) has pursued her curiosity about the social lives of Earth's smallest living organisms. She has been discovering and translating bacterial languages, teaching us about the intricate ways in which bacteria interact and coordinate behavior and demonstrating the profound importance of those interactions in nature.

CALLS FROM THE HALLWAY

The first time I talked to Dr. Bassler, she insisted that I call her Bonnie. Bonnie, who receives fan mail by real post, is remarkably down to earth. Working in Bonnie's lab is fun and always full of surprises. New in her lab, I was pipetting away at my bench when Bonnie shouted my name from her office two doors away. I immediately removed my gloves and ran in her direction as quickly as my short legs could carry me. I wasn't fast enough and she called again, louder. Slightly anxious, I entered her office figuring that I must be in trouble. It turned out that she had been thinking about a recent discussion of ours and wanted to talk about how my project was coming

Figure 1 Professor Bonnie L. Bassler. Credit: Alena Soboleva Photography.

along and go over some new ideas for experiments. I quickly got used to Bonnie's famous hallway calls. No one is ever in trouble, but it is pretty funny to witness the newest lab member when his or her name is first yelled down the hall.

Once, Bonnie came leaping down the hall whooping with joy. It was obvious that something amazing had just happened. We thought perhaps she had won a prize or received some great honor. No, her upcoming seminar trip had been canceled due to bad weather. Bonnie is in demand as a speaker. In this case, she had "won" several days with nothing scheduled on them. The extra time was immediately devoted to her lab gang and to their science. Everybody pipetted a little faster that day.

For Bonnie, her lab group is her family. Her office door stays open when she is in town, and she always stops what she is doing to make time to discuss science, to give career advice, or to talk about life challenges. She also has us to parties at her house to celebrate holidays and graduations. We take days off to go on field trips together. We are an ever-young group of people from all over the world. By now, Bonnie has raised several generations of scientists, and some of her former students' students (myself included) work in her lab. Her goals for us are to have fun, make fantastic discoveries, and get the jobs we most want when we leave her lab.

Bonnie often wakes up in the middle of the night contemplating the next steps in our experiments, wondering about a particular finding, or thinking about how we should deal with reviewers. She answers emails (instantly) until 11 p.m. and then she starts again at 5 a.m., before she heads off to teach her aerobics class. She teaches the aerobics class because that way, she always shows up—she finds it impossible to opt out and not give it her all when people rely on her. She packs her workout clothes when she travels, but without any group expecting her, the gear stays in her suitcase. Bonnie is about the group accomplishing more than the individual—in bacteria, in her lab's approach to research, and in aerobics.

With one single exception, Bonnie is completely focused during lab meetings. Our group gathers weekly around a heavy old wooden table in a small meeting room. Bonnie always leans back in her chair and puts her feet up on the table, wearing Converse sneakers or fancy flats, depending on who she's meeting with that day. It is obvious to us that she couldn't be any happier. She's doing everything she's always wanted. She's on an adventure: thinking deeply, working with smart creative people, together making great discoveries, and publishing findings in excellent journals. She listens carefully, poses questions when anything is unclear, and points out when lab members offer particularly good suggestions for new experiments. Or she says, "Yes, we did that 20 years ago. Does anyone ever read our old papers?" and shakes her head with a disheartened but forgiving sigh. After being awarded the Shaw prize in 2015, Bonnie cut off the lab meeting presenter in mid-sentence: "I didn't listen. Can you repeat that? I was thinking about all the new shoes I'm going to buy with the prize money." She does own an impressive collection of shoes, including a custom-made pair of electric-blue *V. harveyi*-colored sneakers emblazoned with "Queen of Quorum Sensing." Those shoes were a good-bye gift from the very postdoc who presented the lab meeting at which Bonnie was daydreaming!

"I'M JAMES WATSON"

Working in Bonnie's lab brings me joy on a daily basis. It wasn't always like that for Bonnie. Her graduate studies were in a lab that was not particularly supportive and that lacked camaraderie. However, the discoveries she made, investigating how the marine bacterium *Vibrio furnissii* finds and consumes chitin, long sugary chains that make up the exoskeletons of crustaceans, excited her and motivated her to keep going. She thought this was just how science was done: work alone and find private fulfillment. Then she joined Mike Silverman's lab at the Agouron Institute. His approach to science was entirely different—and eye opening for Bonnie. He was an active and supportive mentor and a great role model. Mike cared deeply about Bonnie's well-being and success. He showed her that one could be both a great scientist and a decent human being. Suddenly, being in the lab was fun. Each morning, she would race in to work to look in the incubator for a new surprise on her petri plates. Trained as a biochemist, Bonnie learned the art of genetics from Silverman, and as a postdoc, Bonnie discovered the genes that drive the social lives of bacteria.

At the end of her postdoc training, Bonnie sent out roughly 50 applications for faculty positions. One rejection followed the next. Ultimately, she got two interviews. The first was at the University of Illinois. Bonnie rehearsed her talk to perfection. The interview went great and she was offered a position. The university had an impressive microbiology department working on diverse bacterial systems. Bonnie was ecstatic at the prospect of becoming a professor. A few days later, back at her lab bench, the phone rang. "This is Tom Silhavy," said the Princeton professor, a hero of Bonnie's, famous for discovering how *Escherichia coli* transports proteins through its lipid bilayer. Figuring someone was playing a trick on her, Bonnie replied, "Yeah, and I'm James Watson." Tom then politely persuaded Bonnie that it was indeed he and asked her if she wouldn't like to visit Princeton for an interview. Bonnie, horrified at her blunder, froze, speechless, something that rarely happens to her.

When Princeton made their offer, Bonnie was happy-scared, as she calls it. She was happy at the prospect of working in a department surrounded by scientists investigating how all kinds of organisms—ranging from viruses to bacteria to worms to flies to mice to human cells—convey signals, yet she was scared because she felt woefully underprepared for the job. She knew how to pipette, but she doubted if she could come up with her own research ideas. Teaching classes and writing grant applications were daunting new tasks for which she had no experience. Bonnie, suffering from the imposter syndrome, worried she would fail as a professor. However, Bonnie does not let fear stop her from giving exciting new challenges a shot. She encourages us not to do that either.

Princeton's startup package allowed Bonnie to equip her small new lab in the Department of Molecular Biology. She had four lab benches, just down the hall from her new mentor and friend Tom Silhavy. She bought pipettes and reagents. She poured agar plates with her students. For a long period, she didn't have enough funding to hire a postdoc. Even if she had the money, she worried that she could not provide a postdoc an excellent project to launch a career. She deeply wanted to follow the example of her former mentor Mike Silverman, who generously encouraged her to take her *V. harveyi* project with her when she started her lab. For those reasons, and because the quorum-sensing field had not yet taken off, it was nearly 5 years before the first postdoc joined Bonnie's group.

"JUST TYPE MORE"

Getting research grants was one of Bonnie's biggest challenges in her early career. As a new faculty member at Princeton University, Bonnie worked tirelessly, meticulously crafting application after application to fund her lab. Bonnie received rejection upon rejection. She said to herself, "Oh well, I'll just type more." "I thought it was me," she explained, "I worked on a nonpathogenic bacterium with curious behaviors—the ability to make light and the ability to communicate. I worked in a quirky field that did not yet have traction. It took me a long time to realize that my gender probably played a big role in those rejections. A young woman proposing ideas and experiments that were completely out of the ordinary—that was likely a bad combination." These days, Bonnie considers those rejections as noise in her long and overwhelmingly happy scientific career.

Bonnie's first break came with a National Science Foundation (NSF) Young Investigator award. She had proposed to hunt down the elusive AI-2 autoinducer molecule and uncover how bacteria communicate across species. That NSF grant and, soon after, another small grant from the Office of Naval Research kept Bonnie's little team going for 8 years, enough time for them to make huge breakthroughs and to firmly establish quorum sensing as so very much more than a curiosity.

IT'S BORON

One of Bonnie's most memorable breakthroughs was when she solved the structure of the universal autoinducer that she had named AI-2. Bonnie and her team used the receptor protein LuxP, which binds to AI-2, to fish out the active AI-2 autoinducer from a solution of similar, interchanging molecules. When they gently heated the purified protein, AI-2 would fall out, and when they added this suspension to *V. harveyi* cells, the cells would glow, proving that Bonnie had her hands on the correct molecule. Now they just needed the AI-2 structure. Together with her colleague Fred Hughson, the team produced a beautiful X-ray crystal structure of the LuxP protein with AI-2 sitting inside. They had an impressive 1.5-Å resolution. AI-2 did not resemble any known autoinducer or, for that matter, any other molecule known to humankind. They used mass spectrometry to determine the molecular mass of the molecule within 3 mass units. Together, the data suggested that the AI-2 molecule formed a double-ring structure held together by a carbon bound to four oxygen atoms. However, this central carbon atom puzzled Bonnie and Fred. The chemistry didn't add up: such a molecule would be highly unstable.

One day, as Bonnie and Fred were gazing at the periodic table, their eyes fell on boron. Boron and carbon are indistinguishable in an X-ray structure because their electron density patterns are nearly identical. Boron, 1 mass unit different from carbon, would fit perfectly within the error margin of AI-2's measured molecular mass. Boron would make a stable molecule, easily hanging on to four oxygens. Boron is highly abundant in the ocean, where *V. harveyi* lives. It all seemed to make sense, except that they had never heard of boron having a role in biology. They immediately subjected their AI-2 preparation to boron nuclear magnetic resonance, which revealed an unambiguous boron peak. Bonnie ran straight back across the street from the chemistry department to her lab. Her team squirted a borate solution onto *V. harveyi* cells. The cells started glowing brighter than ever before. At that point, there was no question that boron was the key. "That's the beauty of bacteria," Bonnie explained. "You can generate a hypothesis, test it, and get the unequivocal result the same day." She had finally discovered the molecular underpinning for her "bacterial Esperanto," the universal language that allows bacteria to communicate across species barriers.

Bonnie has spent nearly three decades decoding and translating bacterial languages. Her next goal is to silence that bacterial chatter. She hopes that by preventing disease-causing bacteria from communicating, they will fail to collaborate, which would render them harmless. Her lab is well on its way to successfully proving the merits of this idea. She hopes that her discoveries may someday save lives.

WHITE HOUSE CALLING

For Bonnie, it is frustrating facing gender bias in terms of credibility, positive grant reviews, and the ability to publish in the best journals. Still, Bonnie is grateful for and humbled by the recognitions she has received. In the wake of her AI-2 discovery, at age 40, she received the McArthur "genius" fellowship. That first honor, one she did not apply for, was especially validating of her science and her team's efforts. Other fellowships and prizes followed. Some of her honors include election to the National Academy of Sciences and the National Academy of Medicine, the American Society for Microbiology's Eli Lilly Investigator Award, the Wiley Prize in Biomedical Science, and the Shaw Prize in Life Sciences and Medicine. She is a Howard Hughes Medical Institute Investigator, Squibb Professor, and chair of the Department of Molecular Biology at Princeton University.

One day, Bonnie was sitting in her office, typing away on her computer, when the phone rang. The display read "The White House." "This is going to be a good day," she said to herself, beaming, as she picked up the phone. She was being recruited to be a scientific advisor to President Obama. Bonnie joined the National Science Board (NSB) and was suddenly involved in overseeing the NSF budget and deciding the Foundation's priorities in research and education. She felt that in her NSB position, she could finally give back to the NSF, to which she owed so much, having launched her career. During her term, Bonnie championed NSF funding for early career awardees and for individual investigator grants to ensure that young scientists get a great start and that curiosity-driven science continues to flourish. She helped craft new STEM education initiatives, and she vetted plans for large, shared scientific facilities so scientists have access to state-of-the-art technologies.

I'M NOT SMARTER; I'M JUST OLDER

"When you have your own lab" are the most important words Bonnie ever said to me. The matter-of-fact way she casually tossed off those words in one of our routine conversations, not as a question but as a fact, made me truly start believing in myself. That was one of the transformative moments of my life. Until that day, I had almost not dared to dream that I'd someday have my own lab. I'm so glad I didn't listen when my high school biology teacher told me not to pursue science or when I was told I didn't have the personality to get a Ph.D. I guess when it comes to career advice, unlike in scientific research, cherry picking your sources is the right thing to do!

Bonnie dearly hopes that her cohort of women scientists, using their lives as examples, can successfully break down barriers for the next generation of scientists. She spent her first few years as a young professor at Princeton filled with self-doubt and feeling like an imposter. Later, when each new challenge arose, Bonnie began asking herself, "What's the worst thing that could happen to me if I fail?" She figured she could just try again, try harder, or try something different the next time. Gradually, Bonnie started to realize that her discoveries and her success in science could not just be dumb luck. She now knows that she and her lab gang make a winning combination. The imposter feeling hasn't completely left Bonnie, but she deals with it differently these days: "It never goes away, but it changes. You can tame that voice in your head somewhat, you can use its persistent nagging to inspire you to work harder, and you can do good work in spite of its presence. Scientists your age always think I'm smarter than them and that I'm filled with confidence. I'm not. I've just had more practice overcoming hurdles, accomplishing my goals, and being okay with myself even with my imposter voice rattling around in my head. I'm not smarter than you; I'm just older."

CITATION

Høyland-Kroghsbo NM. 2018. Bonnie L. Bassler: the group accomplishes more than the individual, p 1–8. *In* Whitaker RJ, Barton HA (ed), *Women in Microbiology*. American Society for Microbiology, Washington, DC.

Women in Microbiology
Edited by Rachel J. Whitaker and Hazel A. Barton
© 2018 American Society for Microbiology. All rights reserved.
doi:10.1128/9781555819545.ch2

Antje Boetius: Exploring the Living Infinite

2

Jeffrey Marlow[1]

The sea is everything. It covers seven tenths of the terrestrial globe. Its breath is pure and healthy. It is an immense desert, where man is never lonely, for he feels life stirring on all sides. The sea is only the embodiment of a supernatural and wonderful existence. It is nothing but love and emotion; it is the living infinite.

JULES VERNE, *20,000 Leagues Under the Sea*

Antje Boetius grew up in landlocked Frankfurt, Germany, but a love of the oceans was in her blood. Her father's side of the family was from the Frisian Islands in the North Sea; they were adventurous sorts who "had a lot of seafaring in their genes," Boetius says. "Growing up, my grandfather told me all of these fantastic stories about life at sea, and how great it is—everyone helping each other, fighting the wind and the waves." Boetius's grandfather survived the *Hindenburg* disaster and a torpedo attack on his ship in the Atlantic, so he was familiar with high-stakes situations; in both cases, he shepherded others to safety and was commended for his bravery.

On family vacations to the seaside as a child, Boetius developed a special communion with the ocean, cooling off in the water, diving through the waves, and playing in the mud with her siblings (Fig. 1). She whispered hello to the ocean every morning, and said a terse goodbye upon departure, pouting all the way back to Frankfurt. "The Frisian side of the family had

---

[1]Department of Organismic and Evolutionary Biology, Harvard University, Cambridge, MA 02139

Figure 1 Enjoying the ocean: a seaside family holiday with siblings (from left to right) Meike, Matten, and Antje Boetius. Image courtesy of Antje Boetius.

loads of stories where the ocean was a threat," she explains, "where people fought on their little rocky islands against the sea and the surf. But to me the ocean was mostly comforting—I could swim, and the waves would carry me."

And then there were the underwater exploits of Lotte and Hans Hass, the married adventurers who produced television programs à la Jacques Cousteau. Every evening at 5 p.m., the Boetius family's new color television would unlock 30 minutes of armchair adventure with each new Hass documentary. "You could also see a woman going on expeditions," Boetius recalls, "diving with the sharks and exploring the oceans. I was fascinated with this idea that you could take a ship, put a lot of people on it, and you could explore the seas."

The intellectual fire that was being fanned at home, however, was nearly extinguished at school. "In my time, the ocean didn't exist in school," Boetius explains. "It just simply wasn't part of the curriculum. Biology was about rabbits and trees, and that was rather boring to me." In class, she would daydream of life beneath the waves while the teacher would drill students on seemingly disconnected, trivial facts, like the teeth of a dog or the morphology of flowering plants. But Boetius persevered, working to get good grades not out of interest in the subject matter but as a means to an end, a prerequisite for joining the front lines of research. And after years of putting in her time, she would soon get that chance.

It had been a long journey from Germany to Southern California, but as Boetius lugged her suitcases into the Scripps Institution of Oceanography, there was no time to catch her breath. A secretary whisked Boetius down the hall and into a darkened classroom, where Farooq Azam—a renowned marine microbiologist—was teaching a class on ecological theory. It was a revelation; the jet lag evaporated in a haze of nutrient budgets and biogeochemical models. In an era when "in Germany, you would go to university to listen to an old professor read from their own old books," Azam's passion for the emerging field of marine microbiology shined like a beacon. "It was feeling for the first time that I know what I studied for," Boetius recalls, "what scientific curiosity is and how to turn it to research plans and projects. And that was the beginning of my career as a microbiological oceanographer." The palm trees, sunshine, and soothing metronome of the La Jolla waves didn't do much to dispel the notion of paradise.

In the lab, Boetius started to get her hands dirty with sediments collected from the Clarion-Clipperton Fracture Zone. Straining to see her tiny quarry, she plucked copepods and nematodes from the gooey mess, loving every mud-splattered moment and marveling at how such minuscule seafloor animals could unlock the ocean's secrets. Under the microscope, these alien organisms took on mythic stature, some with bizarre heads and mandibles, others in elegant transparent shells, and a few looking like children's toys. But Boetius sensed that the real architects of these bizarre habitats were even smaller, and as she began to survey the field of deep-sea microbiology, she found an alarming gap. She took a brief foray into researching the pressure-based adaptations of bacteria, and while it was a decent start, "it wasn't quite what I wanted to do because it was all about cultivation. I wanted to work on microbes directly as they come, so I started looking around for labs to work in. But there weren't many options at the time."

Boetius joined scientist Karin Lochte in Victor Smetacek's group at the University of Bremen for her doctoral work, trying all the while to create the field she wanted to study: deep-sea environmental microbiology. It was a steep learning curve. Protocols that worked well in microbial cultures or in upper ocean samples came to a grinding halt when applied to seafloor sediments, where complex food webs, clay particles, and electrostatic interactions presented major obstacles. After years of method development, Boetius presented a thesis demonstrating rate and flux measurements—incorporating cell counts, ATP measurements, and radiolabel tracers—from deep-sea microbial communities.

An equally impressive achievement during Boetius's years as a graduate student (Fig. 2) was her epic appetite for research expeditions, the likes of which would have made even Lotte and Hans Hess queasy. "I was completely expedition-fixated," she recalls. "I went across the seven seas, and did 14 deep-sea expeditions before I even graduated." The idea was to see how deep-sea microbes around the world responded to varying carbon inputs and seasonal climatic events, like the Indian monsoon. By the end of it all, her team was uncovering key links between microbes, the oceans, and the planet's elemental cycles. It was a remarkable effort, a master class of synthesizing discrete observations into a discovery that was more than the sum of its parts. And yet, it wasn't enough for Boetius. "I wanted to do more," she recalls. "I wanted to be able to say exactly which organisms were responsible."

As a postdoc, Boetius joined the fledgling Max Planck Institute for Marine Microbiology in Bremen. Her goal was to learn how to apply methods of fluorescence *in situ* hybridization (FISH), a technique that allows

Figure 2 Academia calling: Boetius receives her Ph.D. in 1996 wearing a boat-shaped mortarboard. Image courtesy of Antje Boetius.

scientists to visualize certain organisms by attaching fluorescent probes to segments of their genetic material, to deep-sea sediments. "I thought this made more sense than only 16S sequencing," she says, "because you can really look at cells and directly see what is relevant. It was more direct, more quantitative than the sequencing." But like the cell counts and rate measurements a few years earlier, applying FISH protocols to environmental samples required some adjustments. Figuring out the right dilution to avoid clogging filters with sediment took time, and developing probes for unknown *Archaea* and *Bacteria* felt like a shot in the dark.

Meanwhile, Boetius's affinity for sea-going research hadn't abated. In 1999, she joined a geophysics expedition to Hydrate Ridge, off the Oregon coast, as the token microbiologist. "I really wanted to work on my big project in the Indian Ocean, but I went on this gas hydrate cruise anyway," she recalls. Earlier reconnaissance had reported white patches on the seafloor: some scientists thought they were methane hydrate ice, while others pointed to hydrate thermodynamic stability diagrams and suggested that the patches were microbial mats. Boetius was enlisted to clarify the situation. At sea, grainy images and cylindrical chunks of the seafloor were recovered each day; she was able to point out the differences between white layers of microbial mats and the hydrate exposed at the seafloor.

When the sediment cores were hauled back on board, something clicked. "I saw the bacterial mats," she explains, "and then I smelled sulfide, and I thought, that's really odd: you have methane in the hydrates, but it smells like sulfide, so what's the link between them?" Recalling recent publications from William Reeburgh (on gaps in the methane budget) and Kai Hinrichs (implicating methanogen-like *Archaea* in anaerobic oxidation of methane), she thought the possible connection between methane and sulfide-producing microbes could be a compelling lead. "So I took a lot of samples."

Back in the lab, the sulfidic hydrate sediments proved to be a bonanza. Each cubic centimeter of sediment had 10 to 100 billion cells—"the densest microbial populations I'd ever seen anywhere in the deep sea," Boetius says. "And they were all in giant consortia, clumps of cells, and I couldn't break them apart by ultrasonication or anything else, so that was very weird." Given the clear olfactory evidence of sulfur metabolism, Boetius started her FISH explorations with fluorescent probes to target sulfate-reducing bacteria (SRB). After probes that encompassed broad groups of sulfate reducers lit up, she strategically narrowed the search by using probes complementary to more specific lineages. Some worked, some didn't, and by the end of the combinatorial effort, Boetius was able to pinpoint the consortium members as close relatives of cold-adapted SRB previously observed in the Arctic by Katrin Knittel, who taught her the secrets of designing probes.

But throughout this effort, only the outer shell of these mysterious cell clumps would light up. Thinking the molecular probes and fluorescent markers weren't able to reach the interior cells, Boetius spent months trying to break the aggregates apart, to access the microbes at the center. Nothing worked. As a last resort, she expanded the search, inspired by a paper from Kai Hinrichs, Ed Delong, and Victoria Orphan: "I thought, why don't I look for the *Archaea*? I made probes for every little branch," and at last, when targeting the newly described anaerobic methanotroph (ANME) lineage, the rest of the cells lit up like a Christmas tree.

Four FISH images of ANME-SRB consortia graced the pages of *Nature* in late 2000, and the consortia are likely still the most iconic of Boetius' scientific discoveries. The unbreakable physical association between the two types of microorganisms led Boetius and her coauthors to propose a coupled metabolism of methane oxidation and sulfate reduction. The expected energetic output hovered right on the edge of biological possibility, but the syntrophy provided an elegant answer to many outstanding questions.

The study is one of the clearest victories for culture-independent, observational microbiology. By carefully collecting geochemical, isotopic, phylogenetic, and spatial clues, Boetius and her Max Planck colleagues broke one of environmental microbiology's biggest mysteries wide open. Nearly two decades later, ANMEs have still not been isolated in pure culture, but they are recognized as an irreplaceable control on global fluxes of methane and critical constituents of the global carbon cycle.

The *Nature* study was the first car of an oncoming train. In quick succession, a number of other researchers published major findings. Victoria Orphan authored seminal studies showing that ANMEs in aggregates had iso-topically light biomass, pointing to methane as a key carbon source. Walter Michaelis and his colleagues showed a connection between anaerobic methane oxidation and carbonate rock formation, revealing reef-like struc-tures in the Black Sea as an alluring study site. Katja Nauhaus led an effort to harness the process in the lab, showing methane-dependent sulfide pro-duction with Hydrate Ridge sediment. These findings—one dramatic result after another—galvanized the field but also changed its tenor. Suddenly, there were prestigious publications to be had, grants to be won, careers to be made. "I noticed there was a bit of the ugly side of science coming through," Boetius recalls. "Everyone was so excited about the anaerobic oxidation of methane [or AOM]. People started fighting about concepts and samples, and there was competition—it was not the way I liked it. There was some barrier between people."

In the midst of this cultural shift, Boetius was on the conference circuit to share her own findings. At a gathering in Brest, France, in 2001, Boetius met Orphan, then a graduate student. Each had been warned by members of their respective labs to be wary of the other as a potential rival; they were cautioned against sharing unpublished data or hinting at current research objectives. But the two women formed a bond built on implicit trust and a common drive to answer big questions rather than claw their way up the career ladder.

The following year, Boetius and Orphan met again at a conference in Davos. While in the Swiss Alps, they went for a hike—"she was wearing some rather non-hike-appropriate footwear and still kicking my ass up the mountain," Orphan recalls—and both spoke openly about their science. Given the preconceptions surrounding their meeting, the refreshing reality was a sigh of relief. "It was really comforting to hear that she was seeing similar things," says Orphan, "and to talk openly and honestly about it." Boetius views the episode as a turning point. "Ever since then, we've shared our knowledge and enjoyed the fun of discussing new ideas and methods, and views on life."

Over the next several years, Boetius was integrally involved in dozens of AOM studies, which collectively helped expose the process as a dominant control on methane fluxes between the subsurface, the water column, and the atmosphere. But then, she took an abrupt turn. "I could never wrap my mind around one topic forever, and only do that," she says. "The AOM work got me many prizes and a professorship and all that, but I also noticed"—blasphemy alert—"I was getting kind of bored restricting myself to under-standing the innermost secrets of a microbe. My mind was also wrapped around the question of climate change and its consequences. At conferences, people found the enigmatic microbe story fascinating, but the big question always was, how do hydrate systems respond to warming oceans? How does the cryosphere more generally respond? How does the entire Earth system cope, including we humans? I can get totally fascinated with Earth systems questions, and it doesn't need to only be the microbiology in it."

A series of expeditions to the Arctic followed, and Boetius's teams examined polar plankton blooms, described the impact of sinking previously ice-bound algae on seafloor systems, and discovered new chemosynthetic habitats in high-latitude waters. Subsequent shifts have led Boetius to engage more deeply with the issues of deep-sea ecosystems, biodiversity, and our vision of how to live with a future ocean. Her most recent projects look at the ecological threats of deep-sea mining. Boetius notes that "locally, it will do damage, that is clear. But land mining is also very destructive. It is a scientific task to test if there are ways to make deep-sea mining somewhat sustainable, for example, by creating a protected area for each exploited area. We're gathering evidence about the connectivity of deep-sea habitats to see if it could work."

Up next is the directorship of the Alfred Wegener Institute, Germany's polar research institute, where Boetius hopes to improve interdisciplinary work, expand scientific dialogue with the public, and develop new observational platforms, especially in rapidly changing polar environments (Fig. 3). "How can we be a bit more brave with the type of science we do?" she asks. "I find that especially in Germany we need to improve the 'big ideas' science, the high-risk, high-reward research. We urgently need to be able to integrate

Figure 3 Boetius at the North Pole during R/V Polarstern Expedition IceArc PS27/3. Image courtesy of AWI.

and use ocean knowledge in different ways, but it's hard to make that happen with the way science and policy-making [are] traditionally done."

This multiact career not only has revealed Boetius as a remarkable scientist who can think deeply across systems but also places her in a broader, increasingly essential category as a synthesizer who can convene and captivate diverse audiences. In some ways, Boetius evokes the German polymath Alexander von Humboldt, given her ability to connect the dots and see the big picture—first with unculturable microbial communities and ultimately with global climate systems—and her boundless curiosity, a voracious drive akin to what Humboldt once described as being "chased by 10,000 pigs." "There are so many great questions out there that are basic," she says, "but so relevant to our big challenge of creating a sustainable Earth."

In a field often driven by depersonalizing forces—the harsh, "what have you done for me lately?" empiricism of the publish-or-perish mindset—Boetius's most impressive achievement may be the community she's cultivated during her career. Tina Treude, now an associate professor of marine geomicrobiology at UCLA, was Boetius's first Ph.D. student. "She is a brilliant and 150% committed scientist," Treude says, "but at the same time she does not forget how to live life." Here is a case in point: "Supposedly someone warned her before hiring me back in 2000 that I party too much. But that did not discourage her; instead, it rather settled her decision to take me!"

Gunter Wegener, a senior scientist at the MPI in Bremen who has worked with Boetius for 15 years, paints a similar work-hard, play-hard portrait. "Now she's too busy of course," he says good-naturedly, "but she used to really be a party queen. We would go out and dance"—favorites include funk and David Bowie—"and then go back to the lab until three or four in the morning."

Underlying the boundless energy is a singular ambition to facilitate meaningful science. "She always wants big results, big stories, new expeditions involving all kinds of people. It drives us all a bit crazy, but it quite often works," Wegener says, with poorly disguised astonishment. As Treude explains, "she is always developing new ideas and brings together interdisciplinary peers to work jointly on big questions."

The symbiotic connections Boetius spent years piecing together through the microscope have played out in her professional relationships as well. "In the end, I think it's the same principles we have in nature," she reflects. "If you cooperate or if you compete, both cost you something, but very often cooperation beats competition if there's trust, if there is return on both sides." Structural features of professionalized research—limited numbers of positions, short-term contracts—can threaten this harmony, but some aspects of the study system are a stabilizing force. Because anaerobic deep-sea microbes can grow extremely slowly (ANMEs double every few months), "we have to build a world where students and postdocs can hop from lab to lab," Boetius says. "It's only by building international connections where you can share and talk and adapt; then we can get a lot further."

From the other side of the Atlantic, Orphan sees this approach bearing fruit. "I really admire her uncompromising support not only for the people in her lab but also for other young researchers in the field," she says. "Antje helps everyone be their best selves, and by letting the excitement of working in this weird niche field spread, you're giving room to let the science blossom."

As a child, Boetius watched in wonder as adventurous documentaries played out on her television. From the comfort of her Frankfurt living room, the creatures of the underwater world seemed impossibly exotic, like denizens of a foreign planet. Now, Boetius is the one pushing those boundaries, merging science and exploration to remarkable effect.

"I think I can make a difference," she says, "supporting deep-sea research at places where no one ever looked, no one ever went. I have this feeling that there's so much out there to discover."

CITATION

Marlow J. 2018. Antje Boetius: exploring the living infinite, p 9–18. *In* Whitaker RJ, Barton HA (ed), *Women in Microbiology*. American Society for Microbiology, Washington, DC.

Women in Microbiology
Edited by Rachel J. Whitaker and Hazel A. Barton
© 2018 American Society for Microbiology. All rights reserved.
doi:10.1128/9781555819545.ch3

Sallie "Penny" Chisholm and Oceans of *Prochlorococcus*

3

Sophie Rowland[1] and Vanja Klepac-Ceraj[1]

FEARLESS AND UNASSUMING

One would think that after researching one organism for over three decades, Professor Sallie "Penny" Chisholm would run out of things to explore. Yet she continues to be dearly devoted to the study of *Prochlorococcus*, the most abundant and the smallest ocean phytoplankton, with a cell diameter of 0.5 to 0.7 µm. Since the discovery of *Prochlorococcus* in 1985 (1), Chisholm's work on this tiniest of all living phototrophs on Earth has revealed a wealth of information on its ecology and evolution. Penny has proven to be adept at turning the immense collection of complex research on this organism into clear and succinct stories for people of all backgrounds to appreciate. Despite its size, *Prochlorococcus* interacts with a vast ecology, and Chisholm's work has shown its effects on a global scale. Chisholm's own story is as unassuming as *Prochlorococcus*, and just as persistent (Fig. 1).

Out of a sea of academics and an ocean of phytoplankton, Penny and *Prochlorococcus* together have changed our understanding of how the oceans work. Over the years, her lab has studied this phytoplankton across many scales, from its effects on the world's oceans to its individual genomes. Currently, their work is focused on the interactions of *Prochlorococcus* with other organisms, especially heterotrophs and viruses. As a phototroph, *Prochlorococcus* helps heterotrophs grow, but the heterotrophs also help *Prochlorococcus* thrive and survive (2). In a similar way, Penny has helped nurture, and continues to teach, new generations of scientists who, in turn, teach her

---

[1]Department of Biological Sciences, Wellesley College, Wellesley MA 02481

Figure 1 Photo of Dr. Chisholm in her lab.

new ways to think about *Prochlorococcus*. It is through her journey with *Prochlorococcus* and the people she has mentored along the way that we can best understand this exceptional woman in microbiology.

A SEA OF POSSIBILITIES

This is not a story of a child who once stared longingly at the mysterious ocean depths as she strolled along the shore. This is a story of a bright individual from the Upper Peninsula of Michigan who followed where her curiosities led. "[People] expect me to say, 'I've been in love with the ocean since I was 5 years old' or something like that, which is not true," said Chisholm. "I didn't even see an ocean until I was in high school." When she first arrived at Skidmore College, Chisholm planned to major in mathematics. She explored a variety of humanities courses, such as philosophy and Asian studies, and if given a second chance, she reflected, she would have explored even more. But an inspiring teacher drew Chisholm away from lifeless numbers and toward lively organisms. She began researching the chemistry of lakes and wrote her undergraduate thesis on the topic. Chisholm ultimately graduated from Skidmore with degrees in biology and chemistry, bringing her one step closer to *Prochlorococcus*.

From the chemical properties of a freshwater system, Chisholm switched her focus to the small autotrophs living in such a system. After college, Chisholm earned her Ph.D. in biology at the State University of New York at Albany with a dissertation on the nutrient uptake of freshwater phytoplankton. It was while working for 2 years as a postdoctoral researcher for Richard Eppley at the Scripps Institution of Oceanography that she began to prefer the organisms of saltier waters. Diatoms, single-celled organisms housed in exquisite microscopic glass shells, were at the time thought to be among the most important photosynthesizers in the sea. While diatoms were fine organisms to study in the lab, Penny's dream was to study an organism as important as diatoms that could be studied both in the lab and in the wild. Soon after, *Prochlorococcus* floated into her life.

In 1986, Penny Chisholm stumbled across *Prochlorococcus* while working with Rob Olson, now retired and emeritus at the Woods Hole Oceanographic Institute, and her postdoc at the time (1). Penny, Rob, and other collaborators were studying the ecology of a different cyanobacterium, *Synechococcus*. They were observing this phytoplankton in natural seawater using flow cytometry and the organism's autofluorescence. When excited by light, *Synechococcus* fluoresces orange due to its photosynthetic pigments. But in addition to *Synechococcus*, there was something else in the samples they were studying. Something smaller. Something glowing red. Upon further investigation, the team discovered this minuscule fluorescent object was another phytoplankton. This bacterium was none other than *Prochlorococcus*. "Not long after that," Chisholm said, "I decided that what I wanted to do for the rest of my career was to really focus on this one model organism and try to understand it."

PERSISTENCE, PERCEPTION, AND PHYTOPLANKTON

In 1976, a young Penny Chisholm landed a position as a professor at the Massachusetts Institute of Technology (MIT) in the Department of Civil and Environmental Engineering, where she works to this day. She began as a professor at a time when less than 10% of MIT's science faculty were women (3). She didn't have an MIT Ph.D. like many of her colleagues. She was the only woman and the only biologist in the entire department. She differed so much from the expected image of an MIT professor that when former lab member Marjorie Aelion, now professor and dean of the School of Public Health and Health Sciences at University of Massachusetts Amherst, first met Chisholm, this was her impression:

"When I went to my interview with Penny to be accepted into the program and work with her on research, she was not there. The secretary, Bea Hanrahan, told me she had just stepped out and to go down the hall and I would find her. I walked up and down the hallway, getting more confused and concerned that I was getting late for this important interview, not understanding why I could not find her. After trying to find Penny several times I went back to Bea and told her I must have misunderstood her. Lo and behold Penny had returned to her office, and Bea introduced us. I had passed Penny several times in the hallway and not known it. She was at the start of her career and looked very young (younger than I looked) and was wearing blue-jean overalls, as I recall. I assumed she was a new student in the department or an undergraduate student helping a graduate student."

How did Professor Chisholm earn the respect of her academic peers when she looked like a female undergraduate student and had a biology degree from a state university? Those details may have made her job harder, but Chisholm had the initiative and innovative thinking that mattered. She had a responsibility to *Prochlorococcus*, to tell its story. Her desire to understand what makes this organism grow and thrive in the ocean kept her focused and moving forward. It is easy to dismiss an inexperienced professor, especially when her mistakes can be discriminatorily blamed on her gender and her education. It is harder to dismiss a new professor when she keeps pushing the boundaries of scientific knowledge with her research.

Keeping her head down and her eye on the future, Chisholm has often been at the forefront of her field and has kept up with emerging technologies. Throughout her career, she has asked questions the technology of the time was not yet able to answer, putting her ahead of the game when the technology eventually evolved. In 2003, soon after genomic sequencing became available, Chisholm and her lab sequenced not one but two strains of *Prochlorococcus*. In comparing the genomes, her lab revealed the integral genetic complexity that allows the organism to survive in a plethora of niches (4). Without Chisholm's foresight, the genetic variability would have otherwise been overlooked. To this day, the Chisholm lab continues to sequence new isolates, with each new genome adding, on average, 160 undiscovered genes to the *Prochlorococcus* pangenome (5). Always mindful of the complete ecological puzzle of *Prochlorococcus*, the Chisholm lab has also sequenced and studied multiple phages and their impact on the cyanobacterium (6). Fueled by burning questions, Penny Chisholm strove to prove the importance of this tiny phytoplankton, and in so doing she also proved her worth as a young female scientist.

After decades of work, Penny Chisholm has finally been recognized for her contributions to the body of scientific knowledge and to the science community. In 2003, she was elected to the National Academy of Sciences, and 7 years later, she became the only female recipient of the Alexander Agassiz Medal for her outstanding work in oceanography. She received the 2011 National Medal of Science from President Obama for her achievements in marine microbiology, and in 2015, she became an Institute Professor at MIT, one of the institution's highest honors, one that has been awarded to a female professor less than 10% of the time.

In the 1990s, before most of these awards, Chisholm used her seniority in the scientific community to push for gender equality. In collaboration with other MIT faculty, she fought long and hard to improve the faculty gender ratio at MIT. For decades, MIT maintained a male-to-female faculty ratio of about 9:1 (3). In 1995, Chisholm became a founding member of a committee on female science faculty consisting of 10 senior faculty, including 7 women from across MIT departments (3). Chisholm served on the First and Second Committees on Women Faculty in the School of Science. The work of these committees revealed that although women may feel supported as junior faculty, senior faculty members became more marginalized as they entered into competition with their male colleagues (3). The committees advocated for real, applicable changes the administration could enact to not only increase the number of women employed at the institution but also improve the number of female employees retained. According to a 2011 report from MIT, their work made a difference, lifting the percentage of female science faculty to 21% (7)—a difference but not a complete solution. There is still more work to be done, and Chisholm continues to fiercely advocate for women in science. More recently she spoke at a workshop for the Society of Women in Marine Science where she discussed challenges still affecting women in science today, like imposter syndrome (7). As a pioneering woman in science, Professor Penny Chisholm is a role model and source of boundless energy and inspiration to many young female scientists.

THE CONTINUING STORY

When trying to determine what to study, Chisholm advises her lab, "Let the organism tell you what's important," as Steven Biller, a research scientist in Penny's lab, recalled. When there is as immense and diverse a collection of data on *Prochlorococcus* as the Chisholm lab has accumulated, it can be difficult to distill the details into a coherent narrative. But in conversing with current and former members of the Chisholm lab, it became clear that Chisholm is adept at stepping back to see the big picture. "I like to think it's out there, floating around, living its life, and we're poking away at it and it's trying to tell us things about what its life is like and how it evolved over eons," said Chisholm. "Taking that disparate information that all my really talented students and postdocs generate and weaving it together into some sort of story is kind of the most fun."

Penny Chisholm's ability to tell stories has helped her to succeed in teaching, too. After working at MIT for 40 years, Chisholm can still be found teaching students introductory biology and the fundamentals of ecology. It struck Chisholm later in her career how her students, having just studied photosynthesis, did not comprehend that the mass of every living thing originated from CO_2 in the atmosphere. This inspired Chisholm to extend her audience outside the scientific community to the general public.

At the start of the 21st century, Chisholm began a collaboration with children's author Molly Bang, author and illustrator of such books as *When Sophie Gets Angry—Really, Really Angry....* Together, the pair has written four books on photosynthesis as part of their *Sunlight Series*. The first book, entitled *Living Sunlight*, explains the power of photosynthesis in shaping our world. The next, *Ocean Sunlight*, covers its importance in the oceans— phytoplankton and the ocean food web. The third installment, *Buried Sunlight*, describes fossil fuels as millions of years of captured sunlight and carbon that humans have released over a couple of centuries. A fourth book, *Rivers of Sunlight: How the Sun Moves Water Around the Earth*, appeared in January 2017 and describes the global water cycle. Teachers across the country use these books to engage their students. The science is rigorous, but the writing style and illustrations make it approachable. And any curious child wanting to know more can find additional resources in the back. Chisholm and Bang wrote these books not just to educate children but also to educate adults about the significance of photosynthesis, interconnectedness of life on our planet, and global change.

PENNY'S PERSONAL TOUCH

Throughout her long, successful career, Penny Chisholm has inspired many other scientists who themselves have gone on to inspire others. "I decided I wanted to work with Penny about two minutes after meeting her. She communicates passion and excitement about her work and has done an amazing job of also bringing this excitement outside of academia to the public," said former lab member Libusha Kelly, assistant professor at the Albert Einstein College of Medicine. Another former student of Chisholm's, Jacob Waldbauer, assistant professor at the University of Chicago, said, "Penny is both extraordinarily perceptive and extraordinarily persistent, and her contributions to the understanding of our planet are a testament to how those virtues can benefit both the scientific enterprise and all of society."

In our interview with Chisholm, she said she most wants to be remembered as a good mentor because "that's what really matters: people." Her research on *Prochlorococcus* may live on in academic literature, but it is her impact on those she has taken the time to help and mentor that will be her lasting legacy. We, the authors of this chapter, have also been directly or indirectly mentored by Penny.

One of the authors of this chapter, Vanja Klepac-Ceraj, is now an assistant professor at Wellesley College. She used to be a graduate student in the lab of Martin Polz in the Department of Civil and Environmental Engineering, working in a lab next to Penny's. Although she was never Penny's graduate student or postdoc, Penny became her unofficial mentor through many spontaneous interactions and discussions in the hallways of the Parsons building. "Working with Penny, I've learned that Penny strongly believes that research and science should be communicated clearly and with passion. She leads by example; her talks are always brilliantly presented with so much excitement about the lab's latest findings on *Prochlorococcus* ecology. She has also encouraged others to do the same by giving seminars to students on what it takes to give a talk. So, when, in 2010, Penny invited me to coteach Fundamentals of Ecology with her, I did not hesitate for a moment. Seeing Penny teach and effortlessly engage students in topics of ecology and climate change has profoundly influenced me as a teacher and scientist. I learned a lot from her way of seeing everything in the context of 'big picture.' She would stress to the students, all future engineers, that the natural world is complex and uncertain and whatever they learn or do, they need to think about the impact across scales. Now, several years later, I still find myself asking, 'How would Penny approach this question? How would she present this concept?' When I teach a classroom full of women at Wellesley College, I often ask myself, 'How would Penny help instill passion for science, critical thinking and thoughtfulness, and a dose of confidence that they too can become successful female scientists?' More than once I catch myself thinking that my daughters and my students will have it much easier than Penny if they choose to become scientists. I am grateful to Penny for her strong involvement in advocating for female scientists and the importance of bringing awareness to gender inequality. While this inequality still persists in many places, I now know that we all, regardless of being male or female, should continue following her footsteps, bringing awareness about gender equity issues and working toward correcting them."

The other author of this chapter, Sophie Rowland, is one of Vanja's students, a grandmentee of Penny Chisholm, if you will. She is another aspiring scientist who has been inspired by this impressively kind and humble woman. "Before interviewing Penny Chisholm, I was not fully aware of who she was. And thank goodness, because I would have been so flustered if I understood the extent of her accomplishments. But she didn't act like this lofty and intimidating figure in science. Instead she took time to talk to me, an undergraduate student she had never met. She was generous enough to not only offer her time, but rescheduled multiple times when conflicts suddenly arose. Chisholm spoke to me with passion for her work. She talked matter-of-factly about her career and her journey, as if everyone were capable of all that she has achieved. Her list of accomplishments is extraordinary, but I am more inspired by her kindness, which I know firsthand she has passed on to her students. To think there are more scientists out there like her makes me stronger in my conviction to pursue microbiology."

Working on this project together, we both had the opportunity to reflect on how we can continue Penny's legacy by treating newer generations with the same kindness, generosity, and respect that she showed us. But for now, Penny told us, "I don't want to retire because it's too interesting; I don't want to miss anything. There's always something new coming around the corner and I don't want to miss out on that." As long as Chisholm is heading her lab, as long as *Prochlorococcus* continues to be a spring of new information, her research will continue. And should Chisholm ever decide to retire, she said she'll "go back to just growing cells and doing simple experiments."

ACKNOWLEDGMENTS

We thank Penny Chisholm (MIT), Libusha Kelly (Albert Einstein School of Medicine), Jacob Waldbauer (University of Chicago), Steven Biller (MIT), and Marjorie Aelion (University of Massachusetts Amherst) for sharing their personal stories with us.

CITATION

Rowland S, Klepac-Ceraj V. 2018. Sallie "Penny" Chisholm and oceans of *Prochlorococcus*, p 19–27. *In* Whitaker RJ, Barton HA (ed), *Women in Microbiology*. American Society for Microbiology, Washington, DC.

References

1. Chisholm SW, Olson RJ, Zettler ER, Goericke R, Waterbury JB, Welschmeyer NA. 1988. A novel free-living prochlorophyte abundant in the oceanic euphotic zone. *Nature* 334:340–343.
2. Biller SJ, Coe A, Chisholm SW. 2016. Torn apart and reunited: impact of a heterotroph on the transcriptome of *Prochlorococcus*. *ISME J* 10:2831–2843.

3. Committee on Women Faculty at Massachusetts Institute of Technology. 1999. *A Study on the Status of Women Faculty in Science at MIT.* Massachusetts Institute of Technology, Cambridge, MA. http://web.mit.edu/fnl/women/women.pdf.

4. Rocap G, Larimer FW, Lamerdin J, Malfatti S, Chain P, Ahlgren NA, Arellano A, Coleman M, Hauser L, Hess WR, Johnson ZI, Land M, Lindell D, Post AF, Regala W, Shah M, Shaw SL, Steglich C, Sullivan MB, Ting CS, Tolonen A, Webb EA, Zinser ER, Chisholm SW. 2003. Genome divergence in two *Prochlorococcus* ecotypes reflects oceanic niche differentiation. *Nature* 424:1042–1047.

5. Biller SJ, Berube PM, Lindell D, Chisholm SW. 2015. *Prochlorococcus*: the structure and function of collective diversity. *Nat Rev Microbiol* 13:13–27.

6. Sullivan MB, Waterbury JB, Chisholm SW. 2003. Cyanophages infecting the oceanic cyanobacterium *Prochlorococcus*. *Nature* 424:1047–1051.

7. Wanucha G. 2014. Women in marine science seize the day. Oceans at MIT. http://oceans.mit.edu/news/featured-stories/women-marine-science. Accessed 10 January 2018.

Women in Microbiology
Edited by Rachel J. Whitaker and Hazel A. Barton
© 2018 American Society for Microbiology. All rights reserved.
doi:10.1128/9781555819545.ch4

Margaret Dayhoff: Catalyst of a Quiet Revolution

4

Jamie E. Henzy[1]

INTRIGUED, IMPRESSED, AND PUZZLED

I have two photos of Margaret Dayhoff near my desk. In one, she is a young, fresh-faced woman with an easy smile and the sparkle of unexplored horizons gleaming in her eyes (Fig. 1, left panel). In the other, she is mature, confident, and somewhat matronly in her sturdy necklace, lapels, and sensible hairstyle (Fig. 1, right panel). In the years between these two photos, roughly 1942 to 1982, she developed methods and approaches that define the field of bioinformatics today. For this, she is widely considered the founder of the field.

Like many researchers, I first became aware of the Dayhoff name as it is referenced in programs that generate multiple sequence alignments. "Dayhoff, M.O." is associated with the point accepted mutation (PAM) amino acid substitution matrix she developed (1), one of the options we're given for making sure the amino acid changes shown in our alignments are the ones most likely to have actually occurred during the sequences' evolution. I admit that I assumed she was a man, since the field of bioinformatics, being a subfield of computer science, has a male, geeky sheen to it. I was not the only one to assume wrongly, as evidenced by a post I found on an online forum on biology. Turbulence Acid (username changed to protect the guilty) says, "Dayhoff is a famous sequence analysis guy *who I cite in my lectures*" (emphasis added). This remark was made in 2002. Hopefully Turbulence Acid has since corrected his error.

---

[1]Department of Biology, Boston College, Chestnut Hill, MA 02467

Figure 1 (Left) Photo of Margaret Dayhoff (then Margaret Oakley) during her high school days. (Right) Photo of Margaret Dayhoff from 1982. Photos courtesy of Ruth Dayhoff and Vincent Brannigan.

When I first learned that the "M" in "Dayhoff, M.O." stood for Margaret, I was intrigued. What was it like to be a woman beginning her scientific career in the 1940s—a time when institutional barriers to a woman's promotion were much greater than they are today, when many universities did not allow women to rise to the rank of full professor or senior scientist? When I read that she is considered the founder of bioinformatics, I was impressed and immediately afterwards puzzled: why was there so little written about this woman, the founder of a scientific field that is growing by the day, becoming more and more relevant to our understanding of life? Our language stumbles when we try to describe her place in history. We find ourselves in the awkward situation of using quotes within quotes: "Margaret Dayhoff, the 'father' of bioinformatics." Of course, "the mother of bioinformatics" has its own problems, suggesting hyperbole, as in "the mother of all storms," and uttered with the local Boston accent, "the *mutha* of bioinformatics" suggests a character you don't want to mess with, who does some impressive bioinformatics. David Lipman, the director of the NCBI, avoided these linguistic stumbling blocks by referring to her as "the mother *and* father of bioinformatics" (2).

Intrigued, impressed, and puzzled that even today, she has not received her due recognition, I've put together a summary of her life. I hope to convince the reader that she deserves more recognition for her role and that her contributions were nothing short of transformative.

HER EARLY LIFE AND EDUCATION

Margaret Belle Oakley was born in Philadelphia, PA, on 11 March 1925, an only child. I'm tempted to guess that the lack of siblings allowed her the space to develop a deep and broad intellect, not having to carve out her own niche. When she was 10 years old, the family moved to the Queens borough of New York City—the perfect environment for leavening her intellect with imagination.

Margaret's college years were framed by WWII. The United States joined the war just as she began studying mathematics at New York University (NYU), which perhaps worked to her advantage, with the war leaving empty seats at universities. When she graduated in 1945 with a B.A. in mathematics, followed by an M.A. in chemistry, the war was coming to a close. She then went uptown, to Columbia University, to do graduate work in quantum chemistry, where she would benefit from the military's rollout of technology and methods developed during the war.

The new Watson IBM Laboratory was founded in early 1945 at Columbia University to provide computer resources for the U.S. military (3). After the war ended, its mission was retooled to that of advancing scientific computing in general, and Columbia researchers were offered training and then access to the computers for their research projects. Dovetailing with the rollout of military computers was another product of the war: an approach to problem solving known as operations research (OR). OR involves the application of mathematical reasoning to complex problems—a sort of precursor to thinking in terms of algorithms. It was first used by Charles Babbage, 19th-century inventor of the first mechanical computer, to figure out the most efficient way to sort mail. OR was further developed during WWI and WWII for solving problems of U.S. military strategy (4). Margaret, with her training in mathematics, became adept at applying OR techniques to her chemical data. As a result, in 1947 she was awarded a fellowship at the Watson Lab. There she learned to write programs on computer punch cards to analyze the resonance energies of small molecules—the subject of her Ph.D. dissertation—setting her on a lifelong path of thinking about ways that computers could be applied to biological data.

DC DAYS

In 1948, at the age of 24, Margaret received her Ph.D., followed shortly by her M.r.s.: she married a physicist, Edward Dayhoff. They had attended the same high school in Queens—Bayside High School, where Margaret had been named valedictorian. As a newlywed, Margaret worked as a research assistant in a chemistry lab for 2 years before the couple moved to Washington, DC, where Edward had been offered a job at the National

Bureau of Standards. In DC, Margaret and Edward collaborated on two projects: their daughters Ruth, born in 1952, and Judith, born in 1955.

Margaret appears to have spent the years between 1952 and 1957 shepherding their two daughters through their early childhood, unaffiliated with any institute or university. This is the point at which Margaret's life could have taken the more traditional route, and you would not be reading about her now. But Edward, as described by his son-in-law, was "endlessly supportive of Margaret's ambitions" (5). Intellectually they were well paired, benefitting from their complementary approaches to science: Margaret was a theorist and Edward an experimentalist, yet their interests converged in their studies of the phenomenon of resonance (Edward worked with magnetic resonance in an effort to determine the fine structure of the hydrogen atom). So in 1957, Margaret plunged back into research, first with a fellowship at the University of Maryland, working with Ellis Lippincott on modeling chemical bonding, and then as a professor of physiology and physics at Georgetown University Medical School.

At around this time, Margaret joined forces with another childhood friend from Queens, Robert Ledley. Margaret's and Ledley's life paths were remarkably intertwined. While Margaret was studying mathematics at NYU, Ledley was uptown at Columbia, studying physics. Afterwards, while Margaret studied quantum chemistry at Columbia, Ledley headed down to NYU to study dentistry. Ledley's real love was still physics, but he was being dutiful to his parents, who had insisted on a field with more assured job opportunities. However, he continued to hoof it uptown for evening classes in physics, and soon after receiving his D.D.S., he earned an M.S. in physics from Columbia. Ledley the physicist married a mathematician, whereas Margaret, trained as a mathematician, married a physicist. Both couples had moved in the early 1950s to the DC area, where Ledley, like Margaret's husband, Edward, worked for the National Bureau of Standards (6).

More importantly, like Margaret, Ledley was keenly interested in the application of computers to biomedical research. In 1960, he chartered the National Biomedical Research Foundation (NBRF) specifically for this purpose. He invited Margaret to join the NBRF, reuniting her with her love of computers. DNA sequencing techniques had not yet been developed, but in 1955, Frederick Sanger had completed the amino acid sequence for bovine insulin. Sanger's degradation method was labor-intensive at both the front and back ends: many copies of the protein were digested, producing a slew of short peptide fragments; the sequences of the fragments partially overlapped and had to be assembled in the correct order to generate the full sequence. In fact, one of Margaret's first projects at the NBRF was to develop a computer program to reassemble such fragments. By 1964, Margaret and Ledley had written FORTRAN programs that could accomplish this task in 5 minutes or less on an IBM 7090 mainframe (7).

At the NBRF Margaret also continued developing programs to analyze biochemical data. For instance, she wrote programs to model the early atmosphere of Earth, gaining insight into the conditions necessary for the origin of life. Her work attracted the interest of a young planetary scientist who collaborated with her to model the atmosphere of his planet of interest, Venus. That scientist was the young Carl Sagan, with whom Margaret published a paper in the journal *Science* in 1967 (8).

THE ATLAS OF PROTEIN SEQUENCE AND STRUCTURE

While Margaret continued sharpening her skills in writing programs to handle large amounts of data, other labs began applying Sanger's technique to their proteins of interest, and sequences started to trickle into the literature, albeit at a snail's pace. It was already becoming quite apparent that proteins related by ancestry (homologs) retained similarity at the sequence level, and biochemists were some of the first to compare sequences to see which regions were most strictly conserved, identifying the likely active sites. Another use of comparing protein sequences involved a major challenge at that time: resolving the genetic code. Ledley was a member of the RNA Tie Club (1954 to 1966), founded by the physicist George Gamov to solve the problem of how a series of nucleotides (adenine, guanine, cytosine, and thymine) could be used to assemble proteins, and they pored over alignments of proteins trying to tease out the code (it was eventually resolved by wet bench methods). At the same time, Emile Zuckerkandl and Linus Pauling were developing the idea that sequence comparisons could tell us something about the rate of evolution and the evolutionary relationships among species (9).

Margaret was better positioned than perhaps anyone to fully appreciate the potential value of converting protein sequences to digital data so that they could be handled *en masse* and compared to reveal insights on how life evolves. She set herself the task of searching the literature for all published sequences, of which there were only ~70 at the time, and organized these into the work she is probably best known for: *The Atlas of Protein Sequence and Structure*, first published in 1965 (10). So began the most productive phase of her career and the inception of the field of bioinformatics.

A LABOR OF LOVE

Each entry in the first edition of the *Atlas* included the name of the protein, how it was obtained, and its sequence in both the three-letter and one-letter amino acid codes (the latter of which Margaret had developed to save computer space and to make the printouts comparing proteins more read-able). This was very useful information for researchers, but the real value lay in the analyses she was able to perform with the digitized data, the results of which were presented in foldout pages in the various editions of the *Atlas* that followed. In addition to PAM matrices, for example, she developed the first methods for computing phylogenetic trees, demonstrating the power of basing phylogenies on protein sequences rather than on morphological data. She was also able to show, by comparing the amino acid sequence of the protein ferrodoxin to itself, that ferrodoxin had evolved by duplication of a shorter sequence motif, thus revealing a method by which proteins could evolve. She became a leading expert on recognizing gene families, which her computer analyses revealed and by which she organized the data in the *Atlas* (9).

The *Atlas* was initially distributed on paper, and later on magnetic tape, to every lab that had contributed a sequence, to the major journals, and to Nobel prize-winning scientists. In general, it was well received; however, there was a tendency to undervalue it, in a way that affected Margaret's career and revealed both gender biases and a dogmatic attitude toward the exper-imental techniques that define the field of molecular biology. When she applied in 1969 for membership to the American Society of Biological Chemists (ASBC), a board member responded, "Personally I believe that you are the kind of person who should become a member of the [Society]. However, the candidate must demonstrate that he or she has done research which is clearly his own. The compilation of the Atlas scarcely fits into this pattern" (11). While this attitude in part likely reflected the perception that women in computing could perform only glorified data entry work, it also reflected a bias against the methods of collecting and comparing data, which were seen as outdated and were associated with musty natural histo-rians, the safari hat set. Another scientist discouraged her application on the grounds that she did not do experimental work, reflecting another bias of the day among wet-lab molecular biologists, who were riding high on the success of their techniques. Experimentation had become sacrosanct to them; any other approach was eyed with suspicion. These attitudes were probably expressed most thornily by James Watson—the poster boy of molecular biology—much to the chagrin of natural historians such as E. O. Wilson (12).

The *Atlas*, however, was ahead of its time, representing a new type of experimental system that the critics failed to grasp as such. The digital data were not associated with a particular research question but could be analyzed from different angles using computer programs, revealing a wealth of new information about life and its evolution (9). It also represented a more passive interaction: nature did not need to be rudely poked and prodded but could reveal secrets when asked politely, through computer programs applied to very large sets of well-organized data. For example, by tabulating the frequencies with which all amino acid substitutions occurred across the entire digitized data set, Margaret was able to estimate the likelihood that any given substitution would be accepted by natural selection (thus, point "accepted" mutation [PAM]). This approach looked at the experiment as done by nature, rather than using the theoretical approach of basing probabilities on how many nucleotide changes would be required for a particular substitution. Thus, it was ironic that her work had been criticized for being too theoretical and speculative, rather than experimental.

Despite others' criticism, Margaret herself fully appreciated the potential insights hidden in this growing collection of sequences, and their value to society, even waxing poetic with a hint of Sagan's dreaminess: "We sift over our fingers the first grains of this great outpouring of information and say to ourselves that the world be helped by it. The *Atlas* is one small link in the chain from biochemistry and mathematics to sociology and medicine" (12). This was bioinformatics, and it was revolutionary.

LEGACY

Margaret died of a heart attack in 1983, a few weeks short of her 58th birthday. While she was not accepted into the ASBC, she did become the first president of the Biophysical Society and a fellow of the American Association for the Advancement of Science, and she served on the editorial boards of several top-tier journals. She also passed along her passion for science to many female students, as well as her daughters: Ruth is a highly respected biomedical informaticist and M.D., and Judith earned a Ph.D. in mathematical biophysics and has written a textbook on neural network architecture (5).

Margaret's work was transformative. She was a woman who was key in converting instruments of war into implements to study life, developing a novel experimental system that continues to unfold and thrive today more than ever. In doing so, she helped to catalyze a sort of new modern synthesis, adding to Darwinian evolution and Mendelian inheritance a revived natural history paradigm that applies a comparative approach, but at the molecular level, to sequences, on a grander, more productive scale. Margaret Dayhoff's story is inspiring in its demonstration of the scale of impact that women can have on science.

CITATION

Henzy JE. 2018. Margaret Dayhoff: catalyst of a quiet revolution, p 29–36. *In* Whitaker RJ, Barton HA (ed), *Women in Microbiology*. American Society for Microbiology, Washington, DC.

References

1. Dayhoff MO, Schwartz RM, Orcutt BC. 1978. *Atlas of Protein Sequence and Structure*, vol 5, p 345–352. National Biomedical Research Foundation, Washington, DC.
2. Wikipedia contributors. 2017. Bioinformatics, on *Wikipedia, The Free Encyclopedia*. https://en.wikipedia.org/wiki/Bioinformatics. Accessed 25 August 2017.
3. Da Cruz F. 2013. Columbia University computing history. http://www.columbia.edu/cu/computinghistory/watsonlab.html. Accessed 25 August 2017.
4. Wikipedia contributors. 2017. Operations research, on *Wikipedia, The Free Encyclopedia*. https://en.wikipedia.org/wiki/Operations_research. Accessed 25 August 2017.
5. Brannigan V. 7 March 2013. Grandma got STEM: Margaret and Ruth Dayhoff. https://ggstem.wordpress.com/2013/03/07/margaret-ruth-dayhoff/. Accessed 25 August 2017.
6. Wikipedia contributors. 2017. Robert Ledley, on *Wikipedia: The Free Encyclopedia*. https://en.wikipedia.org/wiki/Robert_Ledley. Accessed 25 August 2017.
7. Dayhoff MO, Ledley RS. 1962. Comprotein: a computer program to aid primary protein structure determination. *Proceedings of the FJCC*. AFIPS, Santa Monica, CA.
8. Dayhoff MO, Eck RV, Lippincott ER, Sagan C. 1967. Venus: atmospheric evolution. *Science* **155**:556–558.
9. Strasser BJ. 2010. Collecting, comparing, and computing sequences: the making of Margaret O. Dayhoff's *Atlas of Protein Sequence and Structure*, 1954–1965. *J Hist Biol* **43**:623–660.
10. Dayhoff MO, Eck RV, Chang MA, Sochard MR. 1965. *Atlas of Protein Sequence and Structure*. National Biomedical Research Foundation, Silver Spring, MD.
11. Strasser BJ. 2006. Collecting and experimenting: the moral economies of biological research, 1960s–1980s. *Preprints of the Max-Planck Institute for the History of Science* **310**:105–123.
12. Wilson EO. 1994. *Naturalist*. Island Press, Washington, DC.

Women in Microbiology
Edited by Rachel J. Whitaker and Hazel A. Barton
© 2018 American Society for Microbiology. All rights reserved.
doi:10.1128/9781555819545.ch5

Johanna Döbereiner: A Pioneer Among South American Scientists 5

Graciela Brelles-Mariño[1]

There is in every true woman's heart, a spark of heavenly fire, which lies dormant in the broad daylight of prosperity, but which kindles up and beams and blazes in the dark hour of adversity.

> WASHINGTON IRVING, **American writer,**
> **biographer, and historian**

Johanna Döbereiner was a pioneer in the field of plant-microbe interactions. Internationally, she was the most cited female Brazilian scientist and the seventh most cited Brazilian scientist according to the *Folha de S. Paulo* newspaper (cited in reference 1). At least three generations of scientists have been trained under Johanna's guidance. Her former coworkers stated that "She was a woman with a very strong personality, sincere, and dedicated to work. Despite her rigorous treatment of scientific aspects, she behaved always like a mother to all students and trainees in our center. In Brazil she was known as the 'soil doctor', although for those that had the privilege to work with her, we regarded her as our 'Aunt Johanna'" (reprinted from reference 1). This comment shows the supportive, nurturing, almost motherly environment Johanna created in her workplace.

A good picture of Dr. Döbereiner's personal and professional approach is given by Dr. Veronica M. Reis, currently a researcher at Embrapa (Brazilian

---

[1]Center for Research and Development on Industrial Fermentations, Universidad Nacional de La Plata, Facultad de Ciencias Exactas, La Plata, Buenos Aires B1900AJL, Argentina

Figure 1 Dr. Johanna Döbereiner.

Agricultural Research Corporation) Agrobiology Unit and a professor of the graduate program in soil science at the Federal Rural University of Rio de Janeiro. Dr. Reis was Johanna's Ph.D. student, working on the infection and detection methods of the endophyte bacterium *Gluconacetobacter diazotrophicus* (formerly *Acetobacter*) in association with sugarcane. Veronica remembers that Johanna personally guided her students in the laboratory and even checked their petri dishes in the hope of discovering new species of nitrogen-fixing bacteria. At that time, before the advent of molecular biology, bacterial species were described mostly based on morphological and physiological characteristics of the isolates. She loved discussing data and was very good at organizing them. Johanna was very happy in her job and worked until the last days of her life. Dr. Reis says that Johanna was very savvy working in the lab and always taught her students to not waste and to perform their research without using expensive devices or imported reagents. For many years she used an old typewriter, and she started using a computer only at the end of her life. She is deeply missed by all who shared their daily lives with her (V. M. Reis, personal communication, 2016).

Johanna developed a scientific career that lasted about 50 years, starting at a time when it was not common for female scientists to succeed. What was her motivation to overcome such difficulties? Where did she get the strength to build a career at a time in which women were mostly expected to become housewives, especially in South American countries? I believe the quote of Washington Irving at the beginning of this chapter applies very much to Johanna. Maybe the key was her difficult early life.

Figure 2 Dr. Döbereiner in front of the building where she worked for almost 50 years at Embrapa, Seropédica, Brazil. Picture courtesy of Dr. Veronica M. Reis.

EARLY LIFE

Johanna Liesbeth Döbereiner (née Kubelka) was born 28 November 1924 in Ústí nad Labem (or Aussig, in German) in the former Czechoslovakia. The city is presently part of the Czech Republic but used to be in the Austro-Hungarian Empire; it became part of Czechoslovakia in 1918. As the majority of the population was of German origin, the language mostly spoken was German.

During World War II, Johanna was forcibly separated from her family and for about 4 years was able to see her parents and grandparents only occasionally. She worked as a peasant milking cows and spreading manure in farms and orchards. After the war ended, the city of Aussig became again part of restored Czechoslovakia, and most of the German population was forced to leave the country. The family moved to Prague, where Johanna's father, Paul Kubelka, was a physical chemistry professor at a German university. Mr. Kubelka also owned a small factory that produced chemicals for agricultural use. Her mother died in a concentration camp in Prague, and Johanna worked to sustain her grandparents. When her grandparents died, she moved to Germany, where she continued working on farms. She was then accepted to the School of Agriculture at the Ludwig Maximilian University of Munich and graduated in 1950 as an agricultural engineer. At the University of Munich she met a veterinary student, Jürgen Döbereiner, who became her husband in 1950.

FAMILY LIFE AND THE BEGINNING OF
A SCIENTIFIC CAREER IN BRAZIL

Johanna's widowed father and a brother migrated to Brazil in 1948, with the help of foreign teachers located in the country (2). The recently married Jürgen and Johanna Döbereiner followed in 1951, and Johanna was appointed at what is now known as the National Center of Education and Agricultural Research (Embrapa) of the Ministry of Agriculture, in Seropédica, 47 km away from Rio de Janeiro. She became a Brazilian citizen in 1956 and developed a career that lasted until her death at 75 years of age. Johanna deftly balanced family life with an outstanding scientific career. In a television interview, Jürgen, considered a pioneer in the study of toxic plants in Brazil (3), said that Johanna was able to combine work with dedication to her family and that they never had any problem regarding educating their three children, despite both having successful scientific careers.

Dr. Döbereiner's field of research was primarily biological nitrogen fixation (BNF). Nitrogen is an indispensable element for plant growth, and it is typically a limited nutrient in soils. BNF, the second most important process in terrestrial ecosystems after photosynthesis, is carried out only by prokaryotes and involves the reduction of atmospheric nitrogen to ammonia, whereupon it becomes available to plants. In developed countries, nitrogen input for agricultural crops relies on the use of synthetic fertilizers that are environmentally and economically costly since they are derived from petroleum, while the excess of nitrate contaminates subterranean sources of water. Fixation of nitrogen through BNF is less costly and is environmentally friendly, playing a significant role in the production of crops such as sugarcane, sorghum, and rice, where it helps restore impoverished soils. The nitrogen-fixing microbes can exist as free-living organisms or in associations ranging from loose associations and associative symbioses to complex symbiotic associations with legume plants.

Johanna joined the soil microbiology laboratory at (presently) Embrapa and started working on BNF and nitrogen-fixing bacteria, more specifically in associative symbioses. She was not an expert in that field, since she knew about BNF only from her agronomy courses in Munich; however, her work on the soil microorganism *Azospirillum* (formerly *Spirillum*) was pioneering, and she became an advocate for the use of BNF as a natural way to enhance Brazil's soybean production.

In the late 1950s and 1960s, she discovered two new nitrogen-fixing bacteria associated with the rhizosphere of gramineous plants: *Beijerinckia fluminensis* (1958) with sugarcane and *Azotobacter paspali* (1966) with

Paspalum notatum cv. batatais. These studies on free-living and rhizospheric nitrogen-fixing bacteria were carried out mostly between 1950 and 1970, when there were very few reports on the presence of these bacteria in tropical soils (4).

In 1946, Brazil became the first country in Latin America to start producing soybean in large amounts. That year, production was more than 10,000 metric tons; by 1949, the production reached 30,000 metric tons, and it continued growing to 1,057,000 metric tons in 1969 (5). In the 1960s, Johanna's work was influential in the development of the Brazilian Program on Soybean Amelioration, which was based on BNF. This approach was the opposite of that being followed at that time by the United States, whose soybean production heavily relied on the use of chemical fertilizers.

A great deal of effort has been put into finding nitrogen-fixing organisms capable of providing nitrogen to agriculturally significant crops. One of the crops that appear to benefit from their associations with microbes is sugarcane. Johanna and her group realized that sugarcane yields in Brazil were very good in spite of little input of nitrogen fertilizer; some sugarcane varieties could obtain around 60% of their nitrogen through BNF. During the 1980s, her group discovered that sugarcane contained a microorganism that they named *Acetobacter diazotrophicus* (1988), presently *Gluconacetobacter diazotrophicus*. The group also isolated another endophyte (nitrogen-fixing bacterium able to colonize the interior of plant tissues) called *Herbaspirillum seropedicae* from maize, sorghum, and rice. Several other endophytes were discovered by Johanna's group. The term endophyte was first introduced to the area of nitrogen fixation research associated with graminaceous plants in 1992 by Döbereiner (reviewed in reference 6). The term includes all microorganisms that are able to colonize the inner tissues of plants, promoting plant growth without causing any disease symptoms. As a very brief summary, Johanna's group was involved in the discovery of several nitrogen-fixing bacteria, such as the rhizospheric *Beijerinckia fluminensis* and *Azotobacter paspali*, associative *Azospirillum lipoferum*, *Azospirillum brasilense*, and *Azospirillum amazonense*, and the endophytic *Herbaspirillum seropedicae*, *Herbaspirillum rubrisubalbicans*, *Gluconacetobacter diazotrophicus*, *Burkholderia brasiliensis*, and *Burkholderia tropica* (6).

Dr. Vera Lucia Divan Baldani was a Ph.D. student in Johanna's laboratory and is presently a researcher at Embrapa Agrobiology. When asked about the contribution of Johanna to science and in particular to soil microbiology, Dr. Baldani stated, "During her 49 years (1951–2000) devoted to research in biological nitrogen fixation, sustainable production of food, and environmental awareness, Johanna led and contributed to the implementation of the process of BNF in legumes and non-legume plants in Brazil. Her contribution was instrumental in the Brazilian Program on Soybean Amelioration, which started in 1964, and aimed at substituting (synthetic) nitrogenous supplies with biological supplies. The program was successful and now Brazil is the second main soybean producer in the world and saves billions of dollars in nitrogen fertilizers. During the energy crisis in the nineteen seventies, she led and promoted the search of alternatives to nitrogen fertilizers for non-legume plants. This resulted in the discovery of nine nitrogen fixing species, including *Azospirillum* spp., studied and used worldwide as an inoculant for corn and wheat" (V. L. D. Baldani, personal communication, 2016).

THE PERSON BEHIND THE SCIENTIST

We can learn about a person's scientific achievements by reading papers, but the real dimensions of a human being, or at least glimpses of them, come from the memories of those who shared some aspects of his or her life. There are plenty of anecdotes and stories about Johanna, and the ones below are just a sample.

Prof. María Valdés is an award-winning researcher and academician from the National Polytechnic Institute, Mexico, who said, "Johanna was a great woman and a great scientist with whom I developed a very nice friendship. We organized the first course on *Azospirillum* outside Brazil, in my laboratory in February 1979 and sponsored by CONACTY (Mexico) and CNPQ (Brazil). But I met her much earlier during the RELAR (Rhizobiology Latin American Meeting) meetings. Thanks to her, Brazil produces a good amount of ethanol from sugarcane without nitrogen fertilizer and with nitrogen-fixing organisms. When the Nitrogen Fixation Research Center (CIFN; now the Center for Genomic Sciences in Cuernavaca, Mexico) was inaugurated in Cuernavaca, Johanna was the 'big name' invited. When she realized I was not there, she stated: if Maria does not come, I won't be present for the inauguration either. She was the first one working in BNF in Mexico. Then, she put everybody to work to find me" (M. Valdés, personal communication, 2016). This anecdote shows the honesty and unselfishness of Johanna.

Dr. Fábio Bueno dos Reis Junior, presently a researcher at Embrapa Cerrados, was Johanna's student. He said, "She supervised me at the beginning of my scientific career, as an undergraduate and MSc student and she had a great impact on my life, not only as a scientist but in a general

Figure 3 Dr. María Valdés (in dark jacket) kindly greeting Dr. Döbereiner at the former Nitrogen Fixation Research Center in Cuernavaca, Mexico. The other person in the picture is the late researcher Dr. Jesús Caballero-Mellado. Picture courtesy of Prof. María Valdés.

sense. She was a model of leadership, ethics, and enthusiasm. I remember an interesting anecdote that happened at the beginning of my undergraduate research with Dr. Johanna, in 1993. I started working in the laboratory with no fellowship and she asked me to go to her office to talk about it. She asked me if I would like to have a scientific initiation fellowship and I said yes. Then, she picked up the phone and called to Brasília, to the CNPQ (National Council of Technological and Scientific Development) head-quarters and talked with the then-president of the institution just to ask for a fellowship for her new undergraduate student. That's how I got my first fellowship to start working in scientific investigation" (F. B. dos Reis Junior, personal communication, 2016).

Johanna was a strong woman with a rather difficult personality at times, and that was probably the reason she survived her hard early life and succeeded as a scientist in an environment dominated by males; however, she deeply cared about the people who worked with her. A researcher whose name will not be disclosed remembered that when her daughter was 8 months old, her marriage was going through a crisis and she felt like ending it. One afternoon Johanna, who was her advisor, unexpectedly arrived at her home and spoke to her for hours…. She is still married after almost 40 years.

Regarding Dr. Döbereiner's personality, I asked Dr. Vera Baldani if her character was strong. "Strong?" she replied. "It was like iron!" She remem-

Figure 4 Dr. Döbereiner and Dr. Fabio Lopes Olivares during a break in the lab in 1998 celebrating the 100th anniversary of the Vasco da Gama soccer team from Rio de Janeiro. Dr. dos Reis Junior remembers, "Students in the lab were saying that Dr. Johanna was also a supporter of our soccer team, but in fact I think that she was not interested in football at all." Picture courtesy of Dr. Fábio B. dos Reis Junior.

bered, "I started as a master's student without having ever read a scientific publication. Johanna was my advisor and when I went to her office to talk to her, she gave me five papers and asked me to come back in two days. But I didn't understand a word of those papers! I went back to the office and Johanna said we would discuss biological nitrogen fixation in non-legume plants. She asked me several questions that I was unable to respond. Then she said: are you dumb or what? Get out of here and come back once you learned. I started crying right there and went back home to read and learn." Nowadays we would consider Johanna's response politically incorrect, but Dr. Baldani's conclusion is quite different. She said that "through all those things, I have learned that we only fight with those we love. If we don't love, we simply don't care, we ignore."

Avilio Franco and Robert Boddey wrote, "Yet above all, those who understand Dr. Döbereiner's strong personality, prize her friendship, encouragement, and her capacity to face work as happy and enthusiastic as a person going on holiday" (reprinted from reference 4).

AWARDS

Johanna was an outstanding scientist and the recipient of many honors and awards during her long and fruitful career. She was one of the founding

Figure 5 Celebrating Dr. dos Reis Junior's 24th birthday in 1994. Those celebrations were frequent at Embrapa Agrobiology, and Johanna always attended celebrations and interacted with students and younger researchers. Front row (left to right): Dr. Veronica Reis (seated on railing), Dr. Vera Baldani (yellow shirt), Dr. Lucia da Silva (white shirt), and Dr. Johanna Döbereiner. Middle row (left to right): Dr. Leonardo Cruz (white T-shirt), Dr. Jean Araujo (brown T-shirt), Dr. Claudia Martins (gray T-shirt), Dr. Fábio Reis Junior (white T-shirt) and Dr. Elyson Amaral. Back row (left to right): Dr. José Antônio Pereira, Dr. Marcelo da Silva, Dr. Claudio Cunha, and Dr. Fábio Olivares. Picture courtesy of Dr. Fábio B. dos Reis Junior.

members of the Third World Academy of Sciences in 1981 and a member of the New York Academy of Sciences and the Brazilian Academy of Sciences (1977). She was also honored with a membership in the Vatican Pontifical Academy of Sciences (1978). The Pontifical Academicians are 80 members from diverse countries who have made outstanding contributions to science and are nominated by the Pope. Prizes and awards Johanna received include the Embrapa Frederico Menezes Veiga prize (1976), Research Agriculture of Today prize (1977), the Bernard Houssay Prize from the Organization of American States (OAS) (1979), the Prize for Science of UNESCO (1989), and the Mexican Prize for Science and Technology (1992). She was named Doctor "Honoris Causa" by the Federal Rural University of Rio de Janeiro and the University of Florida (7). She was even nominated for the Nobel

Figure 6 Johanna (on the far right) visiting the soil microbiology lab at Embrapa Cerrados in Brasília in 1989. Back row (left to right): Dr. Milton Alexandre Teixeira Vargas, Dr. Ieda de Carvalho Mendes, and Dr. José Roberto Rodrigues Peres (white shirt, black tie), her former student and now the general head at Embrapa Cerrados.

Prize in Chemistry in 1997, and although she did not win it, the nomination garnered a lot of attention from the news and the admiration of her peers, deservedly so.

Johanna's legacy goes far beyond the discovery of bacterial species and the more than 300 papers she wrote. She made tremendous contributions to science over her long career, which sustained the development of science focusing on agricultural applications in Brazil. As a consequence of her findings, several lines of research were further initiated, and other groups working on BNF with nonleguminous plants were established in the country. Currently, besides the Embrapa team, there are groups in the states of Paraná, Rio Grande do Sul, Rio de Janeiro, Minas Gerais, Goiás, Ceará, and Distrito Federal (6).

Some of Johanna's contributions also impacted the development of the Brazilian economy. As mentioned earlier, her contribution was influential for launching the Brazilian Program on Soybean Amelioration, aimed at replacing synthetic fertilizers with biofertilizers based on BNF. Her findings were also instrumental in the development of the bioethanol industry. Brazil is considered an international leader in terms of sustainable production of biofuels. Together Brazil and the United States lead the industrial production of ethanol fuel. The ethanol industry in Brazil uses sugarcane that is produced with no input of synthetic nitrogen fertilizer and relies on some of the endophytes discovered by Johanna. This way of producing ethanol is more efficient and cheaper than that of the U.S. industry based on the use of corn, synthetic fertilizers, and subsidies to farmers.

Johanna Döbereiner passed away on 5 October 2000 and is remembered by all those whose lives she touched, not only for her scientific achievements but also for her leadership, ethics, and enthusiasm.

ACKNOWLEDGMENTS

I am thankful to Drs. Veronica M. Reis, Fábio Bueno dos Reis Junior, María Valdés, and Vera L. D. Baldani for sharing their memories and pictures from their personal collections and Drs. Veronica M. Reis, Fábio Bueno dos Reis Junior, and Eduardo C. Schröder for critically reading the manuscript.

CITATION

Brelles-Mariño G. 2018. Johanna Döbereiner: a pioneer among South American scientists, p 37–47. *In* Whitaker RJ, Barton HA (ed), *Women in Microbiology*. American Society for Microbiology, Washington, DC.

References

1. **Baldani JI, Baldani VLD, Reis VM.** 2001. Johanna Döbereiner: fifty years dedicated to the biological nitrogen fixation research area. Johanna Döbereiner Memorial Lecture, p 2–4 *In* Finan TM, O'Brian MR, Layzell DB, Vessey JK, Newton W (ed), *Nitrogen Fixation: Global Perspective. Proceedings of the 13th International Congress on Nitrogen Fixation.* CABI Publishing, New York, NY.

2. **The Pontifical Academy of Sciences.** Deceased academicians. http://www.casinapioiv. va/content/accademia/en/academicians/deceased.html. Accessed 14 November 2017.

3. **Schield AL.** 2011. Dedication to Jürgen Döbereiner, p xv–xvi. *In* Riet-Correa F, Fister J, Schield AL, Wierenga T (ed), *Poisoning by Plants, Mycotoxins and Related Toxins.* CABI Publishing, Cambridge, MA.

4. **Franco A, Boddey R.** 1997. Dr. Johanna Döbereiner: a brief biography. *Soil Biol Biochem* 29:IX–XI.

5. **Shurtleff W, Aoyahi A.** 2004. *History of Soy in Latin America.* Soyfoods Center, Lafayette, CA. http://www.soyinfocenter.com/HSS/latin_america1.php. Accessed 14 November 2017.

6. **Baldani JI, Baldani VLD.** 2005. History on the biological nitrogen fixation research in graminaceous plants: special emphasis on the Brazilian experience. *An Acad Bras Cienc* 77:549–579.

7. **American Society of Plant Biologists.** Women pioneers in plant biology. http://aspb. org/wipb-pioneer-biographies/. Accessed 14 November 2017.

Johanna Döbereiner passed away on 5 October 2000 and is remembered by all those whose lives she touched, not only for her scientific achievements but also for her leadership, ethics and enthusiasm.

ACKNOWLEDGMENTS

I am thankful to the Vreericke M. Reis, Fábio Sousa, the Bém Jones, Marta Valle, and Vera T. D. Baldani for sharing their memories and pictures from their personal collections and Eric Vreericke M. Reis, Fábio Braveromo Kox Jones, and Eduardo C. Sebba for critically reading the manuscript.

CITATION

Balla-Marino C. 2019. Johanna Döbereiner: a pioneer among South American scientists, p 57–72. *In* Whitaker RJ, Barron HA (ed), *Women in Microbiology*. American Society for Microbiology, Washington, DC.

References

Women in Microbiology
Edited by Rachel J. Whitaker and Hazel A. Barton
© 2018 American Society for Microbiology. All rights reserved.
doi:10.1128/9781555819545.ch6

Diana Downs: A Path of Creativity, Persistence, and Rigorous Testing

6

Norma C. Martinez-Gomez[1]

The excitement of putting things together began in Diana's childhood: she was a very curious and creative child who liked to experiment and observe nature. Her ambition to become a scientist started when she was a college student, and since then she has dedicated her academic career to dissecting metabolic integration. Diana started her research career describing a novel mechanism for induction of a cryptic phage in *Salmonella enterica* serovar Typhimurium (1). Her data seemed to call into question the dogma of the time, and being a student, she initially assumed that her conclusions were wrong. However, after repeating experiments and controls, she embraced the fact that her data were correct and considered the possibility that the dogma needed to be tweaked. When describing these striking observations, Diana mentions an important reflection she had of those times as a young investigator: "…believe your data enough, even if no one has reported or seen it before you. Be confident in your training and experimentation; otherwise, discovery may be blunted." Diana obtained her Ph.D. soon after she published these results, which was the beginning of a research career full of intriguing observations and rigorous testing that often challenge established notions that lack validation.

Joining the Downs laboratory has been the most challenging, engaging, and transformative experience of my academic and personal life. Diana is a researcher who loves genetics, and she teaches and trains all her students with an incredible passion and energy. The positive feedback between her

[1]Department of Microbiology and Molecular Genetics, Michigan State University, East Lansing, MI 48864

high expectations and infectious motivation provided a unique environment for learning and achievement. Before I met Diana, I thought of metabolism in a unidimensional manner, mostly focused on the details of a chemical mechanism of a particular enzyme or pathway. In contrast, her research taught me to see metabolism as a multidimensional, complex, and highly interconnected sum of metabolic pathways. System-level approaches have advanced at such rates that we often find ourselves overwhelmed with large data sets, trying to put seemingly unrelated pieces together to generate a larger story. Diana has a rare gift for translating these complex data sets into testable models. By integrating classical genetic approaches, genomics, and rigorous biochemical studies, her work has identified unpredicted metabolic connections and described robustness of the metabolic network when perturbed (2–4). These discoveries are the result of her drive to understand the "exceptions," or pieces of the metabolic puzzle that do not fit a model. In an era where technology, global techniques, and modeling are abundant, her work reminds us to remain skeptical and rigorously validate new claims.

Identifying the many interconnections among metabolic networks is a challenging task that requires unwavering persistence. Diana is able to motivate students for the challenge by nurturing passion and curiosity. I have always been inspired by the creativity and elegance of the experimental designs that Diana and her team members implement, particularly when using classical genetic approaches. If an experiment is successful, Diana's response is full of a contagious excitement driving students to envision a model that can be rigorously tested. When the results of the experiment are unexpected or target a gene of unknown function, Diana will say, "Good. You just gotta bang your head against the wall," encouraging the student to think about the result from many different angles and design creative experiments to test those hypotheses. If the results suggest that a model is wrong, Diana highlights what we have learned from it, allowing us to quickly move on to design a new model. The emotional highs for intriguing observations inherently allow her students to enjoy science. Consistent with her mentoring style, Diana has a poster of Albert Einstein in her office stating, "Imagination is more important than knowledge." Discussions across lab benches and chalk talks during lab meetings are common. Diana encourages her students to challenge one another while promoting respect for the peer-review process even at early stages of a project. By integrating genetic and phenotypic analyses, her research has defined functional characterization of numerous enzymes (5–11). As a result, she has challenged commonly accepted practices in research that are not always thoroughly tested, including the assumption of enzyme function and chemical mechanisms based on gene annotation (12, 13) and the assumption that conservation of metabolic components can be used to accurately predict network structure and function (14).

One cannot describe Diana's research without mentioning her as a mentor and role model. When asked about her mentoring skills, her students immediately mention phrases such as "endless enthusiasm," "accessibility," and "engagement," but the first and most commonly used word is "passionate." One student commented that "Diana's spirited way of interacting with me encouraged me to throw myself into my project." Another said, "'Get excited, this is awesome!' she sometimes demanded with a huge smile on her face." Another student noted that "...the one characteristic Diana has above everyone else is her enthusiasm for science. Even on a bad day she is always optimistic about your research and will give you the drive to want to do more. I can't explain it very well, but she can motivate you to do a seemingly impossible experiment just for the pure joy of understanding science." This excitement is exemplified by Diana's daily habit of arriving to the lab earlier than anyone else to peek inside the incubator and check all the plates growing there. Diana knew our phenotypic results before we even arrived to lab. She simply could not wait to hear about a result. I remember many times having just taken my coat off and Diana coming to my office with a big smile. I instantly knew our experiment had worked. As motivational as her energy is, high expectations are also part of the Downs lab daily dynamic to develop research projects. Common utterances, particularly with new students, include, "Where are your controls?...Jeepers! This could have been done yesterday!" and "If you do not have time to do it right, how are you going to have time to do it again?" Her passion for logic and problem solving coupled with the freedom to test hypotheses shapes her lab's dynamic: a jovial, bustling research center with a full house of friendly, energetic students who can't stop talking about the perplexities of metabolic integration. Diana's students hold a weekly "Secret Science Hour (SSH!)" meeting to discuss projects in a more casual format. Those meetings are held without Diana and reflect the environment of enthusiasm and ownership over projects that she encourages. Her phrases are used by many of her mentees and have made such an impact in our own mentoring styles that one year for her birthday, her students designed a dichotomous key to guide us to the advice Diana might give, called "What Would Diana Say?" (Fig. 1). Diana's mentoring style promotes independence, curiosity, creativity, and

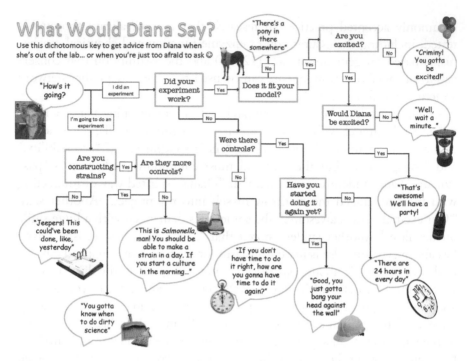

Figure 1 Diagram made by students from the Downs laboratory giving an overview of Diana's mentoring style.

confidence. The last is an important outcome, particularly when women are pursuing research careers. For example, when interviewing to become an assistant professor and having conversations with professors who were experts in areas far different from my own, I found myself in some cases with an initial feeling of insecurity that hindered my participation in conversations. Thanks to my training, I was able to break through these moments of impostor syndrome by taking a step back to look at the bigger picture. Rather than getting overwhelmed by the details that I didn't know, I saw these conversations as doorways into new areas of scientific knowledge. Inevitably excitement replaced my insecurities, and I organically transitioned to critical and logical thinking, as I had done many times before as a member of the Downs lab. By doing so I was able to ask insightful questions and learn from experts about a wide range of topics. When I have prepared manuscripts, the same logical thinking, integration of results, and generation of testable models have been crucial to achieving confidence in my own work and ideas.

Diana empowers students to fulfill their potential by focusing training on critical thinking while letting students broaden their areas of expertise. If a student's passion is teaching, she organizes his/her projects such that they can balance teaching opportunities with research. If their interests are geared towards the biotechnological industry, she encourages applications to funding opportunities that promote short-term internships. Her former mentees include faculty members of biology, geology, biochemistry, and microbiology departments, successful scientists in the biotechnological industry, and excellent teachers at small universities.

Diana's commitment to outstanding mentorship goes beyond graduate training. Diana has provided insights to numerous former students in each career transition they have experienced, including during the transition to a postdoctoral research position, during the tenure process, or when transitioning to industry. She constantly reminds us to look for work environments that provide scientific growth and independence. Importantly, she shares strategies to overcome specific personal challenges, such as starting a family or moving across the country with small children. At the beginning of my postdoctoral career, my daughter, then 5, was struggling with many changes in her life, including new friends and a new school. She was faced with these challenges at the same time that I needed to work the longest hours in the lab and focus intensely on my research. It was a moment in my life when I felt gigantic guilt, as I knew my daughter needed me at home. I called Diana. First she shared with me that early in her career she also dealt with the emotional turmoil of sacrificing the "now" with her young children for her future, and how it tore her heart out on an almost daily basis. I no longer felt alone. Then, she gave me useful advice to reach the balance I needed. She suggested that I set aside one day each week for scheduled mommy-daughter time. Incorporating this dedicated time for my daughter into my weekly agenda was a powerful incentive to effectively organize my time, and it showed my daughter that even though I was working very late hours and weekends, she was still a priority. Those special days with my daughter are unforgettable, and my work advanced as I needed at the same time. I am grateful to Diana for her wise advice allowing me to find the balance my family and I needed during a very trying time in my career.

Diana exemplifies a strong female scientist who essentially "has it all" and demonstrates that scientific excellence and a fulfilling personal life are not mutually exclusive. She is very open about sharing her own experience, recognizing that starting a family and a career simultaneously is hard. Doing both inevitably results in times when you must put science in the backseat. At critical stages in your career this can halt your professional trajectory. Therefore, Diana does not recommend attempting to do so during graduate school. Nonetheless, she provides a welcoming, family-friendly environment in her laboratory and encourages prioritization of family and a healthy lifestyle. At the same time, she guides students into developing a successful academic career during these particularly stressful times. When I became pregnant during the last year of my graduate training, Diana worked with me to develop a plan for both bench work and writing with attainable goals. A key component of that plan was outlining future goals for when I came back from maternity leave. In particular, she guided me in planning my transition back to the program, as she understood that stepping back into the lab would be challenging. Diana was mentoring from experience, as she herself went through this process when she was a postdoc and understood how difficult it can be to regain balance. She provided office space for me to set up a playground and allowed me to bring my daughter as often as needed, especially during the first few weeks after I returned to work. Even when those times were overwhelming, Diana made everything seem doable as long as we worked hard and remained focused. She always defies the "you can't" with a "she did." When starting her assistant professorship, she also had very young children and decided to reach balance in both her personal and professional lives by prioritizing doing good science and publishing it. Everything else during those early times, such as traveling to give seminars or attending conferences, she considered peripheral things. Upon reflection, she recognizes that this likely affected her trajectory, but it provided her the family life she wanted, and therefore, she does not regret a day of it. Including her children in her professional life helped, too: they knew the lab on the weekends, they knew when their mom was writing grants, and they celebrated with her when they were submitted. Diana obtained numerous awards and grants and published many papers very early in her career, and this rapid success allowed her to obtain her tenure earlier. We, as her students, could not help but be inspired, and many of us carry on Diana's legacy as a researcher and mentor with "we did, too."

Overall, Diana's research and mentoring are reminders of how extraordinary curiosity and creativity combined with rigorous testing lead to a successful academic path and outstanding discoveries. Her commitment to mentoring continues to shape numerous scientific minds, and her unique mentoring style shows how providing space for creativity and continuous support can empower women to achieve it all.

ACKNOWLEDGMENTS

I extend special thanks to Diana for giving me permission to highlight her story in this chapter. I also thank Jeff Gralnick, Lauren Palmer, Jannell Bazurto, Mark Koenigsknech, and Jodi Enos-Berlage for contributing to the chapter.

CITATION

Martinez-Gomez NC. 2018. Diana Downs: a path of creativity, persistence, and rigorous testing, p 49–56. *In* Whitaker RJ, Barton HA (ed), *Women in Microbiology*. American Society for Microbiology, Washington, DC.

References

1. **Downs DM, Roth JR.** 1987. A novel P22 prophage in *Salmonella typhimurium*. *Genetics* 117:367–380.
2. **Downs DM, Ernst DC.** 2015. From microbiology to cancer biology: the Rid protein family prevents cellular damage caused by endogenously generated reactive nitrogen species. *Mol Microbiol* 96:211–219.
3. **Koenigsknecht MJ, Downs DM.** 2010. Thiamine biosynthesis can be used to dissect metabolic integration. *Trends Microbiol* 18:240–247.
4. **Downs DM.** 2006. Understanding microbial metabolism. *Annu Rev Microbiol* 60:533–559.
5. **Boyd JM, Pierik AJ, Netz DJ, Lill R, Downs DM.** 2008. Bacterial ApbC can bind and effectively transfer iron-sulfur clusters. *Biochemistry* 47:8195–8202.
6. **Skovran E, Downs DM.** 2003. Lack of the ApbC or ApbE protein results in a defect in Fe-S cluster metabolism in *Salmonella enterica* serovar Typhimurium. *J Bacteriol* 185:98–106.
7. **Beck BJ, Downs DM.** 1998. The apbE gene encodes a lipoprotein involved in thiamine synthesis in *Salmonella typhimurium*. *J Bacteriol* 180:885–891.
8. **Gralnick J, Downs D.** 2001. Protection from superoxide damage associated with an increased level of the YggX protein in *Salmonella enterica*. *Proc Natl Acad Sci U S A* 98:8030–8035.
9. **Boyd JM, Drevland RM, Downs DM, Graham DE.** 2009. Archaeal ApbC/Nbp35 homologs function as iron-sulfur cluster carrier proteins. *J Bacteriol* 191:1490–1497.
10. **Lambrecht JA, Downs DM.** 2013. Anthranilate phosphoribosyl transferase (TrpD) generates phosphoribosylamine for thiamine synthesis from enamines and phosphoribosyl pyrophosphate. *ACS Chem Biol* 8:242–248.
11. **Flynn JM, Christopherson MR, Downs DM.** 2013. Decreased coenzyme A levels in ridA mutant strains of *Salmonella enterica* result from inactivated serine hydroxymethyltransferase. *Mol Microbiol* 89:751–759.
12. **Martinez-Gomez NC, Downs DM.** 2008. ThiC is an [Fe-S] cluster protein that requires AdoMet to generate the 4-amino-5-hydroxymethyl-2-methylpyrimidine moiety in thiamin synthesis. *Biochemistry* 47:9054–9056.

13. Martinez-Gomez NC, Palmer LD, Vivas E, Roach PL, Downs DM. 2011. The rhodanese domain of ThiI is both necessary and sufficient for synthesis of the thiazole moiety of thiamine in *Salmonella enterica*. *J Bacteriol* **193**:4582–4587.

14. Bazurto JV, Downs DM. 2016. Metabolic network structure and function in bacteria goes beyond conserved enzyme components. *Microb Cell* **3**:260–262.

Women in Microbiology
Edited by Rachel J. Whitaker and Hazel A. Barton
© 2018 American Society for Microbiology. All rights reserved.
doi:10.1128/9781555819545.ch7

Nicole Dubilier: A Force of Nature 7

Elizabeth G. Wilbanks[1]

It was the summer of 1970 in Wiesbaden, Germany, when Nicole Dubilier received her acceptance to train with the Stuttgart Ballet. Under the direction of John Cranko, a charismatic South African with big dreams for the little-known group, the dance company was on a meteoric rise. Just the year before, Cranko led their New York debut with "revolutionary" performances that rocketed the ensemble from relative obscurity to international fame. The local critics had remarked that from a city better known for producing Mercedes-Benz than ballet, Cranko had assembled "a ballet company second only to the New York City Ballet"—high praise indeed from New Yorkers (1).

"My mother says I liked to dance before I could crawl," recalls Dubilier. But for the 13-year-old, ballet was more than just a childhood fantasy. Born in New York City to a German emigrée and American industrialist father, the child ballerina once performed at Lincoln Center with the Royal Ballet of London in their production of the Nutcracker. Cast as a rat, "yes, a rat!," she scampered at the feet of the legendary Rudolf Nureyev, a man hailed as one of ballet's most gifted dancers—a true lord of the dance, or in Dubilier's words: "Total asshole. But impressive."

Newly arrived in Germany, she auditioned for Stuttgart "to be with the best of the best." But when the offer arrived, Dubilier balked. Divorce had splintered her family across the Atlantic earlier that year. Leaving her father in New York, she had boarded the *S.S. France* for her mother's native Germany. Dancing with Stuttgart would have meant another departure—

---

[1]Department of Ecology, Evolution, and Marine Biology, University of California, Santa Barbara, Santa Barbara, CA 93106

this time for boarding school, leaving her mother and three younger brothers behind in Wiesbaden. Instead, she stayed and set off on a path very different but with no less star power.

Clad not in toe shoes but in a black and white cocktail dress with a piano key skirt, Dubilier walked confidently onstage in a packed, 4,000-seat auditorium in New Orleans. As chair of the American Society for Microbiology's (ASM's) annual meeting, she welcomed the crowd to ASM Microbe 2017, with an opening session highlighting discoveries from the cell biology of leprosy to the physics of bacterial cell division to bioenergetics at the origin of life. Dubilier is no stranger to science's international stage. She is a director at the distinguished Max Planck Institute for Marine Microbiology in Bremen, Germany, where she heads the Department of Symbiosis, and a professor at the University of Bremen. In 2014, Dubilier, who is a member of both the European and American Academies of Microbiology, was awarded the Leibniz Prize, Germany's highest research honor.

Originally trained as a zoologist, Dubilier sidestepped into the microbial realm, chasing mouthless, gutless worms in the shallow, sandy sediments of enviable research stations like Bermuda, Lizard Island on the Great Barrier Reef, and the Mediterranean isle of Elba. These beautiful, ghostly creatures, more reminiscent of tiny paper straw wrappers than animals, eschew normal meals in favor of far more exotic fare. Like the chemosynthetic symbioses discovered at deep-sea hydrothermal vents, these shallow-water worms partner with unusual bacteria to thrive in habitats with scarce organic nutrients. Beneath their cuticle, these worms harbor chemoautotrophic bacteria capable of feeding their hosts by harnessing the chemical energy from sulfide oxidation to fix carbon dioxide into organic carbon.

Dubilier and her research group have explored the symbioses of marine invertebrates from shallow sediments to deep-sea hydrothermal vents and cold seeps. For years, chemosynthetic symbioses were thought to be a *prix fixe* menu of sulfide or methane, oxidized by a single symbiont species. But work led by Dubilier found that it can be more of an *à la carte* affair. The worm from Elba, *Olavius algarvensis*, hosts no fewer than five different bacterial species, which assemble an interdependent food web to provision their host. Dubilier and her collaborators also discovered symbionts capable of harvesting energy from unexpected sources: the carbon monoxide leeching from beds of decomposing seagrass, the hydrogen bubbling out of hydrothermal vents at mid-ocean ridges, or even propane gas erupting from asphalt volcanoes deep in the Gulf of Mexico.

Dubilier's research resonates with today's medical microbiome mania and the dawning realization that all animals develop intricate relationships with bacterial partners. What once were academic curiosities—sea monsters—are now seen as extreme examples on a continuum of bacterial dependency.

In the case of my own career, I had never found the nutrient cycles, sketched out in textbooks over a planetary scale, inspiring. But Dubilier's experiments managed to find hidden intermediates and track the sulfur cycle—all in the space beneath a worm's skin! The scale captured my imagination. What *else* might we be missing with our myopic, macrobial perspective? What undiscovered worlds await if we can see life from a microbe's view?

Her research animals may be gutless and spineless, but the same surely cannot be said of Dubilier. A small, tenacious woman with an easy laugh and irreverent sense of humor, she is someone colleagues describe as "a force to be reckoned with" (Fig. 1). Dubilier calls it like she sees it with no-nonsense candor—defiant of (but not oblivious to) any ruffling of feathers. During a recent row with leadership over a motion to disband ASM's Junior Advisory Group for ASM Microbe, "my heart was *pounding*," Dubilier said, "but I had to fight for what I thought was right." Former postdoc adviser Colleen Cavanaugh recalls an outspoken, young Dubilier so engaged with her science that her questions after lectures trended more towards passionate interrogation than staid inquiry.

As a newly minted Ph.D. at a small meeting, I remember asking a similar "question," where I essentially told a distinguished professor his interpretations were way out on a limb from his data. Maybe my voice shook; my hands certainly did, but I had spent months thinking about these papers, and I'd be damned if I sat by when I had something to say. If I was unsure how things had gone, some senior colleagues cleared it up for me afterwards,

Figure 1 Dr. Nicole Dubilier, a force to be reckoned with.

joking that I sure had my "panties in a twist"—a snipe that my gallant postdoc mentor, Victoria Orphan, shot down without hesitation.

I was still crushed. I evidently did not understand how to navigate things. Clearly, I had done something wrong. But that feeling faded as I watched Nicole throughout that meeting. She was blunt and bold. With confidence and humor, she suffered no fools but held few grudges—a quality, she would later tell me, she attributes to growing up as the eldest of four and the only girl of her family. "I was used to…having my way is the wrong way of putting it. I was used to feeling confident about myself," she laughed. Wherever it came from, her confidence was infectious.

After she declined Stuttgart's offer, Dubilier continued dancing closer to home, performing often with the Hessian State Ballet in Wiesbaden. With time, however, the adolescent ballerina began to rail against ballet's body fetish. She found that neither her own body nor her conceptions of female beauty conformed with the impossibly wisp-like proportions of the Balanchine ideal. More importantly, though, Dubilier describes this as time when "I was starting to discover my *mind*!" She was thriving as a student at the all-girls Helene-Lange-Schule, where she described her classmates as "brilliant." Continuing professionally in ballet meant leaving school at 15. To her surprise, Dubilier realized, this was no longer something she was willing to do.

Dubilier's plan B was born of her summers spent with her family in Fire Island Pines, a sandy spit of land 60 miles east of Manhattan. These were months awash with what she calls "just that pleasure of the ocean." Though her only connection with marine animals was eating clams fresh from the bay, teenage Dubilier dreamed of a future as a marine biologist. She laughs ruefully as she recalls planning for a career where she could combine physical and mental work: scuba diving in the mornings with the beautiful men of Jacques Cousteau videos, working a bit in the lab during the afternoons. Besides, smiled Dubilier, thinking of Cherry Grove and Fire Island Pines, glamorous gay enclaves in their post-Stonewall heyday, "I thought all men were as beautiful as the guys there."

Biology, though, held no interest for her in school. It was politics, with its drama and debates, that fired her up. Dubilier recalls discussions, as war broke out in the Middle East, over the German press's one-sided portrayals of the Arab-Israeli conflict. Domestically, in a divided Germany, she was fascinated by the politics of then-Chancellor Willy Brandt, a lady's man, social democrat, and reformer, who won the 1971 Nobel Peace Prize for his efforts to ease tensions between East and West. Years later, while in graduate school, Dubilier worked a side job in governmental public relations, where she hosted cultural dignitaries and took part in one of the first West German envoys to East Germany after the fall of the Berlin Wall.

The nudge towards a career in science came at a tipping point after high school, when Dubilier went to Helgoland, an island in Germany's North Sea. There she found a "basic, emotional satisfaction" in her work cleaning tanks at a marine station, a closeness with the ocean and marine organisms (2). It was then that she decided to pursue zoology at the University of Hamburg. The connection, though, between her basic loves—for learning and for the ocean—"remained very vague"—frustratingly so, she says, for many years.

Dubilier's Deep Sea Screwdriver: mix one part vodka with two parts orange juice. Load into a Niskin bottle aboard a CTD rosette and seal closed. Chill by lowering to 1,000-meter water depth. Upon recovery, garnish with orange slice and serve shipboard.

This might well be an apocryphal story, reveries from thirsty American oceanography where alcohol is banned aboard research vessels (unlike German ships). If it isn't true, though, don't tell me. Veracity and tasting notes aside, I love this recipe for its legend: Dubilier, the adventurer and *bon vivant*. Talking with Dubilier, one gets this sense of her irrepressible *joie de vivre*. She may have been classically trained in ballet, but afterwards her tastes wandered, first to jazz and then farther afield to less structured booty-shaking. Dubilier's procrastination of choice while writing or answering emails? Streaming YouTube dance videos, particularly those she classifies as "tits & ass." Her eyes light up describing what draws her to dance: "It's joyous. It's vibrant. It's sexual. It's just… there's so much in it that is *alive!*"

"She exemplifies the humanity that underlies great science," says collaborator Pete Girguis, who considers Dubilier something of an academic "big sister." He mentions her idiosyncrasies fondly and describes how her collaborative generosity set an example for him in his own career, a symbiosis of symbiosis researchers: "To me, her being a role model isn't an idealized depiction of a scientist, […] mainly because that does her a disservice!"

Dubilier remembers her graduate research as a struggle. "I needed discipline as a ballet dancer," she says, recalling that she felt ill-suited to the free-form structure of independent research. "I would wander off and then think about which bar we can go to at night," Dubilier laughs, because "there was a lot outside science that was very interesting!" Her research on the sulfide adaptation of the worm *Tubificoides benedii* involved ample fieldwork. But the freezing temperatures and boot-sucking mud in the Wadden Sea (whose name translates literally as *mud*) provided a rude awakening to her Jacques Cousteau dreams. Dubilier swore her next animals would inhabit more appealing climes (2).

Hardest, though, was that she just didn't feel passionate about her work. Envious of those who got their best ideas in the shower, Dubilier says, outside of the lab she thought about everything *but* science. It seemed everyone loved their research but her, and she wanted to quit her Ph.D. "6 million times." She remembers the advice passed along by her partner, Christian, from his father, a man she deeply respected: "Tell her to finish it, because she'll never forgive herself for not." That such a stark and simple truth inspired perseverance is perhaps unsurprising for Dubilier, an aficionado of the gritty realism and snappy dialogs best depicted by American crime novelists like Elmore Leonard. She forged on to finish.

Ph.D. in hand, Dubilier left Hamburg in 1992 for Santa Catalina Island and a University of Southern California summer course that would prove a turning point in her career. She describes herself as "clueless" about DNA at the time ("You mean I was supposed to *keep* the supernatant?!"). "It could have been *Molecular Biology for Dummies* for all I knew!" said Dubilier of the course. But it most certainly was not. Taught by Donal Manahan, the course ran at a blistering pace, from 8 a.m. to midnight many days, and featured guest lectures from such rising stars in symbiosis and microbial ecology as Margaret McFall-Ngai, Ned Rudy, Ed Delong, and Steve Giovannoni.

Dubilier was fascinated by the power of PCR, an invention less than 10 years old at the time and for which the Nobel Prize in Chemistry would be awarded the following year. New assistant professors and former postdocs from Norm Pace's group, Ed DeLong and Steve Giovanonni, taught students how to PCR amplify 16S rRNA genes from environmental samples, a new approach that was revolutionizing the field. Dubilier thrilled at the instant gratification of looking at a sequencing result, after years repeating tedious physiology experiments: "We were discoverers! We just sequenced stuff and we saw the coolest things! The only thing we had to be able to do was look."

Well, it was a little bit harder than that. Giovanonni showed them how to pour and load the slab gels to sequence their PCR product base by base, lane by lane. "High throughput" sequencing of the day was accomplished by pouring a final gel at 3 a.m. on their way home from the bar (though it infuriated Giovanonni that the students often found late night sequencing less exciting than roaming down to the harbor to feed the moray eels that would slither ashore, called up by a knocking on the rocks).

With this introduction to molecular biology, Dubilier was hooked. The frontier sensibilities of the field at that time certainly suited her, and in her work, one can see this love of discovery. First with electron microscopy (in her very first and single-author publication on epibionts of *Tubificoides bendii*) and then with 16S rRNA gene sequencing and fluorescence microscopy *in situ* hybridization, Dubilier peered into the strange inner lives of worms to find their unseen bacterial partners. A pioneer with metagenomic sequencing of symbionts and now metaproteomics, she describes the appeal of this discovery-based approach: "We're sailing into *terra nova*!"

Dubilier fell in love with symbiosis as a postdoc working at Harvard with Colleen Cavanaugh, a pioneer in the study of chemosynthetic life at hydrothermal vents. To Cambridge, she brought with her a year of independent funding, her newly acquired molecular biology skills, and a package full of the gutless worms *Phallodrilus leukodermatus*, which were discovered in Bermuda's sandy sediments by her Ph.D. advisor Olav Giere. The combination of symbiosis and molecular biology research was, for Dubilier, "a total game changer." She had hit upon a field that really excited her, though precisely why, she says, is hard to describe. The idea of synergy between such different organisms just fascinated her. Perhaps it was the combination of timing, technology, and her creatures' idyllic habitats, or maybe it was an echo of her love for politics and international relations—biological diplomacy in an alliance wrought over evolutionary time. Whatever the reason, Dubilier knew she'd found her groove.

With the funding of an NSF grant she wrote with Cavanaugh, resources for more fieldwork, it would seem nearly time to run the credits on Dubilier's success story against a backdrop of tropical research stations. But later that year she was diagnosed with breast cancer, ductal carcinoma *in situ*. Treatments and operations upended her plans. Not one to be sidelined, she still ventured far afield, diving for worms in Australia, Belize, and Bermuda, but lamented that she "wasn't very productive" on their project. Then after 2 years in Boston, when her husband Christian finished his fellowship in orthopedic surgery, came more transitions: a return to Hamburg, the birth of their son, and starting new jobs.

After a short stint back in Hamburg, Dubilier found her way to the Max Planck Institute for Marine Microbiology in Bremen, where she has been ever since. The door of opportunity, though, did not simply swing open: her career there began with just a 3-month postdoctoral contract. "I was persistent to the point that, they later told me, they were worried I was going to be this super-annoying person once I arrived," she recounts (2). For nearly the next decade, Dubilier thrived working in Rudi Amann's Molecular Ecology Department, first as a postdoctoral fellow and later as a research associate.

Eventually, it was pride in her work and the spark of competition, as she watched colleagues move past her into more senior positions, that nudged Dubilier out of her early career Neverland. "I thought it would be perfect to be a postdoc forever. [...] I never ever wanted to grow up," Dubilier says. "Actually, I did want to grow up. I just didn't want to be a grown-up. There's a big difference." Her confident, Sandbergian career advice, "Don't put your foot on the brake," is a bellwether of her hard-won success, but underneath lies a sense of doubt and vulnerability, a reflection on a humanity as complex as the symbioses she studies. "Don't kid yourself. Don't think you can stay somewhere in the middle," she advises—as much to herself as to anyone listening.

"I think it's so important to say," Dubilier pauses, sipping her Corona, "my career has been everything but straightforward, *everything but straightforward*, and I nearly lost my way multiple times." As we drink our beers during happy hour in the sprawling exposition hall at the ASM meeting, she reflects, "the only thing I'm superproud of is my persistence." More than just an antidote to a success story's caricature, her sentiments echo 2017's feminist battle cry: "Nevertheless, she persisted."

CITATION

Wilbanks EG. 2018. Nicole Dubilier: a force of nature, p 57–64. *In* Whitaker RJ, Barton HA (ed), *Women in Microbiology*. American Society for Microbiology, Washington, DC.

References

1. Saal H. 18 July 1971. Who ever heard of a German ballet company? *New York Times*, New York, NY. https://nyti.ms/1GSTT3L.
2. Azvolinsky A. 1 July 2015. Sold on symbiosis. *The Scientist*, Ontario, Canada. http://www.the-scientist.com/?articles.view/articleNo/43337/title/Sold-on-Symbiosis/

Women in Microbiology
Edited by Rachel J. Whitaker and Hazel A. Barton
© 2018 American Society for Microbiology. All rights reserved.
doi:10.1128/9781555819545.ch8

Katrina J. Edwards: A Force in the World of Environmental Microbiology

8

John R. Spear[1]

Katrina J. Edwards was a force in the environmental microbiology world. She was a woman of diverse training in the fields of geology, isotope geochemistry, and geomicrobiology, which allowed her to do novel work in little-understood places. With time, she came to focus on the subsurface life and processes that sustain our planet. With the National Science Foundation, Katrina founded the Center for Dark Energy Biosphere Investigations (C-DEBI), which has led to the production of more than 300 publications and counting, adding to the knowledge of what biological, hydrological, and geochemical processes occur in the subsurface of the Earth. Katrina did all of this as an incredibly effectual mentor to the many students and postdocs who moved through her lab, as well as those of her C-DEBI collaborators and colleagues. Mirroring her own diverse educational background, Katrina drew together people of diverse skill sets in science and engineering to design and implement novel tools to escape the restrictions of laboratory experiments and better understand the world in an *in situ* fashion, the way the world needs to be understood. Katrina did all of this by blending science intimately into her personal world in what ultimately became a destructive work/life balance. But the lessons she imparted to us all remain: build a solid foundation for your career, work hard and sweat the details in your love of science, and make your local world better by being a warm and caring person of keen wit to all of whom you meet!

---

[1]Department of Civil and Environmental Engineering, Colorado School of Mines, Golden, CO 80401

THE WOMAN: MOTIVATION AND PATH

If there is one word to describe the motivation of Katrina Jane Edwards, it is Passion, with a capital P: Passion for life, Passion for love, Passion for family, Passion for science, Passion to tell a story. She became known as the original Mistress of the Dark World. Like for most creative people, Katrina's path in life was not unlike that of a coyote; though it takes longer to get from point A to point B by a zig-zag path, the rewards in doing so can be rich and fruitful. The experiences acquired "along the way" often become transformational and provide points of seemingly random connection to future experiences, which themselves become the radial-frame threads in the web of one's life. My and Katrina's threads intersected when I was a postdoc in Colorado and she an associate scientist at the Woods Hole Oceanographic Institution (WHOI) in Woods Hole, MA. We had both been on nonlinear paths in life, weaving our own complex webs, and I believe that we are better educators and citizens of the world for our students, our peers, our colleagues, our friends, and our families by taking such a path—less defined and less traveled. We both undertook different kinds of occupations between high school, undergraduate education, and graduate school. We completed our doctoral degrees in 1999, worked together in various environments, pursued successful careers in academics, and helped in weaving together a new field of science in geobiology.

Katrina grew up as an exceptionally bright child in Columbus, OH, and graduated from the Columbus Alternative High School. She worked with friends and family at the nearby Delaware, OH, municipal airport, was a passionate pilot, loved to fly, and worked her way up to become a flight instructor; this perhaps served as the root of her educational nature. Concurrently, she pursued a degree in geology at The Ohio State University, earning her bachelor's degree with honors in 1994. Next, she went on to the University of Wisconsin, Madison, for a master's degree focused on isotope geochemistry and a Ph.D. advised by Dr. Jillian Banfield in geomicrobiology, the first such degree awarded by the university. Geomicrobiology (a subset of the broader field of geobiology) had been considered a respectable subfield of microbiology for decades. But it was not until the late 1990s and early 2000s that the field gained critical mass from people like Jill, herself well known in the field, and Katrina, who were cross-disciplinarily trained— for her, geology to isotope geochemistry to microbiology.

From 1998 through 2000, Katrina started to publish both the primary and ancillary works of her Ph.D. thesis in a number of papers important for both her and her coauthors, including her Ph.D. advisor. One considered the distribution of specific microorganisms and their implications for the production of acid mine drainage (1), another the oxygen isotope ratios in the crystal of ultramafic minerals (2), another the microbial oxidation of pyrite (3), another the broader context of pyrite dissolution (4), and still another the seasonal variations of microbial communities in an extreme acid mine drainage environment (5).

In March of 2000, Katrina published a landmark cover story for the journal *Science*, entitled "An Archaeal Iron-Oxidizing Extreme Acidophile Important in Acid Mine Drainage" (6). This seminal paper described the acid-tolerant slime streamers that contained a new species of iron-oxidizing [taking Fe(II) to Fe(III) as an energy source] *Archaea* that grew at a pH of 0.5 and at a temperature of ~40°C at Iron Mountain, an acid mine drainage site in Northern California. This paper remains her most cited and continues to be frequently referenced to this day. Importantly, from these first six papers, it becomes clear that Katrina's primary motivation as a scientist was the pursuit and dissemination of new knowledge in a field that she helped to pioneer. For her, this was cemented in the wonderful works of her Ph.D. experience, which laid a solid foundation from which her impactful career was born.

This, then, serves as an excellent piece of advice for all who want to be successful in art, or science, or whatever: a strong foundation of work is one which you can stand upon for the rest of your life, no matter what your own passion may be. That foundation can underpin your future with the knowledge and satisfaction of doing well, while at the same time advancing science and/or enriching the human experience, which no technology can ever replace. Ultimately, such a foundation can be a comfort as you move through life. But of course, bear in mind the words of the adage, "Comfort is the great adventure thief!"

In 1999, Katrina started as an associate scientist at WHOI and started a geomicrobiology laboratory that focused on the microbial "degradation" (I would argue the weathering and cycling) of rocks, minerals, and organic matter by microbiota, specifically microbiota that "rust the crust." Two important articles from her time at WHOI appeared in the American Geophysical Union's *Eos*, entitled "Energy in the Dark: Fuel for Life in the Deep Ocean and Beyond" and "Iron and Sulfide Oxidation within the Basaltic Ocean Crust: Implications for Chemolithoautotrophic Microbial Biomass Production" (7, 8). These papers considered how chemosynthetic life, dominated by chemolithoautotrophs (microorganisms that get their energy from inorganic elements like hydrogen and carbon from CO_2; life fueled by chemistry rather than photosynthesis), is more encompassing of the kinds of life on the planet than that which tends to surround us in the form of photosynthetic life. To an extent, this paper set the stage for her transition in 2006 to the University of Southern California (USC) as a professor of biological and earth sciences, where Katrina was a key member of a six-person cluster hired to jump-start geobiology at USC. There she began an 8-year legacy of mentoring many undergraduate, graduate, and postdoctoral fellows who had the commonality of all working at the rock-microbe interface, primarily in subsurface environments and primarily with Katrina's laser focus on life-sustaining energy in the deep oceanic subsurface.

In her first years at USC, Katrina really began to think about life in the subsurface of the Earth and the fact that there was little understanding of the "intraterrestrials"—the microbiota that live miles below the Earth's crust or the deep ocean sediments. This led Katrina to write a monumental grant to the National Science Foundation, for which she received 29 million dollars to establish the C-DEBI (http://www.darkenergybiosphere.org) at USC in order to better understand the little-studied microbiotas that live in the deep subsurface beneath the floor of the oceans. With collaborators from a number of national laboratories and universities (self-included), C-DEBI's primary aim was to research the reciprocal interactions between microbiota, rocks, minerals, and geochemical fluids in the Earth's crust beneath the ocean's floor and how these interactions affect global biogeochemical processes. All the affiliated people of C-DEBI came to be known to the greater microbiological community with their black t-shirts emblazoned with "Mainly Microbe" across the front and the C-DEBI light and dark microbial cell's yin-yang logo on the back. This was a whole tribe, and the shirts remain pervasive at scientific meetings and on field works and research cruises.

In 2010, Katrina commented to the *USC College (Dornsife) Magazine* that the C-DEBI work considered "what lies in-between—hundreds of square kilometers in aerial extent, down kilometers below the ocean bottom, lies an active, living intraterrestrial ecosystem—this is what I think about almost all of the time" (9). Katrina continued, "This is our space; the deep biosphere is the new moon for this team of researchers. The series of expeditions that we are putting together are our *Apollo* missions, and this new center provides us with exactly the launch pad we need to accomplish our ambitious goals" (9). In yet another well-made statement that flaunted Katrina's motivation, she observed, "It's shocking. You go below the very surface of the ocean and basically fall off the edge of our knowledge about this planet. In the present day we know much, much more about space and the surface of other planetary bodies than we do about the inner space of our world" (9). The work of C-DEBI and the large team of researchers involved with the program became Katrina's primary path to fulfill her motivation to come to know the unknown of one of Earth's most important ecosystems, the deep subsurface.

CHALLENGES AND RESPONSES

Katrina faced many challenges in her career. I did not know her until our life threads intersected in 1999, but by the time we met, she kind of had this legend status: her nature, her intellect, her curiosity, her charm, and her scientific interests preceded her into every room she entered. I'm sure that this was a bit of burden growing up; I know that it became one over the course of her career. Katrina was particularly good at balancing the many aspects of her life. She raised three children (Ania, Katya, and Nakita Webb) while starting her career, more than once taking one or more of them into a meeting of fellow scientists, many of whom did the same thing. Babies and little kids can add a lot to a meeting—by imparting what's really important in life! In science, you never know when the ideas are going to come, or when a thought stream could lead to some promising research idea or question answered or circumstance to happen. To balance our science lives, many of us interweave life, family, and children with our work, any of which may be at hand at a given moment. It can be a delicate act to successfully merge a career in science with your life's life. But with warmth, love, humor, and friendship, Katrina was always able to accomplish this. Jason Sylvan, a former postdoc in Katrina's lab and now a faculty member at Texas A&M University, recalled, "My fondest memories of Katrina are the parties she regularly threw at her home because she made her lab feel like it was a big family. There was a real warmth there that I think people outside of her lab did not experience, but it was real. I knew that I always had her support, which meant a lot to me."

As her career heated up with the production of more than 100 publications (and still climbing), which brought more collaborators, more attention, and more kinds of science now possible, the success also brought the burdens of management and less time—to just be, which in and of itself is something that we all need. Katrina had high standards for herself and looked to the high standards of her mentors as a model. Katrina, in turn, mentored with those same high standards and had a number of successful mentees, graduate students, and postdocs move through her lab into prolific careers of their own (e.g., Jason Sylvan, Cara Santelli, Brandy Toner, Beth Orcutt, Clara Chan, Ann Pearson, and Olivier Rouxel) as they were able to bear the load of what C-DEBI and Katrina's other projects set out to accomplish.

Keeping a large team of collaborators managed and running smoothly is a huge challenge. Most academics are not trained to manage large numbers of people, large amounts of funding, or the production of multiple scientific stories in fixed amounts of time to coax out the insights that lie in the shadows and which the creative mind must find. This is a notable fault of the academic preparation process and nature of funding agencies that have shifted toward larger-team projects that can be detrimental to effectivity. Going back to her childhood, Katrina was a woman who gave much of her heart and mind to every project she encountered. Katrina completely loved the C-DEBI world of people, projects, science, and engineering! Her passion was there, despite also the nerves and stress of the large C-DEBI program, travel, and many research cruises to field sites. Katrina eventually passed away after a long illness, much too early at 46 years of age.

A challenge of our technological age is to maintain our work/life balance. Early in her career Katrina was a master of this, but as time wore on that balance appeared more difficult. A work/life balance becomes more difficult in an era of being always "on," always "connected," always "engaged" in ever-extending social networks of connectivity that can tweak a human psyche. Successful people tend to always be "in the game," for not being so can impart a feeling of lost opportunity. However, this often comes at an incalculable cost. The number one thing to remember is that we get one chance at this life and to be in this world, and life, and a good-quality one, must always win. Some of us lose that in the fast-paced world of discovery science. To jump in and characterize the black hole of the unknown and put forth a new discovery to make an incremental step or a major leap forward in science can become an addiction. But the necessity of the work/life balance can never be lost in the name of a rich life to be fulfilled only by what may be possible with the pursuit of the unknown in a scientific endeavor. We all need to make the most out of our lives, always; ultimately, this is what makes the world a better place. Katrina was making the world a better place until the end, but there was so much more that could have been ahead.

HER LIFE'S INSPIRATION AND ENCOURAGEMENT

Katrina's inspiration and encouragement to us all live on in the personality of the organization she founded: C-DEBI. C-DEBI's mission "is to understand the extent, function, dynamics and implication for the existence of a deep biosphere on Earth" (9). C-DEBI has been enormously successful at the establishment of a solid community of diverse and young investigators who work in a wide expanse of environments from the sediment world of microbial communities on the floors of the oceans to the cracked and fractured subseafloor ocean in areas such as North Pond (mid-Atlantic Ocean) and the Juan de Fuca Ridge of the northeast Pacific and the South Pacific Gyre. All of the people who have been involved with C-DEBI carry forth its foundational personality into their own futures. With her infectious and rambunctious energy, drive, curious intellect, and warm personality, Katrina and the colleagues she drew into the Center created something indescribably special, and that something is now spreading to new labs, to newly mentored students, and to new ideas that will feed the science of tomorrow.

Part of Katrina's inspirational quality was her ability to understand that the observations of those who preceded her were critical to better understand what was in front of her. In the 1930s the Scripps Institution of Oceanography's Claude ZoBell and his Ph.D. student David Updegraff (one of my Ph.D. mentors) found what they thought was evidence of life everywhere they looked, from the oceanic water columns to the sediments to the deep subseafloor rock and geochemical fluids. ZoBell thought that the radiolytic splitting of water could produce molecular hydrogen that fueled life in any number of environments. Thomas Gold followed up on that with his important paper and book *The Deep Hot Biosphere*, works that we have just served retrospection upon (10–13). As a thorough scientist, Katrina understood that it is just as important to look to the past, even the deep past, to stand on the shoulders of giants so that you may see into the future. Or, as Carl Woese commented, "If I could see further than those before me, it is only because I was looking in the right direction." Katrina might well have said the same. As Katrina commented in starting C-DEBI, "This is the decade of the intraterrestrials—we have definitive, measureable deliverables that contribute to our knowledge of really big questions: Is there a consequence for all that life down there, or is there none?" (9). That was Katrina echoing the sentiments of ZoBell and Updegraff in pointing C-DEBI to its future.

As a graduate student, Katrina wanted to understand the natural world on its own terms, and by looking at things differently and in different ways, her designs of *in situ* experimentation methods, tools, and practices are an inspiration to us all. Her thoughtfully crafted field experimentation to study the microbiota of Iron Mountain was top notch. Her outstanding field skills and "lab hands" were critical in the success of all of her many research cruises. Katrina always had a strong, unique vision for what she set out to accomplish, and she always had incredible drive to see it through. She was one of the first ocean biogeochemists who understood the power of drill-hole

Figure 1 Katrina Edwards, at work crushing rocks. Photo courtesy of Ann Close.

technologies and then married geomicrobiology with engineering in a trendsetting fashion. A legacy denoting this clear vision and drive is CORK (Circulation Obviation Retrofit Kit) technology. Katrina brought people together from different science and engineering backgrounds to refine CORK for better observatories in subseafloor investigations. Installation of a CORK into a drilled borehole allowed for the *in situ*, isolated examination of subsurface processes such as geochemical fluid flow and sampling of biological material to consider hydrological, geochemical, and biological processes in igneous oceanic crust (14). Some of this work was captured in an award-winning feature-length documentary that she executive produced about the North Pond ocean drilling project (15).

HOW SHE INSPIRED ME

A refrain that several of us who worked with Katrina have is that she was never about hierarchy or aristocracy; she was always highly engaged with whom she was with at the moment; she was accessible and on the same level; you were her teammate. She was this way with everyone in both informal and formal sessions. Katrina never had a bias in talking with people about science—every field of science mattered, as if to say that all fields are linked, which indeed they are. Systems behave as a whole because of the sum of their parts. Katrina understood and appreciated this: understanding what an ecosystem is and the elemental cycles that make that ecosystem run requires knowledge beyond your own field. Katrina taught her students what it was to see something, an idea, that isn't there yet and how to turn that idea into something that exists, is successful, and serves as a beacon to others who are excited about that idea. Katrina's loving dedication was instrumental for the ignition of that beacon that has become the amazing field of geobiology. For example, in thinking about mineral crystallography, geochemistry, and microbial community, Katrina was able to visualize, articulate, and contribute to what has become geobiology.

HOW SHE SHOULD INSPIRE YOU

- Build a solid foundation upon which to stand for your life and career!
- Be a great human to all of those who surround you!
- The poet Hilaire Belloc (16) penned the poem *The Microbe* in 1897 (later quoted by Willy Wonka) with words applicable to Katrina: "Oh! Let us never, never doubt what nobody is sure about!"
- Trust yourself and embrace your unique abilities!
- Work hard and sweat the details and do so for your innate love of science!

Katrina Jane Edwards, 15 March 1968–26 October 2014
It was not supposed to end this way.

ACKNOWLEDGMENTS

Thanks to Will Berelson, Matthew Schrenk, Jason Sylvan, Emily Kraus, and Ann Close for helpful thoughts and reviews.

J.R.S. is supported by the NASA Astrobiology Institute Rock Powered Life project and the Zink Sunnyside Family Fund.

CITATION

Spear JR. 2018. Katrina J. Edwards: a force in the world of environmental microbiology, p 65–74. *In* Whitaker RJ, Barton HA (ed), *Women in Microbiology*. American Society for Microbiology, Washington, DC.

References

1. Schrenk MO, Edwards KJ, Goodman RM, Hamers RJ, Banfield JF. 1998. Distribution of *Thiobacillus ferrooxidans* and *Leptospirillum ferrooxidans*: implications for generation of acid mine drainage. *Science* 279:1519–1522.
2. Edwards KJ, Valley JW. 1998. Oxygen isotope diffusion and zoning in diopside: the importance of water fugacity during cooling. *Geochim Cosmochim Acta* 62:2265–2277.
3. Edwards KJ, Schrenk MO, Hamers R, Banfield JF. 1998. Microbial oxidation of pyrite: experiments using microorganisms from an extreme acidic environment. *Am Mineral* 83:1444–1453.

4. Edwards KJ, Goebel BM, Rodgers TM, Schrenk MO, Gihring TM, Cardona MM, McGuire MM, Hamers RJ, Pace NR, Banfield JF. 1999. Geomicrobiology of pyrite (FeS$_2$) dissolution: case study at Iron Mountain, California. *Geomicrobiol J* **16**:155–179.
5. Edwards KJ, Gihring TM, Banfield JF. 1999. Seasonal variations in microbial populations and environmental conditions in an extreme acid mine drainage environment. *Appl Environ Microbiol* **65**:3627–3632.
6. Edwards KJ, Bond PL, Gihring TM, Banfield JF. 2000. An archaeal iron-oxidizing extreme acidophile important in acid mine drainage. *Science* **287**:1796–1799.
7. Bach W, Edwards KJ, Hayes JM, Sievert S, Huber JA, Sogin ML. 2006. Energy in the dark: fuel for life in the deep ocean and beyond. *Eos (Washington DC)* **87**:14.
8. Bach W, Edwards KJ. 2003. Iron and sulfide oxidation within the basaltic ocean crust: implications for chemolithoautotrophic microbial biomass production. *Geochim Cosmochim Acta* **67**:3871–3887.
9. Andrews S. Spring/summer 2010. The power and promise of the ocean. *USC College Magazine.* University of California, Los Angeles, CA.
10. Gold T. 1992. The deep, hot biosphere. *Proc Natl Acad Sci USA* **89**:6045–6049.
11. Gold T. 1999. *The Deep Hot Biosphere.* Springer, New York, NY.
12. Colman DR, Poudel S, Stamps BW, Boyd ES, Spear JR. 2017. The deep, hot biosphere: twenty-five years of retrospection. *Proc Natl Acad Sci USA* **114**:6895–6903.
13. Spear JR, Walker JJ, McCollom TM, Pace NR. 2005. Hydrogen and bioenergetics in the Yellowstone geothermal ecosystem. *Proc Natl Acad Sci USA* **102**:2555–2560.
14. Edwards KJ, Wheat CG, Orcutt BN, Hulme S, Becker K, Jannasch H, Haddad A, Pettigrew T, Rhinehart W, Grigar K, Bach W, Kirkwood W, Klaus A. 2012. Design and deployment of borehole observatories and experiments during iodp expedition 336, mid-Atlantic ridge flank at North Pond. *Proc IODP* **336**:Tokyo (Integrated Ocean Drilling Program Management International, Inc).
15. Oki Productions, C-DEBI. 2014. *North Pond: The Search for Intraterrestrials.* https://vimeo.com/117447690.
16. Belloc H. 1897. *More Beasts for Worse Children: The Microbe.* Duckworth and Company, London, United Kingdom.

Women in Microbiology
Edited by Rachel J. Whitaker and Hazel A. Barton
© 2018 American Society for Microbiology. All rights reserved.
doi:10.1128/9781555819545.ch9

Alice Catherine Evans: The Shoulders Upon Which So Many Stand

9

Lorraine A. Findlay[1]

It can be stated that Alice C. Evans (1881–1975) was the first professionally successful female microbiologist (Fig. 1, left panel) and that all succeeding female microbiologists "stand on her shoulders." She was the first woman to hold a permanent appointment as a bacteriologist at the U.S. Department of Agriculture (USDA), she was the first woman to hold a senior appointment in the U.S. federal government, and she was the first woman president of what is now known as the American Society for Microbiology (ASM). She accomplished all this in the early decades of the 20th century, a time during which female scientists were dismissed as unimportant and not professionally respected.

I admit that prior to my becoming involved with the ASM Committee on the Status of Women in Microbiology (CSWM), Alice Evans was unknown to me. The CSWM had initiated the ASM Alice C. Evans Award in 1983, annually recognizing a person most advancing the role of women in microbiology, and I questioned why this award was named after Evans. As chairperson of the CSWM from 2005 through 2013, I had the opportunity to delve into the life of Alice Evans. Permit me to fondly address this great microbiologist simply as "Alice."

Alice was born on 29 January 1881 on a farm in Bradford County, PA. Her family was Welsh and exhibited qualities that would mold Alice's life. They were poor and thrifty, worked hard, were religious, emphasized education, and spent free time in study. Her primary education was in a country schoolhouse. Her family's emphasis on education enabled her to attend a

---

[1]Department of Allied Health Sciences, Nassau County College, Garden City, NY 11530

FIGURE 1 (Left) Alice C. Evans, 1881–1975. Courtesy of Center for the History of Microbiology (CHOMA)/ASM Archives at University of Maryland, Baltimore County (UMBC). (Right) Women's basketball team, the Susquehanna Collegiate Institute, 1896. Alice is on the left. Courtesy National Library of Medicine (NLM).

secondary school, the Susquehanna Collegiate Institute. At this school, Alice played basketball (Fig. 1, right panel), a game very new for women. People were shocked; the girls showed their legs. At a game, an elderly doctor refused treatment for Alice's finger that became dislocated. Alice's reaction was, "It did not bother us" (1).

As these institutes disappeared around 1900 with the advent of community high schools, her class of seven was the last to graduate in 1901. Upon graduation, Alice expressed a philosophy, later emphasized in her memoirs (1), which would continually influence her choices in life: "I always stepped into the only suitable opening I could see on my horizon." As teaching was the only profession available to women, Alice taught grades 1 through 4 in her own country schoolhouse. But she yearned for "a way to escape." The College of Agriculture at Cornell in Ithaca, NY, was giving a 2-year course to teachers to inspire in their students a love of science and nature. The course was free to rural teachers. Alice was poor, and so following her philosophy, she jumped at the opportunity. However, when the course was complete, Alice no longer wanted the certificate. Her interest in sciences had been whetted; Alice wished to remain at Cornell and study science.

Bacteriology was a young science at the time (1900 to 1910), and Cornell offered the science free of tuition. Alice again addressed her poverty and followed her philosophy. "I had to take Bacteriology and that was perfectly satisfactory to me." It appears that bacteriology chose Alice and not the reverse. At Cornell, Alice lived on campus with 30 women (Fig. 2). Following her family's habit of devoting free time to study, her light was the last to go out at night. She also worked part-time doing housekeeping and clerical work in the alumni library. Her schoolmates called her "the grind" (1). At 28, she completed a bachelor of science in agriculture in 1909.

In her senior year at Cornell, Alice's professor, Professor Comstock, recommended Alice for a bacteriology scholarship at the University of Wisconsin. This scholarship had never before been held by a woman! A door to a career had opened unexpectedly, and Alice again stepped into the only suitable opening available. She received her master of science degree from Madison but declined to pursue a doctoral degree, as she did not see its value for her. It was 1910, and at 29 years of age, society proclaimed that she was an old maid.

FIGURE 2 Evans with her "sisters" at Cornell University. Alice is in the forefront, on the left. Courtesy NLM.

Alice's professor at the University of Wisconsin was able to appoint her as a bacteriologist at the USDA. Alice became a federal civil service employee as a bacteriologist on 1 July 1910, the first woman to do so. Until the USDA built its new wing in Washington, DC, her research was done at the Cooperative State Facility in the University of Wisconsin Dairy Division. Alice was pleased to be able to remain in Wisconsin. Her project was to investigate better means of cheesemaking, an important industry in Wisconsin. She published a number of papers specifically on the bacteriology of improving the flavor of cheddar cheese. Finally, in 1913, the USDA East Wing was completed in Washington, DC. It was a white marble building north of Independence Avenue, between 12th and 14th Streets, NW. After 3 years at the state facility in Wisconsin, Alice regretted having to leave Madison but accepted the only suitable opening. On her way to Washington, Alice visited the University of Chicago. A professor there stated, "I was told the USDA did not want any women scientists." Reflecting on this memory, Alice commented, "I was on my way, where I had not wanted to go, and where I was not wanted." However, she did "not let this bother her." It had not occurred to USDA officials that a woman might have been chosen for the position. When news broke that a woman scientist would be joining, the USDA officials "almost fell off their chairs" (1).

At the USDA, Alice studied the bacteria of freshly drawn milk from cows. She first studied streptococci and, in 1916, published a paper authored by three future presidents, including herself, of the Society of American Bacteriologists (SAB) (2). The SAB was the former name for ASM. Eventually, Alice focused on the causal organism of bovine contagious abortion; early reports warned that the organism might be dangerous to human health. Bernhard Bang had discovered the causal organism of bovine contagious abortion in 1897, identified it as rod shaped, and placed it in the bacilli as *Bacillus abortus*. It was also known that milk from healthy goats carries the bacteria of human undulant/Malta fever; in 1887, Lord David Bruce identified the organism causing this disease as spherical and placed it in the cocci as *Micrococcus melitensis* (3). Alice discovered that the two bacteria, *B. abortus* (bovine) and *M. melitensis* (caprine), were alike in culture and reported this finding at the annual meeting of the SAB in Washington, DC, in December of 1917.

Alice's subsequent publications on the finding in the *Journal of Infectious Diseases* in July 1918 (4–6) were viewed as controversial. She wrote, "Considering the close relationship between the two organisms and the frequency of *B. abortus* in cow's milk, it is remarkable that we do not have disease resembling Malta Undulant Fever in the US. Are we sure cases of abortion may not sometimes occur among human subjects in this country as a result of drinking raw cow's milk?" (6). Alice made a strong call for the pasteurization of cow's milk. Reaction to her papers was skeptical; if the organisms were related, some other bacteriologist (implied a male) would have noticed it. Alice realized that this was not valid criticism; it did "not bother her" (1). The organisms behind undulant/Malta fever and contagious abortion (previously identified as *B. abortus* and *M. melitensis*) were established by Alice to be one and the same species. Today the organism is named after Lord Bruce, *Brucella abortus*, and the disease is known as brucellosis.

Tuberculosis (TB) was a common disease transmitted via milk from cattle. The dairy industry was deeply concerned about the economic loss caused by TB in cattle, before any human concerns became known. The Commission on Milk Standards certified milk; grade A milk may be raw but must be drawn and bottled with strict cleanliness, and free from disease via testing for TB in the animals. In 1918, the belief was that certified milk was safe. Alice's study was unacceptable to dairymen, who had invested in expensive equipment for certified milk. They opposed pasteurization and accused her of collaborating with the manufacturers of pasteurizing equipment. Dairy lobbyists moved to have her fired. The industry failed to consider that milk, although certified and produced by healthy animals, might cause human disease.

A respected bacteriologist, Theobald Smith of Rockefeller University, rejected Alice's work and her call for pasteurization. A man of Smith's stature could delay the truth for years, while mothers drinking raw milk were still miscarrying. Smith's opposition was not surprising. The 19th Amendment, which acknowledged the right of women to vote, was not ratified until 1920; Alice's papers were published in 1918. Smith was not accustomed to considering scientific ideas from a woman, and the idea of sexual discrimination was not accepted at that time. Alice held steadfast to her conclusions. She wrote to a fellow scientist, Henry Welch of Johns Hopkins, to intercede with Smith, although Smith remained unyielding.

When World War I drafted most male scientists, Alice was able to resign from the Dairy Division, and in April 1918, she joined the Hygienic Laboratory, which later was known as the Public Health Service. In 1930, with a staff of only 133, it became the National Institute of Health. Working at the Hygienic Laboratory allowed Alice's interest in *Brucella* to expand.

Unfortunately, the "Spanish flu" influenza pandemic reached Washington, DC, in early October 1918. In a city filled with dislocated people, primarily women doing wartime work, who were crowded in garbage-strewn, slum boarding houses, thousands died. Alice was asked to stop her *Brucella* research and work on the flu, which she herself eventually contracted, but from which she subsequently recovered. After World War I finally ended, the returning veterans suffered horribly; most were destitute and moved into shanties and tents in the fields near the Capitol Building. Because they were ill clothed, hungry, living under crowded conditions, and vulnerable to disease, *Streptococcus* spread among them. Streptococcal disease was important to Alice, who had survived scarlet fever as a child. Alice was again asked to temporarily change the focus of her research, this time to epidemic streptococcus (Fig. 3).

Alice eventually returned to her research on *Brucella* and her call for pasteurization. Brucellosis is a respiratory airborne infection, and in 1922, Alice became infected, remaining in Johns Hopkins Hospital in Baltimore for more than 10 weeks. This was her first of five experiences in five hospitals over the next 9 years. Her longest stay was 14 months in duration. The tendency of this disease to incapacitate patients for long periods and its

FIGURE 3 Evans (left) with Rebecca Lancefield at the 1964 ASM meeting. The first two women presidents of ASM both studied the various species of *Streptococcus*. Courtesy of CHOMA/ASM Archives at UMBC.

potential use as a biological warfare agent resulted in the inclusion of *Brucella* on the Select Agents and Toxins List. The disease caused joint pain, fever, chills, night sweats, and crushing fatigue; internal lesions from *Brucella* were found during surgery in 1928. In spite of her suffering from brucellosis, Alice continued to work at the Hygienic Laboratory, where she always wore her immaculate white starched uniform.

In 1926, Alice was invited to serve as a member of the Committee on Infectious Abortion, of the National Research Council, Federal Department of Agriculture. This appointment was under Theobald Smith, who subsequently refused to serve as chairman when he found out that Alice would be a member. Again Henry Welch interceded and Alice served "without memorable incident" (1). Then, momentously in 1928, the 9 January issue of *Time Magazine* prominently announced the national news that "Alice C. Evans was elected President of the Society of American Bacteriologists at Rochester, New York" (1). Alice, at the age of 47, became the first female president of ASM (Fig. 4), although she could not attend the society meeting, as she was bedridden with brucellosis.

A fortuitous event occurred in 1929 that supported Alice's fight for pasteurization. Paul de Kruif, a well-known author who had published the book *The Microbe Hunters* in 1926 (7), had previously interviewed Alice several times. de Kruif published the article "Before You Drink a Glass of

FIGURE 4 Meeting of the Society of American Bacteriologists (now ASM) at the Jefferson Hotel, Richmond, VA, December 1928. Evans as president is seated center. Courtesy of CHOMA/ASM Archives at UMBC.

Milk" in the *Ladies' Home Journal* magazine in September 1929 (Fig. 5), which further inflamed the controversy on pasteurization (8). In addition, other authors (9) were accumulating evidence that supported Alice's conclusions regarding the spread of brucellosis in raw milk. Alice was even quoted as saying, "The fat was in the fire" (1).

In 1934, it appeared that Alice had finally won her fight, as the Federal Program for the Eradication of *Brucella* began. The program mandated that cattle be tested for the presence of *Brucella* and, if infected, be slaughtered. By 1947, most of the dairy cattle in the United States were certified "*Brucella*-free." Eventually, between the years 1964 and 1967, all the U.S. state governments voted to adhere to the Federal Pasteurized Milk Ordinance, Act 233, requiring all commercial milk to be pasteurized.

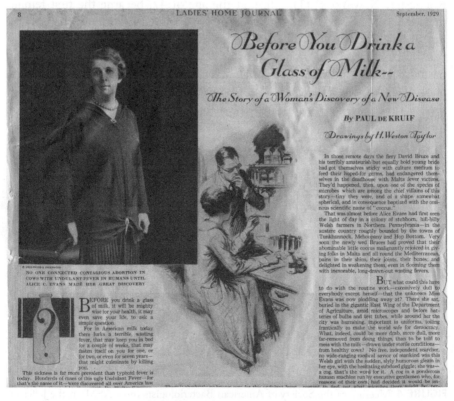

FIGURE 5 Paul de Kruif's article on Evans in the 1929 *Ladies' Home Journal*. Courtesy of CHOMA/ASM Archives at UMBC.

Today, brucellosis remains a significant disease of humans. In 1936, Alice estimated that the actual number of cases in the United States was at least 10 times the number of reported cases. During the 1960s, *Brucella* outnumbered all other agents as the cause of laboratory-acquired infections.

Alice retired at 64 years old in 1945 when WWII had just ended. Retirement meant a decent pension, and she lived independently until 1975. In 1966, when Alice applied for Medicare, the forms required swearing that one had no affiliation with the Communist Party. Alice refused, citing that it denied her a constitutional right. Interestingly, her application was processed without her signature. Despite her last episode of brucellosis in 1943, 21 years after her initial date of infection, Alice, in her retirement, followed the plight of chronic brucellosis patients. In 1961, Alice authored an article published in the *Journal of the American Medical Women's Association* campaigning for the better diagnosis of the chronic disease (10).

Alice was a worldly and modern woman. As early as 1928, Alice took her first airplane trip from Croydon, England, to Holland, just one mere year after Charles Lindbergh made his famous Atlantic crossing. Alice later traveled several times to Europe, visiting Wales, Scotland, London, Holland, and France, even though at that time the federal government did not pay for employees to attend meetings. Her niece remembered Alice as fashionably dressed, wearing a blue knee-length dress with a matching sailor hat and a carefully selected necklace and brooch, and with her brown hair lightened and cut short with style (Fig. 6, left side) (11). Alice also did extensive lecturing on the subject of females entering male-dominated careers. She is quoted as saying, "Women have proved that their mental capacity for scientific achievement is equal to that of men. Women do not receive the same recognition as those of men" (1). Alice established in 1969, through the American Association of University Women, a scholarship fund at the Federal City College.

In 1930, Alice was selected as the delegate to the First International Congress of Microbiology in Paris, and she was the only woman included in the book *Men Against Death*, authored by Paul de Kruif in 1932 (12). In 1934, she was bestowed an honorary medical degree from the Women's Medical College, now the Medical College of Pennsylvania. At that time, Alice received and cherished a letter from Marie Dressler, who won the Best Actress Oscar for her performance in the 1930 film *Min and Bill*.

Known as the Mother of Pasteurization in the United States, Alice received numerous awards and recognition. In 1936, she was awarded honorary doctor of science degrees from Wilson College in Chambersburg,

Figure 6 (Left) A later photo of Evans. Courtesy of CHOMA/ASM Archives at UMBC. (Right) Cover of the *SGM Quarterly* magazine, August 1995, designed by artist Mark Duffin. Reprinted with special permission from the publisher.

PA, and the University of Wisconsin. In 1936, Alice was again selected as delegate to the Second International Congress of Microbiology. In 1941, she was appointed delegate to the Committee on Brucellosis Meeting in Mexico and was elected President of the First Congress of the Inter-American Committee on Brucellosis in 1946. She continued to serve as honorary president of the Inter-American Committee on Brucellosis until 1957.

In her 80s, Alice authored her memoirs. She was included in the 1985 biographical book *Profiles of Pioneer Women Scientists* (13) and, at 89 years of age in 1970, was elected to the prestigious National Academy of Science. Posthumously, she was elected to the National Women's Hall of Fame in 1993 and especially honored as one of the four "Mount Rushmore" faces of microbiology on the cover of the *SGM Quarterly* magazine in August 1995 (Fig. 6, right side).

Alice never chose to be a role model, but her legacy is that she is the first professional female microbiologist to become one. She was a pioneer in leading other scientists, essentially all of whom at the time were male, to recognize and accept the validity of scientific research conducted by women. The research that she conducted, correctly identifying the causative organism of brucellosis, which resulted in changes to federal laws involving the pasteurization of milk, was outstanding, original, and independent. Alice's special inspirational qualities were her industriousness, perseverance, steadfastness, courage, and support of women. Markedly, she never declined an opportunity and "always stepped into the only suitable opening on her horizon." This is a thoughtful philosophy for women scientists to adopt today.

ACKNOWLEDGMENT
I thank Jeff Karr, Center for the History of Microbiology (CHOMA)/ ASM Archives at University of Maryland, Baltimore County, MD, for making available several documents.

CITATION
Findlay LA. 2018. Alice Catherine Evans: the shoulders upon which so many stand, p 75–85. *In* Whitaker RJ, Barton HA (ed), *Women in Microbiology*. American Society for Microbiology, Washington, DC.

References
1. Evans AC. 1963/1969. *Memoirs. ASM Corporate Archives and Collections at Library at University of Maryland, Baltimore County*. UMBC, Baltimore, MD.
2. Rogers LA, Clark WM, Evans AC. 1916. Colon bacteria and *Streptococci* and their significance in milk. *Am J Public Health (N Y)* **6**:374–380.
3. Bruce D. 1889. Observations on Malta fever. *Br Med J* 1(1481):1101–1105.
4. Evans AC. 1918. Further studies on *Bacterium abortus* and related bacteria I: the pathogenicity of *Bacterium lipolyticus* for guinea-pigs. *J Infect Dis* **22**:576–579.
5. Evans AC. 1918. Further studies on *Bacterium abortus* and related bacteria II: a comparison of *Bacterium abortus* with *Bacterium bronchisepticus* and with the organism which causes Malta fever. *J Infect Dis* **22**:580–593.
6. Evans AC. 1918. Further studies on *Bacterium abortus* and related bacteria III: *Bacterium abortus* and related bacteria in cow's milk. *J Infect Dis* **23**:354–372.
7. de Kruif P. 1926. *The Microbe Hunters*. Harcourt, Brace and Co, New York, NY.
8. de Kruif P. 1929. Before you drink a glass of milk. *Ladies Home J* **9**:8–9.
9. Hardy AV, Jordan CF, Borts IH, Campbell-Hardy GE. 1931. Undulant fever with special reference to a study of *Brucella* infection in Iowa. *Natl Inst Health Bull* **45**:158.
10. Evans AC. 1961. Chronic brucellosis: the unreliability of diagnostic tests. *J Am Med Womens Assoc* **16**:942–945.
11. Burns VL. 1993. *Gentle Hunter*. Enterprise Press, Englewood, NJ.
12. de Kruif P. 1932. *Men Against Death*. Harcourt, Brace and Co, New York, NY.
13. O'Hern EM. 1985. *Profiles of Pioneer Women Scientists*. Acropolis Books, Washington, DC.

Women in Microbiology
Edited by Rachel J. Whitaker and Hazel A. Barton
© 2018 American Society for Microbiology. All rights reserved.
doi:10.1128/9781555819545.ch10

Mary K. Firestone: Groundbreaking Journey of a Microbial Matriarch

10

Jennifer Pett-Ridge[1]

When Mary Firestone was little, she used to help her dad plant flowers, and she liked to sit on the ground with her hands in the dirt. That crumbly Oklahoma City, OK, topsoil was a world away from the heavy Puerto Rican rainforest clay from which she would eventually sort roots or the hard-as-a-brick pedons of Hopland, CA, where she and her graduate students studied the great CO_2 exhale that happens annually with the autumn's first rains. All that was in the future; to 4-year-old Mary Kathryn, it just felt good to push her fingers into the soil. By the time she was 12, Mary already pictured herself in a white lab coat and working as a scientist, probably because her father let her weigh and pour things in his water and soil testing laboratory at the Oklahoma Gas and Electric Company. She grew up watching her technically fearless mom repairing home appliances. Not surprisingly, she got into trouble in 7th-grade sewing class for taking apart the sewing machine; Mary didn't like the way it was stitching. But later, when Mary's graduate advisor, Jim Tiedje, came upon her in a similar situation, sitting cross-legged on the floor in the lab with a piece of equipment disassembled into many pieces, he simply said, "I'm not even going to ask!" and walked out. It may be the reason Mary survived in Jim's lab, first as a technician and later as a graduate student—she wasn't afraid of analytical instruments and could take them apart and put them back together.

Today, Mary K. Firestone (Fig. 1) is a professor of soil microbiology in the Department of Environmental Science, Policy, and Management at the University of California, Berkeley (UCB). During her 40 years on the

[1]Physical and Life Sciences Directorate, Lawrence Livermore National Lab, Livermore, CA 94550

Figure 1 Mary Firestone, professor of soil microbial ecology, Department of Environmental Science, Policy, and Management, UCB.

faculty, Mary has been active in university governance, serving in many roles, including Vice Chair and Chair of UCB's Academic Senate. She continues to enjoy taking things apart and reconstructing them, but now she more often applies this skill to complex microbial interactions in soils. A fellow of the American Geophysical Union, Ecological Society of America, Soil Science Society of America, and American Academy of Microbiology, Mary was surprised and pleased to be elected to the National Academy of Sciences in 2017. She has received numerous honors and awards throughout her career, notably the Berkeley College of Natural Resources Career Achievement Award (2013) and, most recently, the Natural Resource Ecology Laboratory Award of Excellence in Ecosystem Science (2017).

A particularly meaningful accolade came early in Mary's career, the Emil Truog Soil Science Award, which Mary received in honor of her dissertation. Emil Truog was an early-20th-century soil scientist, and the award in his name is given by the Soil Science Society of America to recognize "a recent Ph.D. degree recipient who has made an outstanding contribution to soil science as evidenced by his/her Ph.D. thesis or dissertation." In 1979, the members of Mary's department at Michigan State University nominated a young woman "because she had done this cool ^{13}N study" using a radionuclide with a half-life of 10 min to study the gaseous products of denitrification in soil following the onset of anoxic conditions (1–3). She showed with soils and cultures of *Flavobacterium* and *Pseudomonas* that sequential production of denitrification enzymes occurred as oxygen became depleted, resulting initially in N_2O, and later, primarily N_2 gas. Mary was rightfully proud of her award, and while out to dinner with her advisor and his wife, Linda Beth Tiedje, she showed them the certificate, which read, "To Mary Firestone, for his excellent research in soil science." Well. Linda Beth was having none of that. The next day, Jim asked, "Mary, can you send me the

Truog Award certificate, because I need to show it to everybody in the department." He showed it to the faculty who had nominated Mary, and they wrote a joint letter to the Soil Science Society of America, questioning the clear expectation that all Truog Award winners would be male. Soon afterward, Mary received a revised certificate.

Mary Firestone started working in Jim Tiedje's lab as an undergraduate summer worker; at the time, Jim was a young assistant professor 7 years her senior. Mary came with a solid foundation in math, physics, and biochemistry and convinced Jim to hire her despite a somewhat checkered academic record, which reflected the substantial time she had spent in antiwar demonstrations. She stayed on in the Tiedje lab as a technician after she finished her B.S. in microbiology. It was a great choice…and ultimately proved to be an excellent career move. Jim Tiedje is one of the most prolific and influential soil microbiologists of the past half-century and has mentored a generation of successful microbial ecologists. Among many topics, he has long been fascinated by the ecology, physiology, and biochemistry of denitrification. Mary's 1977 M.S. topic was the isolation and characterization of a tertiary amine monooxygenase (4), and it led her back into microbiology and, from there, soils. Apparently, working with soils resonated and perhaps linked her to the little girl who had sat in her backyard, digging holes. Whatever it was, her work in the Tiedje lab ignited a lifelong passion. When she started getting a little antsy after her master's, Jim asked, "Why don't you do a Ph.D.?" Though some faculty in her department questioned her on-farm agronomy credentials, Mary had long been a sounding board for the other members of the lab (she edited the thesis of Tiedje's first Ph.D. student [Box 1]) and had most of the coursework already taken. Her husband, Rick, a physicist, worked with her to make ^{13}N tracers, and she finished her dissertation in short order (1979). In the end, Mary received a great deal of support from her all-male soil science colleagues, support that lasted well into her career.

Box 1 An early mentoring experience

Mary was an undergrad during the Vietnam War era, which was very disruptive on campuses and engaged many students (including Mary). The first Ph.D. student of her advisor, Jim Tiedje, was from South Vietnam, supported by USAID. Mary volunteered to edit the first draft of his thesis as her contribution to their struggle of the times. That student went on to become a significant science and education leader in Vietnam, was elected Rector of Can Tho University and director of their first biotech center.

As Mary was completing her dissertation, she had a conversation with a prominent soil geochemist from Cornell, and he asked what kind of career path she was planning: "Are you going to apply for faculty positions?" She replied, "Do you know of any woman on a soils faculty in the country?" He did not say a word for 5 minutes, finally noting a young woman working as a postdoc in his own lab. Shortly after graduating, Mary accepted a temporary lecturer position at UCB. The job was not a tenure-track position, but she had been told before accepting it that the department would soon be hiring a tenure-track soil microbiologist, and she hoped that the Berkeley location held promise for a second job in nuclear chemistry for her husband. A few months later, in a faculty meeting, Mary was surprised to learn that the soil microbiology position was to be a full professor and departmental chair. She had assumed that the future hire would be an assistant professorship—one for which she could apply. Later, she mentioned to a few selected faculty, "What really galls me is there are people in this country who are going to think I did something wrong in this position!" Needless to say, UCB did eventually post a tenure-track assistant professor position, and Mary was hired into it. Was she the only woman in her department? "Of course! It felt like I was the only woman…in all of soil science. In the country," she recalls.

Being a successful junior faculty member meant being an excellent instructor and sometimes a little assertive (Box 2). In the classroom, Mary originally taught Soil as a Medium for Plant Growth, which had previously been led by Professor Kenneth Babcock, who was known for his rigorous infusion of physical chemistry into soil science. Apparently, some of the older graduate students initially doubted that the young, female newcomer could meet their high expectations. But Mary's version of the course was a great success; it won over skeptics, and for at least one of those early students (Tetsu Tokunaga), it sparked a career-shaping interest in diffusion and redox processes. In the years that followed, Mary was assigned to an office in an out-of-the-way space in Giannini Hall, but she really wanted to be in Hilgard Hall, where all the other soil science faculty and staff were housed.

Box 2 Mary, on being the only woman in the room

For years, I'd be in meetings with male colleagues and they'd be debating something that clearly couldn't work, because, let's say, the sky is blue. So I'd say, "But that wouldn't work because the sky is blue." There'd be a moment of silence, and the debates would continue. Five minutes later, a male colleague would say, "But the sky is blue," and everyone around the table would stop and say "Of course! What a brilliant insight! We're totally changing our thinking because of it!"

When an empty office became available in the basement, she asked the department chair if she could have it. The answer was an emphatic "No." And so she began asking every week, but the answer was always the same. Finally, frustrated, Mary just moved into the vacant office. The chair was angry, but he saw that she had already moved over all her things to the new space, and he finally said, "OK, I want you to ask me one more time." When she did, he finally responded, "Yes. Now it's OK."

When UCB hired Mary Firestone, she received no start-up funds. In retrospect, Mary notes, "They thought they were doing me a favor." She realized quickly that the way to survive was to build her own group of people with whom to interact. So, she put in her first grant proposal to the USDA, a project on denitrification, and it was funded. That allowed her to begin to equip her lab and supported her first master's student projects on denitrification. Josh Schimel then joined the lab as her first Ph.D. student, working on "belowground competition for nitrogen—plants versus microbes." Soon thereafter, a large cohort of graduate students and postdocs joined the lab, including Jenny Norton, Tom Keift, Ken Killham, and Steve Hart and, later, John Stark, Trish Holden, and Eric Davidson, all of whom have made profound impacts on the field. They worked on the microbiology of soil wet-up (5), soil nitrogen cycling (6), degradation of pollutants (7), carbon flow in the rhizosphere (8), and the soil effects of acid rain (9). They identified heterotrophic nitrification (10) and used ^{15}N pool dilution to measure the gross rates of nitrogen transformation between soil organic matter, NH_4^+, NO_3^-, and microbial pools (11). It must have been a remarkably stimulating time: three of the Firestone lab's most highly cited papers stem from this period (11–13).

When Mary and her then-postdoc Eric Davidson decided to write an NSF grant proposal on nitrous oxide, they discussed the idea that the rate of N_2O production was dependent on the rate of the processes that produced it, as well as biological and physical controls on the loss of N_2O. Mary had been trying to get this idea across for some years: "But you'd say that to people and their eyes would glaze over—nobody ever understood!" One day, she and Eric were sitting in her office, talking round and round on the topic when Eric finally said, "Oh, do you mean like this?" And he drew a pipe and put holes in it (Fig. 2). "There's N coming through the pipe, and some of it escapes?" "Yeah, that's it!" "Huh, that's interesting—now I understand." Mary insisted on putting the little drawing in the grant proposal, which got funded and later led to a highly cited and well-known conference paper. Mary has always been a fan of finding ways to simplify and humanize

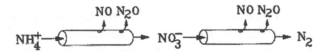

Figure 2 From "Microbiological Basis of NO and N_2O Production and Consumption" (12). The original caption is, "A conceptual model of the two levels of regulation of N trace gas production via nitrification and denitrification: (*a*) flux of N through the process 'pipes' and (*b*) holes in the pipes through which trace N-gases 'leak.'"

complex topics, and this became the first of many conceptual diagrams to come.

Unlike many faculty, Mary frequently encourages her students and technical staff to participate in writing proposals. Indeed, most leave her group having written at least 1 or 2 federal grant applications, some for millions of dollars. She sees enormous value in the process of slowly constructing a good hypothesis, imagining the experimental details, and finding a way to persuade the potential reviewer. As she likes to say, "You've gotta put all the sex and violence on the first page!" For Mary, writing a grant proposal is an inherently worthy intellectual process, independent of whether the project gets funded. Still, many of her students recall occasions when Mary pulled the plug on something they had worked very hard on—sometimes only days before submission. Though hard to swallow, her philosophy is an oft-shared piece of advice: "If you put garbage in front of reviewers, you get typecast. You need to think about how many people will review that proposal, and that they are forming long-term opinions of your science." This philosophy is even truer for that most intimate of proposals—the job interview. Says a former advisee, "Of course I'd expect an advisor to give feedback on my chapters, my talks, etc., but after that was done, she'd ask, 'So what are you going wear?' This stunned me at first. I had no idea how important it was, particularly for a young female, to dress a certain way to project a certain image. You don't want them to see you as a student, but as a potential colleague."

In 1983, Mary was pregnant. No one in her department knew, because she was up for tenure and she knew that if she didn't get approved on the first round, she was in trouble. She was still one of only a few women science faculty on campus. It was a scary time. She didn't want anybody voting on her tenure case to know she was pregnant and wanted to make sure her case got out of the department and the college before she was showing. Luckily, the timing worked out, and once she was 'showing', Mary went to the dean's office to say, "I'd like to take maternity leave." But at that time, there was no such thing as maternity leave at UCB. She was told to take a sabbatical, even though she protested, "But that's not what sabbaticals are for!" The chair of her department let it be known that (to him) maternity leave consisted of the period of confinement, i.e., the amount of time one spent in the hospital. But in the end, the Dean's office offered a compromise: "Write up what you would like and give it to us." So, Mary wrote (and was given) a fairly liberal allowance, with a semester off, no teaching, while remaining responsible for her graduate students and lab. Five years later, Mary got a call from the UCB Academic Senate, saying, "Mary, we've heard you codified a maternity leave policy. Would you please meet with the chair of our committee, who is trying to write a policy for the campus?" Mary passed along the policy she'd written, and a similar version still stands as the university maternity leave policy.

Like many young women, Mary faced the pull of caring for a child versus pursuing a career. But she was lucky to have a husband who was willing to split the day, with Mary working the morning shift and Rick going into the office in the afternoon. Still, there were tough times, particularly when two more babies came along: family demands meant that she had to say no to some offers that came her way. Some of those missed opportunities were hard to swallow. But in the long run, Mary got good advice: "Nobody ever gets to the end of their life and says 'I wish I'd spent more time at work.'" She recounts one particular occasion when "An older soil scientist called me up to do something, and I said, 'I can't, I've got a 1-year-old and a 3-year-old.' There was a long pause, and he replied, 'Oh my, I really wish I'd said that when I was your age.'" Now when she is talking to young women in her group about managing expectations after a baby comes, she counsels, "There is no good time to have kids, only impossible times. Just tell everyone that you're not planning on doing any work while on maternity leave. That way there is no pressure for you to do anything at all, and if you actually decide or are able to do something, then people will be pleasantly surprised."

At UCB, Mary served as chair of the Soil Science Department, the first chair of the Department of Environmental Science, Policy and Management, and later, chair of the UCB Academic Senate. In the mid-1990s, she helped shepherd the faculty through a difficult structural transition, forming what is now a forward-looking, multidisciplinary unit in environmental sciences that enjoys one of the highest academic reputations in the world. In 2012, the campus awarded her the Berkeley Faculty Service Award for her "outstanding and dedicated service to the campus, and whose activities as a faculty member have significantly enhanced the quality of the campus as an educational institution and community of scholars."

Somehow, among the demands of family, faculty service, and scholarships, Mary has always put her students first, particularly when it comes time for thesis writing. Many of her students fondly recall long sessions devoted to crafting a single hypothesis statement or the title for a manuscript. Says Josh Schimel, "Mary devoted an immense amount of time, energy, and patience dealing with the wonderful writing an inexperienced, sleep-deprived student was producing at 3 a.m. in the morning, day after day." For others, who might be loath to leave out any of their precious insights, she warns that crafting a story is as much about what you don't include as what you do include—and that some conclusions are better left stored safely on one's hard drive. Others have even more visceral memories about their thesis writing process: one says, "I will never forget the feeling of walking into Mary's office, and she'd turn her chair around, clear her desk, get a pen and a blank piece of scrap paper, and then look at you and you knew that she was ready to listen and to work. The feeling was of being really listened to, of being truly mentored."

Today, Mary Firestone holds the Betty and Isaac Barshad Chair in Soil Science, an endowed position named after a student of the legendary professor Eugene Hilgard, a founder of the field of soil science. She maintains an active lab group, and when one looks around for leaders in the field of soil microbial ecology, Mary Firestone, the people she trained, and increasingly (now) the people they trained form a critical nucleus. It is a telling comment about Mary's positive influence as a mentor that she has remained in close contact with nearly every graduate student and postdoc who has come through her laboratory, about 60 persons all told. It is also no accident that Mary's former advisees and the extended "Firestone Lab Family" have for decades gathered at conferences, to share a meal and laugh over good memories (Fig. 3). Her former students describe their time in her lab as a "transformative experience" and credit her with much of their career success (for at least one, this included facilitating their marriage!). Many speak of how she mentored every step of their career process over several years, using a variety of approaches, "…including kicking me under the table when needed." "She modeled the kind of faculty member I want to be: active and engaged in the department, committed to her students for the long haul, and, most importantly, cognizant that we are all human and must have a whole life as well as a research life." Mary has fostered this atmosphere of collegiality and respect in her lab for nearly 20 years, and it is a tribute to her as a scientist and mentor that such a broad network of ecologists and microbial ecologists has developed under her guidance (Fig. 4). In the words

Figure 3 Alumni of the Firestone group gather regularly at national conferences and Mary's annual potluck holiday party. Above, Firestoners at the Ecological Society of America Annual Meeting in 2002 and 2015.

of Mary's first Ph.D. student, Josh Schimel, now a professor at UC Santa Barbara, "I consider my decision to do my Ph.D. with her as the best professional decision I ever made. I do not believe that there is a better scientist or graduate mentor in the business."

ACKNOWLEDGMENTS

Thanks to Mary and Rick Firestone for interviews related to this chapter. Thanks also to Jim Tiedje, Tetsu Tokunaga, Josh Schimel, Jenny Norton,

Figure 4 Mary Firestone and Professor Jizhong Zhou in Ardmore, OK, embarking on a new project on carbon cycling and sustainable biofuels.

Eric Davidson, Tom Kieft, Don Herman, Teri Balser, Whendee Silver, Egbert Schwartz, Valerie Eviner, Christine Hawkes, Eoin Brodie, Pam Templer, Kristen DeAngelis, Jim Prosser, Dorthe Petersen, Romain Barnard, Dara Goodheart, Katerina Estera-Molina, and other current and prior members of the Firestone lab for sharing memories and input.

CITATION

Pett-Ridge J. 2018. Mary K. Firestone: groundbreaking journey of a microbial matriarch, p 87–97. *In* Whitaker RJ, Barton HA (ed), *Women in Microbiology*. American Society for Microbiology, Washington, DC.

References

1. Firestone MK, Tiedje JM. 1979. Temporal change in nitrous oxide and dinitrogen from denitrification following onset of anaerobiosis. *Appl Environ Microbiol* **38**:673–679.
2. Firestone MK, Firestone RB, Tiedje JM. 1979. Nitric oxide as an intermediate in denitrification: evidence from nitrogen-13 isotope exchange. *Biochem Biophys Res Commun* **91**:10–16.
3. Firestone MK, Firestone RB, Tiedje JM. 1980. Nitrous oxide from soil denitrification: factors controlling its biological production. *Science* **208**:749–751.
4. Firestone MK, Tiedje JM. 1975. Biodegradation of metal-nitrilotriacetate complexes by a *Pseudomonas* species: mechanism of reaction. *Appl Microbiol* **29**:758–764.

5. Kieft TL, Soroker E, Firestone MK. 1987. Microbial biomass response to a rapid increase in water potential when dry soil is wetted. *Soil Biol Biochem* **19**:119–126.
6. Schimel JP, Firestone MK. 1989. Nitrogen incorporation and flow through a coniferous forest soil profile. *Soil Sci Soc Am J* **53**:779–784.
7. Hunt JR, Holden PA, Firestone MK. 1995. Coupling transport and biodegradation of VOCs in surface and subsurface soils. *Environ Health Perspect* **103**:75.
8. Norton JM, Smith JL, Firestone MK. 1990. Carbon flow in the rhizosphere of ponderosa pine seedlings. *Soil Biol Biochem* **22**:449–455.
9. Killham K, Firestone M, McColl J. 1983. Acid rain and soil microbial activity: effects and their mechanisms. *J Environ Qual* **12**:133–137.
10. Schimel JP, Firestone MK, Killham KS. 1984. Identification of heterotrophic nitrification in a Sierran forest soil. *Appl Environ Microbiol* **48**:802–806.
11. Davidson E, Hart SC, Shanks CA, Firestone MK. 1991. Measuring gross nitrogen mineralization, and nitrification by 15 N isotopic pool dilution in intact soil cores. *Eur J Soil Sci* **42**:335–349.
12. Firestone MK, Davidson EA. 1989. Microbiological basis of NO and N₂O production and consumption in soil, p 7–21. *In* Andreae MO, Schimel DS (ed), *Exchange of Trace Gases between Terrestrial Ecosystems and the Atmosphere*. John Wiley & Sons, New York, NY.
13. Hart SC, Stark JM, Davison EA, Firestone MK. 1994. Nitrogen mineralization, immobilization and nitrification, p 985–1017. *In* Weaver RW, Angle S, Bottomley P, Bezdiecek D, Smith S, Tabatabai A, Wollum A, Mickelson SH, Bigham JM (ed), *Methods of Soil Analysis, Part 2. Microbiological and Biochemical Properties*. Soil Science Society of America, Madison, WI.

Women in Microbiology
Edited by Rachel J. Whitaker and Hazel A. Barton
© 2018 American Society for Microbiology. All rights reserved.
doi:10.1128/9781555819545.ch11

Lady Amalia Fleming: Turbulence and Triumph

11

Joudeh B. Freij[1,4] and Bishara J. Freij[2,3,4]

During preparatory work for a presentation on Sir Almroth Wright for the American Society for Microbiology Microbe 2016 meeting in Boston, MA, we came across the name of Dr. Amalia Voureka, a Greek physician in training at the Wright-Fleming Institute who was later to marry Sir Alexander Fleming. This piqued our interest for a variety of reasons, including our own Greek roots on one side of the family, Wright's resistance to women working at his Inoculation Department, and the 31-year age difference between Amalia and Alexander Fleming. Upon further investigation, we learned of Amalia's academic contributions, dedication to the cause of the politically persecuted, anti-Nazi wartime activities, jail stints, and eventual service as a member of the Greek Parliament. We were hooked!

EARLY LIFE AND CAREER

The second Lady Fleming was born Amalia Coutsouris to Greek parents in Constantinople, Turkey, in either 1909 or 1912. Her father was a dermatologist who practiced in Turkey until 1914, when his house and laboratory were confiscated, and the family fled to Greece.

Amalia's inclination was to study philosophy and become a writer, but her father wanted her to become a physician. She changed to medicine because she did not think that she would enjoy teaching philosophy after becoming a

[1]Department of Microbiology and Immunology, Johns Hopkins Bloomberg School of Public Health, Baltimore, MD 21205
[2]Division of Infectious Diseases, Beaumont Children's Hospital, Royal Oak, MI 48073
[3]Oakland University William Beaumont School of Medicine, Rochester, MI 48309
[4]Wayne State University School of Medicine, Detroit, MI 48201

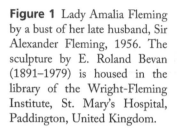

Figure 1 Lady Amalia Fleming by a bust of her late husband, Sir Alexander Fleming, 1956. The sculpture by E. Roland Bevan (1891–1979) is housed in the library of the Wright-Fleming Institute, St. Mary's Hospital, Paddington, United Kingdom.

professor; her father's dream of Amalia becoming a physician was thus realized. She studied medicine at the University of Athens, specializing in bacteriology, and graduated in 1938, after which she took a job at the city hospital of Athens. She also married her brother's friend, architect Manoli Voureka.

After the invasion of Greece by Axis forces in World War II, Amalia became involved in the Greek resistance. With the help of police, her wartime activities included helping to secure fake identification cards and ways of escape for Greek resistance fighters, Allied officers and soldiers, Jews destined for the concentration camp at Bergen-Belsen, Germany, and even Italian soldiers who did not wish to fight with the Germans. She used a house inherited from her maternal aunt as a hiding place for escapees before leaving the country. Other resistance activities included carrying radio transmitters around in fake potato sacks, and with money she made working at a small hospital in Nea Ionia, Amalia bought wool to make socks for soldiers.

Eventually, Amalia and 40 others were betrayed by a colleague who had been tortured by Italian forces. She was arrested in November 1941 and spent 6 months in Averoff Prison, where she was repeatedly threatened with execution during interrogations but was not physically tortured. Upon release by the Allied forces, she found her house destroyed and her estranged husband's studio in ruins. At the end of the war, Amalia found herself behind the times in terms of scientific advances. To remedy this, she applied for and won a 6-month British Council scholarship to study in London, beating out 44 other applicants (1, 2).

YEARS AT ST. MARY'S HOSPITAL

Amalia joined the Inoculation Department at St. Mary's Hospital in October 1946 as its first female researcher. She initially wanted to study viruses, but there were no available openings, so Sir Alexander Fleming agreed to have her work in his laboratory. The past chair of the department, Sir Almroth Wright, had made the Inoculation Department inhospitable to women, and Amalia was not allowed to eat with other staff.

The 6-month grant became 5 years, helped by money donated by philanthropist Ben May of Mobile, AL. Mr. May owned Gulf Lumber Co. and in 1947 wrote to Sir Alexander Fleming offering to help provide funding for "service of things that are worthwhile" (3). The result was the purchase of a phase-contrast microscope, which allowed Amalia to pursue research on *Proteus vulgaris* and other bacteria.

Amalia eventually left Sir Fleming's laboratory, albeit for a brief period of time, to become head of the bacteriology department at Evangelismos Hospital, Athens, in December 1951. During her years in Sir Alexander Fleming's laboratory, she published nine papers in three journals, four as the single author and three as the first author, although only three were with Sir Fleming (and published as Amalia Voureka) (4–12).

When his first wife died on 29 October 1949, Sir Fleming became despondent. Although Amalia was still legally married, she had been separated from her estranged husband for many years, and her fluency in many languages made her very helpful as an interpreter and translator, so she escorted Sir Fleming to numerous social functions. Through this closeness, Sir Fleming gradually began to emerge from his despondency, and he eventually proposed to Amalia during a trip to Athens in November 1952, despite a 31-year age difference. They were married on 9 April 1953.

RESEARCH AT ST. MARY'S HOSPITAL

Amalia's research at St. Mary's Hospital focused on bacteria and antimicrobial agents. This was cutting-edge research for its time, and she was highly productive. Her major research findings, in order of their appearance in the published literature, included demonstration of variable endpoints for streptomycin bacteriostatic power depending on inoculum size, solution type (e.g., broth, saline, or distilled water), and anaerobic conditions (4), and showing that penicillin sensitivity of some resistant bacteria, such as penicillinase-producing *Staphylococcus aureus*, can be restored by mixing them in culture with other sensitive bacteria (5); she also studied the frequency of penicillin resistance among staphylococci recovered from the nostrils of asthmatic patients (6).

The phase-contrast microscope acquired through the philanthropy of Ben May was a powerful tool in Amalia's work: she described a method of growing bacteria in slide culture for examination with this technology

(7), reported on the effects of penicillin in various concentrations on the morphology, motility, and growth characteristics of *P. vulgaris*, and demonstrated that flagellar movement responded to radiant heat (8). She subsequently reported on the recovery of bacteria with altered morphological and cultural characteristics with either transient or permanent loss of virulence in patients treated with chloramphenicol (9). Amalia expanded these observations with *in vitro* studies on coliforms exposed to chloramphenicol with or without specific antiserum, and she demonstrated the emergence of atypical bacterial colonies that could revert back to normal after multiple subcultures (10).

Amalia's work in her later research years focused on *S. aureus*. She exposed this bacterium to a variety of injurious agents, such as chloramphenicol, terramycin, nitrogen mustard, and hydrogen peroxide, and demonstrated that new bacterial variants emerged with altered biochemical properties that were transmissible to their progeny (11). This suggested to her that the alterations induced by these chemicals were genetic in nature. In her last published work, she compared virulent mouse *S. aureus* strains to ones rendered avirulent by exposure to chemicals and showed the difference to be due to inability to produce α-hemolysin, dermonecrotic toxin, and Panton-Valentine leucocidin (12).

RETURN TO LONDON

After Amalia's marriage to Sir Alexander Fleming, they spent their honeymoon in Cuba and the United States while he was on a lecture tour. Unfortunately, their marriage was not to last long. Sir Fleming died of a heart attack on 11 March 1955, after less than 2 years of marriage. Lady Amalia Fleming remained in London after her husband's death, becoming a "professional widow" devoted to perpetuating his memory and achievements. She was described as "deeply suicidal" after Sir Fleming's death and spent minimal time in the research laboratory, and her scientific output ceased (2, 13). Lady Fleming commissioned a biography of her late husband by André Maurois.

RETURN TO GREECE

Lady Fleming began spending more and more time back in Greece after her husband's death, and she decided to move back permanently in 1967. However, 5 weeks after she returned to Greece, a military coup on 21 April 1967 established the rule of the right-wing Regime of the Colonels. Lady Fleming became an active supporter of the politically persecuted and thus became an opponent of the junta, especially because political prisoners were being tortured and their families reduced to poverty.

In 1970, she testified in the defense of 34 intellectuals accused of plotting to overthrow the government. Lady Fleming was arrested on 31 August 1971 with three others (a Greek American divorcé from Minneapolis, MN, a theology student from Pennsylvania studying at Athens University, and a prison guard) for her involvement in an escape plan for a political prisoner, Alexandros Panagoulis, sentenced to death for the attempted assassination of Georgios Papadopoulos, who was the head of the military coup. She was kept in jail for a month and suffered bullying interrogations, but she was released because of poor health (she had developed diabetes and cardiac disease). Her trial, on 28 September 1971, was brief, and she was given a 16-month prison sentence; however, Amalia was in jail for only a brief period. Lady Fleming was deported to London on 14 November 1971 and her Greek citizenship revoked. She had to be forcibly put on an airplane from Athens to London and initially refused to disembark after her arrival to England. Lady Fleming published *A Piece of Truth* in 1973, a book about the struggles against the Greek military junta, hers and those of others.

FINAL RETURN TO GREECE

Lady Fleming returned to Greece from exile in 1974, after the fall of the junta. She joined PASOK (the Panhellenic Socialist Movement), later became head of the Greek Committee for Amnesty International, and was active in several other human rights organizations. Lady Fleming ran for the Greek Parliament and was elected to office in 1977, 1981, and 1985. She helped establish the Athens Biomedical Sciences Research Centre Alexander Fleming, which began operation in 1997, and a suburban Athens hospital in Melissia is named after her. Lady Fleming passed away on 26 February 1986.

FINAL THOUGHTS

In *A Piece of Truth*, Lady Fleming described herself as "by nature un-believably lazy—or, to tell the truth, an unbelievably lazy person who has always had to work very hard throughout her life; and this is perfectly appalling" (14). We doubt that many would agree with that assessment.

CITATION

Freij JB, Freij BJ. 2018. Lady Amalia Fleming: turbulence and triumph, p 99–104. *In* Whitaker RJ, Barton HA (ed), *Women in Microbiology*. American Society for Microbiology, Washington, DC.

References

1. Uglow JS (ed). 1998. *Dictionary of Women's Biography*, 3rd ed, p 205–206. Northeastern University Press, Boston, MA.
2. Maurois A. 1959. *The Life of Sir Alexander Fleming: Discoverer of Penicillin*. E P Dutton & Co, Inc, New York, NY. (Translated from the French by Gerard Hopkins.)
3. Hoffman R. 2001. Ben May: the quiet philanthropist, p 288–295. *In* Hoffman R (ed), *Back Home: Journeys Through Mobile*. University of Alabama Press, Tuscaloosa, AL.
4. May JR, Voureka AE, Fleming A. 1947. Some problems in the titration of streptomycin. *BMJ* 1:627–630.
5. Voureka A. 1948. Sensitization of penicillin-resistant bacteria. *Lancet* i:62–65.
6. Voureka A, Hughes WH. 1949. Frequency of penicillin-resistant staphylococci. *BMJ* 1:395.
7. Voureka A, Fleming A. 1949. Staining of flagella. *J Gen Microbiol* 3:xxiii.
8. Fleming A, Voureka A, Kramer IR, Hughes WH. 1950. The morphology and motility of *Proteus vulgaris* and other organisms cultured in the presence of penicillin. *J Gen Microbiol* 4:257–269.
9. Voureka A. 1951. Bacterial variants in patients treated with chloramphenicol. *Lancet* i:27–28.
10. Voureka A. 1951. Production of bacterial variants *in vitro* with chloramphenicol and specific antiserum. *Lancet* i:29–31.
11. Voureka A. 1952. Induced variations in a penicillin-resistant *Staphylococcus*. *J Gen Microbiol* 6:352–360.
12. Voureka A, Ogilvie AC. 1952. Toxin production in two strains of *Staphylococcus* and their variants. *J Gen Microbiol* 7:48–53.
13. Brown K. 2004. *Penicillin Man: Alexander Fleming and the Antibiotic Revolution*. Sutton Publishers, Phoenix Mill, United Kingdom.
14. Fleming A. 1973. *A Piece of Truth*. Houghton Mifflin, Boston, MA.

Women in Microbiology
Edited by Rachel J. Whitaker and Hazel A. Barton
© 2018 American Society for Microbiology. All rights reserved.
doi:10.1128/9781555819545.ch12

Katrina T. Forest: A Renaissance Woman in Microbiology

12

Katherine McMahon[1]

Dr. Katrina Forest is a full professor in the Department of Bacteriology at the University of Wisconsin-Madison (UW-Madison), and she probably identifies more as a biophysicist than as a microbiologist (Fig. 1). Her formal training in applied biology and molecular biology was done at MIT and Princeton, respectively. However, she fell in love with microbes in college and has been embedded in a microbiology department for 20 years. She joined the faculty at Madison as a new assistant professor in January 1998. I first had the opportunity to work closely with her in 2009 when we codirected a residential college for women in science and engineering, and more recently have collaborated with her on a few research projects. In addition to being my colleague, she is one of my dearest friends, and I was pleased to interview her multiple times while preparing this chapter. I was curious about the advice she would have for me and other women embarking on a career in microbiology, but I was also excited to have an excuse to learn more about her own winding path.

Katrina (Katy) is an outstanding example of a highly accomplished woman who came to love microbes by an unconventional route. She teaches a course that is likely completely unique among microbiology programs: Microbiology at Atomic Resolution. A true renaissance woman, Katy reads and thinks broadly, is not afraid to try new things, loves languages and living abroad, and is almost always formidably energetic.

---

[1]Department of Bacteriology, University of Wisconsin-Madison, Madison, WI 53706

FIGURE 1 Photo of Dr. Katrina Forest.

When I asked Katy to highlight some "life lessons" that would help to guide and inspire other women in microbiology, she came up with these four things:

1. Don't rush. It is easy to be in a hurry to move onto the next thing, but often it's better to stay on pace rather than to sprint past the next milestone ahead of schedule, and risk not savoring the outcome of the current one.
2. Do not underestimate the value (both short-term and long-term) of self-care and life balance. Pushing your limits leads to burnout, which ultimately will hurt your career.
3. Read broadly and attend seminars/talks that are far outside your specialty. You never know what kinds of opportunities or inspiration will strike you. Keep your eyes and ears wide open for new ideas.
4. Don't be afraid to reinvent yourself over time. Your Ph.D. and postdoc topics do not define you, and neither do the first 10 years of your faculty position!

Her life story shows how she developed this wisdom. Katy Thompson graduated from Oregon High School in Wisconsin in 1984 but has never considered herself to be a bona fide Midwesterner. She spent most of her youth in Maryland and spent a year abroad in Germany during high school. Katy was eager to return to the East Coast when applying for entry to college. She was considering majoring in either languages and linguistics at Georgetown University or physics at MIT. Curiously, it was a German class that she sat in on while visiting MIT (plus an inviting picture of a sailboat on the cover of their recruiting brochure) that sealed the deal on attending MIT.

When Katy started at MIT as a freshman, she briefly considered architecture as a major but declared physics. However, she was smitten with a love for the biological sciences after taking biology as an elective course, and she switched her path. This decision to switch to biology was further reinforced when she learned about structural biology in a biophysical chemistry course and, in an "a-ha!" moment, realized that it had grabbed her attention. Although her path in biology continued from there, she credits her early training in the physical sciences with much of her unique perspective on microbiology. She was fortunate to get involved in an undergraduate research project early on, with Daniel Wang in the Department of Chemical Engineering. There she worked on increasing the production of xanthan gum in *Xanthomonas*. Katy is notably proud of the fact that her first experience working with living things in a research context did not involve a traditional model organism like *Escherichia coli*!

Katy met her husband, Cary Forest, when she was finishing up high school in Oregon, WI. Cary is 2 years older and was attending UW-Madison as a physics major when they started dating. They took a short break while Katy was at MIT, but they got back together again in time for Katy to follow Cary to Princeton, where they both earned Ph.D.s. She notes that this was probably the first example of her being in maybe too much of a hurry: instead of savoring her time at MIT she finished her B.S. 1 year early, in 1987, in order to follow Cary.

At Princeton, Katy studied with Clarence Schutt and worked on solving the structure of pertussis toxin; she earned a Ph.D. in molecular biology in 1993. Katy feels fortunate to have trained as a crystallographer at a time when the field of crystallography was small and tight-knit. Schutt's mentoring style was grounded in believing that his mentees were smart and capable of great things; he was largely hands-off but supportive when students came to him for guidance.

One of her defining memories during this time is of a first-year course led by Professor Tom Silhavy, in which they read seminal papers in the broad field of molecular biology. She remembers initially being terrified in class but eventually building an incredibly valuable sense of self-confidence in dissecting and critiquing work, even outside her field of direct expertise.

In a similar vein, it is noteworthy that Katy initially started her Ph.D. research trying to solve the structure of a viral RNA but switched projects to pertussis toxin more than 2 years into her time at Princeton (an example of not being afraid to reinvent oneself as required). She remembers clearly the moment when a postdoc in the lab suggested trying electron microscopy to tackle the pertussis toxin structure, which she then vigorously pursued through a collaboration at Rutgers University. The crystal structure of pertussis toxin was eventually published by another lab, effectively scooping her. However, Katy doesn't remember being particularly upset or set back by this. She proudly reflects on the fact that the techniques she developed during her Ph.D. work were ultimately used to solve the structure of the Lac repressor. When pressed, she admits to being disappointed that she didn't try to publish her work on pertussis toxin, or more that her advisor did not push her to do so. However, the fact that she was able to pick up and dust off, move forward, and stay enthusiastic about her work is testament to the kind of tenacity required to be successful in academia. As her time in graduate school came to an end, she was still undecided about being a faculty member, since she was genuinely interested in entering the nascent biotechnology industry. In any case, she decided to pursue a postdoctoral position and think about it more.

Katy is forthcoming with some very personal details about her time as a trainee. She and Cary very much wanted a baby while in graduate school but were unable to have one. They rushed on to postdoc positions in San Diego in 1993, with Katy going to the Scripps Research Institute. Katy had a fellowship and could set her own research agenda while solving the first crystal structure of a type IV pilin. She took time to have her daughter Tess in 1994 and son Gabriel in 1996. When asked about how her principal investigator responded to her pregnancies, she said he didn't even have a chance to object. She recounted taking a very proactive approach, meeting with him to explain how she planned to work from home until she could come back to the lab. She noted that there was no leave policy for postdocs, so she created her own plan. Katy says her overwhelming memories of "science-ing" with babies are filled with joy and with being very, very tired. She wishes that she could have slowed down more and spent more time with her babies.

Late in 1996, Cary decided to start his search for a faculty position. With a 2-year-old and an infant, Katy wasn't ready. But Cary charged on and secured job offers at multiple universities. Katy interviewed at each and ultimately secured her position in the Department of Bacteriology at UW-Madison. The family moved to Madison in 1997, and Katy spent 6 months writing and setting up her lab before officially joining the faculty in January 1998.

Katy remembers being so grateful that her nascent colleague, collaborator, and friend Heidi Goodrich-Blair also arrived at UW-Madison in 1997. They were both young female faculty with small children in a department composed (mainly) of men. They supported one another and took risks together. In 2000 they proposed a cluster hire with the theme of symbiosis and were amazed to have it approved by the department and university. It was interesting to Katy that the other department faculty were willing to let two very junior faculty take on such an intensive project, which ultimately distracted greatly from their research programs. Together they ran the searches that brought multiple high-profile faculty to campus, creating a legacy that persists today, even after Heidi's recent move to chair the Microbiology Department at the University of Tennessee. When asked if it was worth it, Katy says, "Of course!"

As a pretenure faculty member, Katy remembers early successes (her first NIH grant), teaching a brand new and unique course of which she is very proud (Microbiology at Atomic Resolution), and having trouble recruiting graduate students as a biophysicist in a bacteriology program. She remembers lots of sleep deprivation, working (too) hard, and straining to raise two small children with little support. When asked what makes her most proud about her time at Madison, she provides a curious story. While a first-year assistant professor, Katy was attending a symposium featuring summer undergraduate research projects from a broad collection of disciplines. She remembers an animal sciences student presenting on her efforts to develop a cryopreservation method for embryos. Katy was quick to see the connection to her own skills: crystallographers are wizards of cryopreservation. She teamed up with a postdoc in the animal science lab and developed a technique that was wildly successful. The work led to a patent and a crossroads: Katy was tempted to leave her faculty position to start a company to bring the technology to market, but she ultimately opted to stay a professor. Today, the technology is used by virtually every clinic providing *in vitro* fertilization for couples eager to have their own biological children. This kind of success, measured not by number of citations or even royalties, very much defines Katy. She is not afraid to step outside her discipline to pursue something that interests her. The outcomes can have wide impact, but not very many people may know about it.

As a more senior faculty member, Katy continues to pursue her passions and reinvent herself. She speaks of how rewarding it is to collaborate with faculty in divergent disciplines. Scientifically, she is probably most well-known for her work on type IV pili and bacterial phytochromes. But lately she has joined teams studying light-harvesting proteins in freshwater bacteria, proteins that allow bacteria to specifically colonize their nematode hosts, and lectins. She was recently the driving force behind a major new initiative at UW-Madison to actively support research computing. A few years ago Katy was part of the "What is Human?" initiative through the

Center for the Humanities at UW-Madison, which explored how human ideas are transforming as a result of recent advances in science and technology. She is also always looking for ways to blend her science with art, including hosting "artists in residence" in her lab from time to time (Fig. 2).

In my conversations with Katy about this chapter, and in so many other conversations over coffee or bourbon, Katy has emphasized the need to slow down and take care of yourself. It is easy to be in a rush to move on to the next step, be it finishing your Ph.D. (even with unfinished papers languishing) or your postdoc work (with unfinished projects or papers languishing) or through your pretenure years. It is a challenge to figure out when to pounce on an opportunity that may disappear with time and when to pass in order to finish something else. But don't feel like you have to jump at every single opportunity. Katy always qualifies this outstanding advice by noting that emphasis on self-care is not meant to imply that science is a 9-to-5 job. Being successful requires that you are willing to burn the candle at both ends when needed to meet a grant deadline or push through a pile of commitments. Science has always been, and will always be, hard work. But don't try to subsist on 4 hours of sleep day in and day out. You might feel invincible as a postdoc or junior faculty member, but it will catch up with you. Self-care and pacing should be part of your career plan, lest you burn out

FIGURE 2 A work created by artist-in-residence, Douglas Bosley, titled "LD:4334.001." An observer sifting through a crystal trial looking for suitable crystals. A compression gradient has created crystals of varying size. "LD:4334.001," Doug Bosley, accessed February 6, 2018, http://www.dougbosley.com/items/show/8. Image ©Douglas Bosley (2014), reprinted with permission.

later. Lack of sleep and too much stress take a real and measurable toll on your brain and body.

When I asked her about the biggest challenges she has faced, or continues to face, as an academic, Katy cited many of the problems burdening modern academics in general: an erosion of administrative support that distracts from the work that they were trained to do, the (thankfully rare) collaborator with an outsized ego, and mentees experiencing more and more pressure to succeed combined with mismatched expectations about what it takes to be a successful scientist. On the whole, however, she is relentlessly positive. More than anything, Katy loves her work. She loves microbes and she loves structural biology. She adores her children and UW-Madison. She will not hesitate to drop everything and help a friend who is struggling, and I have been the beneficiary of this generosity many times. I feel very fortunate to have her as a colleague, friend, and role model and sincerely hope that this chapter has in some way introduced you, the reader, to this great scientist.

CITATION

McMahon K. 2018. Katrina T. Forest: a renaissance woman in microbiology, p 105–111. *In* Whitaker RJ, Barton HA (ed), *Women in Microbiology*. American Society for Microbiology, Washington, DC.

Women in Microbiology
Edited by Rachel J. Whitaker and Hazel A. Barton
© 2018 American Society for Microbiology, Washington, DC
doi:10.1128/9781555819545.ch13

Elodie Ghedin: Unlocking the Genetic Code of Emerging Outbreaks

13

Tamara Lewis Johnson[1]

As a scientist, innovator, and trendsetter, Dr. Elodie Ghedin stands out in the field of genomics. From her start as a curious young girl to her early field work in West Africa to her years at the National Institutes of Health and the Institute for Genomic Research (TIGR). Ghedin stays on the cutting edge of genomics research by honing her intellectual genius to track parasites and viruses during emerging outbreaks. In this chapter, I explore the intrinsic motivations of a scientist who seeks to understand how and why parasites and viruses infect humans. Through her use of genomics, Ghedin has collaborated with many scientists and clinicians to address the most pressing public health issues of our day. Her brilliance in deciphering major and minor influenza virus strains may one day provide a roadmap for developing safer and more effective influenza vaccines. Through her intricate and breathtaking genomics research, scientists have been able to detect, track, and monitor emerging infectious disease outbreaks in real time and to prepare for pandemics in the future. A MacArthur Genius and Kavli Science Fellow, Ghedin leverages the power of large-scale data and mathematical modeling to track biomarkers as a means to improve point-of-care testing for infectious diseases. In addition to her scientific achievements, Ghedin is also an outstanding mentor of the next generation of computational biologists. Her unique blend of skills and talents enables her to be a highly valued innovator in the burgeoning field of genomics.

I first met Elodie Ghedin at the 2012 NIAID International Women's Day Lecture held in Wilson Hall on the main campus of the National

---

[1]Office for Research on Disparities and Global Mental Health, NIH/NIMH, Rockville, MD 20892

Figure 1 Elodie Ghedin, Ph.D., Director of the Center for Genomics and Systems Biology at New York University.

Institutes of Health. The Women's Health Research Working Group of the National Institute of Allergy and Infectious Diseases (NIAID) invited her to give the keynote address, which was entitled, "Genomics and Neglected Tropical Diseases: The View from My Half of the Human Race" (1). The purpose of the lecture was to highlight the work of women scientists whose infectious disease research advances the health of women and girls. At the podium, she spoke passionately about the health of women and girls who had developed neglected tropical diseases. Her passion derived from somewhere deep within and was brought to the surface by an intense immersion experience she had had several years prior.

In 1989, Ghedin completed a bachelor's degree in biology from McGill University. While pursuing field research for her master's degree in environmental sciences, she collected data on the water quality of wells in rural villages of Senegal and Guinea. When collecting water samples in the Lac de Guiers in Senegal, she observed children with distended bellies playing in the water in a nearby lake. In the heat of the day, mothers approached her seeking access to drugs for diseases their children had developed. Although malnutrition was a public health concern in West Africa, she suspected that the children were also being exposed to parasites, such as *Schistosoma* in the snails swimming in the water or *Leishmania* from the bites of infected sandflies, leading to enlarged spleens and distended bellies (2). At the time, Ghedin was working in collaboration with the Pasteur Institute and the Canadian Development Agency and her research focus was environmental science, not public health, but remedies to address parasite-related diseases were sorely needed in this region of the world. At that moment, her research interests shifted to infectious diseases.

After she received her master's degree in environmental sciences from the University of Quebec in Montreal in 1993, Ghedin returned to McGill University to get a doctorate in molecular parasitology. As a graduate student, she worked in Dr. Greg Matlashewski's lab at the Institute of Parasitology, where she studied leishmaniasis. After Ghedin completed her Ph.D. in molecular parasitology at McGill University in 1998, she went to the National Institutes of Health to further her research skills in molecular parasitology as a postdoctoral fellow in the Laboratory of Parasitic Diseases. While working in the intramural lab of Dr. Dennis Dwyer at NIH, she interacted with Dr. Michael Gottlieb, an extramural program officer at NIAID's Division of Microbiology and Infectious Diseases. Before working at NIAID, Gottlieb operated a lab at Johns Hopkins University in Baltimore, MD. One of the principal investigators, whose grant Gottlieb administered, was decoding the genomes of the parasites that cause African sleeping sickness (*Trypanosoma brucei*) and Chagas disease (*Trypanosoma cruzi*) at TIGR, now known as the J. Craig Venter Institute (http://www.jcvi.org). Gottlieb suggested that Ghedin conduct her research at TIGR, and he recommended her for a postdoctoral fellowship position.

At TIGR, Ghedin worked on the *Trypanosoma* genome project. She worked closely with her mentor, Najib El-Sayed, to hone her skills in bioinformatics. She was then promoted to principal investigator and became a faculty member of the Parasite Genomics Group. The research environment at TIGR supported risk taking, in which Ghedin was able to explore an area of research independently but still seek mentoring support from her colleagues. Claire Fraser, the director of TIGR, took interest in Ghedin's research and mentored her. Ghedin collaborated with other genomics experts to sequence the genomes of three parasites (*Trypanosoma cruzi*, *Trypanosoma brucei*, and *Leishmania major*). Their findings were published in 2005 (3). Her mentors describe her as a "tremendously creative scientist. Her genius lies in her ability to apply genomic analyses to investigate a broad range of infectious diseases" (4). While at TIGR, Ghedin developed a project, with collaborators, to investigate the evolution of influenza virus on an unprecedented scale. In 2005, Ghedin and colleagues published the first large-scale U.S. study of the evolution of influenza virus, showing that new strains could emerge each flu season, with no cross-seasonal persistence. Elodie designed a special-purpose pipeline capable of processing thousands of flu virus samples. She and her team came up with innovative approaches to decode the genomes of unknown flu virus samples. The technique she developed became the foundation of several large-scale government-initiated early-warning programs used today (4).

When her endocrinologist husband got a faculty position at the University of Pittsburgh's School of Medicine, Ghedin joined the faculty as an assistant professor in the Division of Infectious Diseases from 2006 until 2009. These were some of her most productive years. She published more than 30 articles describing her genomic research on influenza virus and on parasites such as *Leishmania* and *Brugia malayi* in leading research journals. In 2010, Ghedin received the Chancellor's Distinguished Research Award and moved to the Department of Computational and Systems Biology—where she was one of only two women on the faculty—and the Center for Vaccine Research at the University of Pittsburgh.

Ghedin's work demands a wide range of skills and knowledge, but she states that "researchers who mine genomic data are still undervalued" (4). Today the bioinformatics field is flourishing. Interdisciplinary collaboration is clearly the way forward. In 2011, Ghedin received a MacArthur Fellowship. This program awards unrestricted fellowships to talented individuals who have shown extraordinary originality and dedication in their creative pursuits and a marked capacity for self-direction. MacArthur Fellows may use their fellowship to advance their expertise, engage in bold new work, or, if they wish, change fields or alter the direction of their career. The MacArthur Fellows program gave Ghedin access to a cadre of innovators in different fields and exposure of her talents to larger audiences (https://www.macfound.org). In 2012, Ghedin was promoted to the rank of associate professor with tenure at the University of Pittsburgh's Department of Computational and Systems Biology. That same year, she also became a Kavli Frontiers of Science Fellow. The Kavli Foundation and the National Academy of Sciences select scientists in the fields of astrophysics, nanoscience, biology, neuroscience, and theoretical physics and bring them together at an interdisciplinary symposium to promote cross-discipline relationships. The symposia, which are both national and international in scope, enable emerging young scientific leaders to become acquainted with their counterparts in a broad range of disciplines and to stimulate long-term relationships with their peers, as well as to become acquainted with colleagues in other nations (http://www.kavlifoundation.org). Through these programs, she developed a rich network of relationships with both scientists and nonscientists.

When asked about her approach to mentoring young investigators, Ghedin shared a number of astute strategies:

- **Be willing to fail.** More experiments will fail than succeed. This means you are asking the hard questions. The important thing is to keep asking the questions and testing hypotheses.

- **Be a good communicator.** Improve oratory and written skills. Ghedin puts effort in critiquing the writing of young investigators and in teaching them how to present their research to an audience.
- **Try a hands-off approach to mentoring.** She allows her students to work on their own and often on things she is not familiar with. This enables her to learn from her students.
- **Send students to scientific meetings to get exposure to cutting-edge science and to network.** She also encourages women to join women in science networking groups, which enable students to connect with other scientists outside of the lab.

I recently interviewed Laura Tipton, a graduate student whom Ghedin mentors. From the beginning, she found Ghedin to be enthusiastic and encouraging of her pursuit of a Ph.D. in computational biology. Tipton started her Ph.D. studies in 2012 and joined Ghedin's research lab at the University of Pittsburgh in 2013. "Ghedin has always been supportive about allowing students to pick their own research project," she remarked. Tipton's work is focused on the human lung microbiome. Tipton also stated that Ghedin "has a broad network of colleagues" and often connects her graduate students to her network of scientists as a resource for exploring new areas of research or research tools. From Ghedin, Tipton has learned that "effective mentors are supportive of the people they mentor. Good mentors stand out by tailoring the mentoring experience to the needs of the person being mentored."

On 1 April 2014, Ghedin was appointed professor of biology and public health at the Center for Genomics and Systems Biology and the College of Global Public Health at New York University (NYU). Almost a year before her new appointment at NYU, she competed for and received a substantial research grant from the NIH. Her research focuses on developing computational models that integrate large-scale omics data collected over the course of influenza A virus infections that predict severe disease outcome. Her goal is to leverage the power of large-scale data and mathematical modeling to identify risk-stratifying prognostic biomarkers that could be used in the development of point-of-care testing (5).

Ghedin was listed in Thomson Reuters Highly Cited Researchers for 2014 and 2015 (https://www.thomsonreuters.com). This study recognizes nearly 3,000 scientists who published the greatest number of articles ranking among the top 1% by citations received in their respective fields in each paper's year of publication. Analysts assessed more than 120,000 papers indexed between 2003 and 2013 throughout each area of study.

As I reflect on Dr. Ghedin's scientific career, I note that she has taken a nontraditional path through very different research environments. Working first as a young investigator at TIGR, she conducted her research at a private research institute that was on the cutting edge of the genomics field. TIGR's research environment was set up to allow an abundance of creative freedom at this early phase of her career, which enabled her to learn a tremendous amount about the coding potential of infectious agents. At the University of Pittsburgh, she interacted with physician scientists and clinicians at the School of Medicine, which compelled her to think more about the translational aspects of her research and to incorporate the human side of the host-pathogen interactions into her research. At NYU, she states, "I am in a truly multidisciplinary and interdisciplinary scientific environment, and it has reignited the science fire in my belly (and how I felt when I was at TIGR). I feel like I can do anything. My colleagues are extraordinary."

When I asked her to consider a watershed moment since the MacArthur award, she stated emphatically, "The second wave of genomics enabled by new technology is allowing us to answer questions about pathogens and infections that we could not answer with previous tools. [This new technology] is empowering." In a recent interview on her unique set of research skills, she mentioned that she learned early on to not focus too keenly on only one aspect of a problem, but to look at it from different perspectives. This skill has enabled her to be exceedingly effective at being a supporter of the research of others who could bring in the required expertise to better understand the problem. Today, her approach to research enables her to work collaboratively with scientists in other parts of the world and across disciplines. For example, she recently completed genomic analyses of myxomatosis in collaboration with Australian scientists. Myxomatosis is a highly infectious viral disease of rabbits artificially introduced in Australia to reduce the rabbit population. Over time, the rabbits developed resistance to the virus, which is only 50% as effective in reducing the rabbit population as it had been in the past. Her research is focused on exploring the evolution of virulence of this disease at both the phenotypic and genomic scales to explain the evolutionary pressures acting on the pathogen (6).

Ghedin recently coauthored a paper in *Nature Genetics*, in collaboration with a team of scientists, about the role that minor flu virus strains play in transmitting diverse strains of influenza virus among humans (7). In a recent NYU press release, Ghedin explains "that a flu virus infection is not a homogenous mix of viruses but, rather, a mix of strains that get transmitted as a swarm in a population." This team of scientists examined flu virus samples collected in Hong Kong during the first wave of the 2009 flu pandemic to determine how many viral particles are transmitted from an infected donor and able to replicate in a newly infected recipient. To investigate this phenomenon, Ghedin and her colleagues performed whole-genome deep sequencing of flu virus samples and found that most individuals carry not only the dominant strain but also minor strains and variants of major and minor strains. The results of their research enabled the team to look at the variants and establish networks of transmission (8). The future of Ghedin's research—where it will go and how it will continue to evolve—is still emerging. However, one thing is certain: this trailblazing scientist with a giving heart will continue to use her carefully honed, valuable skillset to lead her lab and collaborate with interdisciplinary teams to address important infectious disease questions.

CITATION

Lewis Johnson T. 2018. Elodie Ghedin: unlocking the genetic code of emerging outbreaks, p 113–120. *In* Whitaker RJ, Barton HA (ed), *Women in Microbiology*. American Society for Microbiology, Washington, DC.

References

1. Ghedin E. 7 March 2012. Genomics and neglected tropical diseases: the view from my half of the human race. *The NIH Catalyst.* National Institutes of Health, Bethesda, MD.
2. Rouvalis C. Winter 2012. The genius of Pitt. *Pittsburgh Quarterly.* https://pittsburghquarterly. com/pq-health-science/pq-technology/item/494-the-genius-of-pitt.html.
3. Berriman M, Ghedin E, Hertz-Fowler C, Blandin G, Renauld H, Bartholomeu DC, Lennard NJ, Caler E, Hamlin NE, Haas B, Böhme U, Hannick L, Aslett MA, Shallom J, Marcello L, Hou L, Wickstead B, Alsmark UC, Arrowsmith C, Atkin RJ, Barron AJ, Bringaud F, Brooks K, Carrington M, Cherevach I, Chillingworth TJ, Churcher C, Clark LN, Corton CH, Cronin A, Davies RM, Doggett J, Djikeng A, Feldblyum T, Field MC, Fraser A, Goodhead I, Hance Z, Harper D, Harris BR, Hauser H, Hostetler J, Ivens A, Jagels K, Johnson D, Johnson J, Jones K, Kerhornou AX, Koo H, Larke N, et al. 2005. The genome of the African trypanosome *Trypanosoma brucei. Science* 309:416–422.
4. Shetty P. 2012. Elodie Ghedin: intrepid gene hunter. *Lancet* 379:113.
5. Ghedin E. 2013–2017. *Omics-based predictive modeling of age-dependent outcome to influenza infection.* NIH project number 5U01AI111598-06.
6. Read AF. 2012–2016. *Genomic analyses of the canonical cases of virulence evolution: myxomatosis in AU.* NIH project number 4R01AI093804-05.
7. Poon LL, Song T, Rosenfeld R, Lin X, Rogers MB, Zhou B, Sebra R, Halpin RA, Guan Y, Twaddle A, DePasse JV, Stockwell TB, Wentworth DE, Holmes EC, Greenbaum B, Peiris JS, Cowling BJ, Ghedin E. 2016. Quantifying influenza virus diversity and transmission in humans. *Nat Genet* 48:195–200.

8. Devitt J. 4 January 2016. *Scientists find minor flu strains pack bigger punch*. New York University, New York, NY. https://www.nyu.edu/about/news-publications/news/2016/january/scientists-find-minor-flu-strains-pack-bigger-punch.html.

Women in Microbiology
Edited by Rachel J. Whitaker and Hazel A. Barton
© 2018 American Society for Microbiology. All rights reserved.
doi:10.1128/9781555819545.ch14

Jane Gibson: A Woman of Grace and Acerbic Wit

14

Caroline S. Harwood[1]

Jane Gibson was a member of the first small cadre of female professors of microbiology in America, and as such, she had a major influence on my development as a scientist. Jane died in 2008 at the age of 83 due to leukemia, and I sought her advice and counsel until the end. She was an internationally respected researcher in microbial physiology and biochemistry and a mentor to scores of undergraduates and many graduate students and postdoctoral fellows at Cornell University. I hope that this chapter conveys why Jane was such an inspiration to me and to other scientists.

Jane was born Audrey Jane Pinsent in 1924 in Paris, France, and grew up in Devon, England, and in Switzerland, where she learned to speak German. She was educated at Newnham College, University of Cambridge, and received her Ph.D. from the Lister Institute, London. While at Cambridge Jane attended lectures and was influenced by Marjory Stephenson, who in 1930 wrote *Bacterial Metabolism*, a textbook that was used by generations of students. Gibson's work for her doctorate was mostly on trace elements, and she was the first to discover a role for selenium in bacterial growth, which she published on in 1954 (1). After completing her doctorate, she moved to California to study at the Hopkins Marine Station of Stanford with C. B. Van Niel, which led to an interest in photosynthetic microorganisms. On her return to Britain in 1950, she worked at the University of Sheffield with S. R. Elsden part-time as a volunteer while she devoted herself to the care of her four children and her husband, the biochemist Quentin Gibson. In 1961, Jane worked as a postdoc with I. C. Gunsalus at the University of Illinois,

---

[1]Department of Microbiology, University of Washington School of Medicine, Seattle, WA 98195

Champaign-Urbana, where Quentin was a visiting professor in the chemistry department. Jane always said that "Gunny" was the fairest boss that she ever had and that he did everything to make it possible for her to succeed while working part-time. In 1966, after also spending time at the University of Pennsylvania, Jane and Quentin Gibson moved to Cornell University. Jane started as a part-time assistant professor in the Department of Microbiology, eventually moving to full-time and up the chain of promotion to the rank of professor. Although Jane in many ways had a traditional role as a wife and mother, and the norms of society in the 1960s dictated that this was her full time job, she never gave up on science and kept at it despite having very little outside help for the care of her family. She never spoke of this other than to mention that the period when the youngest of her four children was 2 (shortly before she started at Cornell) was the hardest in her life.

I met Jane Gibson in 1976, when I was a student in the microbial diversity course at the Marine Biological Laboratory (MBL) in Woods Hole, MA. Jane was a faculty member in the course and since she was a woman, I was fascinated to know who she was. I was also terrified of her. Jane was gruff and she claimed to know nothing. She especially knew nothing about Sippewissett Marsh, a major study site for the course. I doubted that this was the case. I followed Jane around from a distance and noticed that she was the only one of the five course faculty members who did experiments. But she would never tell me what she was doing because she was "just messing around and it probably [wouldn't] come to anything." Jane always worked in the lab throughout her career. As might be expected from her history of part-time work while she raised her family, she was incredibly efficient in the lab. She was always interested in and critical of the techniques that people used, and the first section she read in a research paper was the Materials and Methods section. In fact, Gibson was a dedicated teacher of laboratory sciences to undergraduates at Cornell. Her cell biology laboratory course was a campus legend, and she was honored with the Edith Edgerton Career Teaching Award in 1994. During the semesters when she taught the cell biology laboratory, Jane spent at least 10 hours a day setting up experiments for the students and making sure they worked. She logged many hours walking from her lab in Wing Hall to her teaching lab in Emerson Hall. Walking was one of Jane's few extracurricular activities.

Gibson held her science to the highest standard and was a critical and fair reviewer of manuscripts both as an editor and as a primary reviewer. She served as a long-time editor of *Applied and Environmental Microbiology* and then generously recommended me as her replacement. She worked to advance the careers of trainees and colleagues in any way she could. At Cornell, Jane was interested in small metabolite pools of photosynthetic bacteria and in the transport and utilization of organic compounds by all the major groups of phototrophs. She was an expert in the care and feeding of these somewhat difficult organisms. She was especially proud of her isolation and description of *Chloroherpeton thalassium*, a flexing and gliding green sulfur bacterium that she isolated from marine sediments near Wood Hole (2). In 1986, I moved to Cornell for a brief stint of 4 years and reconnected with Jane. I asked her if she would help me figure out how bacteria degrade benzene rings under anaerobic conditions. There were reports in the literature that the purple nonsulfur phototroph *Rhodopseudomonas palustris* had this capability, but I didn't know where to get *R. palustris* or how to grow it. Jane rummaged around in her freezer and resuscitated a strain of *R. palustris* that she had isolated from Beebe Lake in Ithaca, NY. We named it CGA009. The "CG" stood for Cornell Gibson. This was the start of a happy and productive collaboration that ended formally on Jane's retirement in 1996 but continued informally thereafter.

In her retirement Gibson was a visiting scientist in several laboratories, working on *Myxococcus xanthus* at the University of Texas Houston (where Quentin was again a visiting professor) in Heidi Kaplan's laboratory, and at Dartmouth Medical School on *Pseudomonas aeruginosa* in the laboratory of Deborah Hogan. She and Quentin settled in this location to be close to their daughter Ursula and her family. Another daughter, Emma, lived in Woods Hole, and the microbial diversity course was fortunate to have Jane at the MBL in the summers of 2002 and 2003. Jane was called into action to help with young grandchildren as well as with the enrichment of phototrophic bacteria of all types. She was a valued member of microbial genome annotation teams—especially the *Rhodopseudomonas palustris* annotation team, due to her encyclopedic knowledge of microbial physiology.

Gibson was an amazingly hardworking and energetic scientist. I think of her frequently and how she would have handled this or that situation, and I still laugh thinking of her many comments: "That might be enough for a British Ph.D. thesis BUT...," "No one cares how YOUR mind works," and "Oh, he likes to take the saturation bombing approach to science." Her work-life balance was "all work" and "all family."

When she retired, Jane gave me a leather-bound copy of *Microbiologie du Sol, Problems et Methods, Cinquante ans de Recherces* (microbiology of soil, problems and methods, 50 years of research), by S. Winogradsky. She won

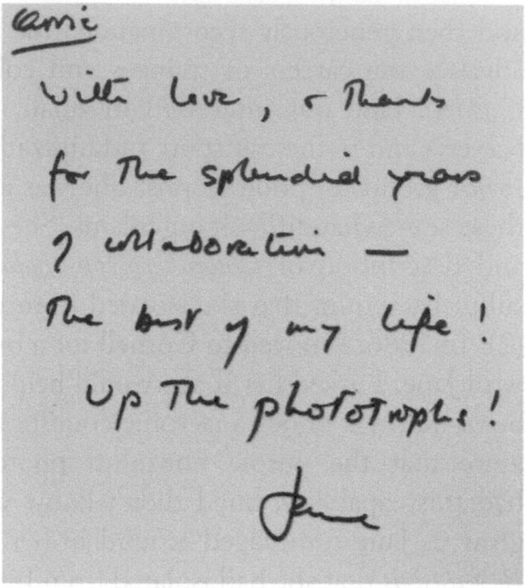

FIGURE 1 Inscription from Jane Gibson.

this this book as a prize in school. Inside, she inscripted "Carrie, with love, and thanks for the splendid years of collaboration—the best of my life! Up the phototrophs! Jane" (Fig. 1). This is my most prized possession, a tangible reminder of everything I learned from and with Jane, a portion of which I hope I've been able to relate here.

CITATION

Harwood CS. 2018. Jane Gibson: a woman of grace and acerbic wit, p 121–124. *In* Whitaker RJ, Barton HA (ed), *Women in Microbiology*. American Society for Microbiology, Washington, DC.

References

1. **Pinsent J.** 1954. The need for selenite and molybdate in the formation of formic dehydrogenase by members of the coli-aerogenes group of bacteria. *Biochem J* 57:10–16.
2. **Gibson J, Pfennig N, Waterbury JB.** 1984. *Chloroherpeton thalassium* gen. nov. et spec. nov., a non-filamentous, flexing and gliding green sulfur bacterium. *Arch Microbiol* 138:96–101.

Women in Microbiology
Edited by Rachel J. Whitaker and Hazel A. Barton
© 2018 American Society for Microbiology. All rights reserved.
doi:10.1128/9781555819545.ch15

Millicent C. Goldschmidt: Scarred Pioneer and Protector of the Biosphere

15

Hazel A. Barton[1]

At one time astrobiology was the stuff of science fiction. Early researchers struggled to convince the media of the legitimate scientific objectives in the field—it wasn't their goal to look for little green men (1). I first became aware of astrobiology in 1999 as a postdoc working in an environmental lab. By that time, the Viking landings were in the distant past and NASA was embarking on an era of near-continuous robotic exploration on Mars. The level of funding for such exploration demonstrated that astrobiology had evolved into a significant discipline, supported by work at numerous institutes at the most fundamental levels of chemistry, physics, and biology (2). Microbiology has proved critical to the advancement of the field, allowing us to understand the myriad of chemical and physical conditions that permit life and expanding the potential habitats for life in our solar system, which now includes Mars, the moons Europa, Titan, and Enceladus, and potentially the binary dwarf planet system of Pluto/Charon. The fundamental question of how to identify life when the traditional tools of microbiology might not work is a stimulating thought experiment, allowing one to explore what new and nonobvious means could be used to determine if life does exist elsewhere (3). All disciplines have a beginning, but I did not know what that beginning looked like or the fact that I routinely sit next to one of its first practitioners.

The early efforts to study the potential of life elsewhere included the work of Millicent E. Goldschmidt, known affectionately as "Mimi" to her friends. I first became aware of Mimi in 2007, when she was appointed to the

---

[1]Department of Biology, University of Akron, Akron, OH 44325

American Society for Microbiology's (ASM's) Committee on the Status of Women in Microbiology (CSWM). At that time, Mimi was the feisty octogenarian with a head of brightly dyed hair and always ready with an opinion. After 60 years as a microbiologist and 40 years serving within ASM, she knew everyone and forgot little—a skill that she would use with blistering effect. If you underestimated and messed with Mimi, you were going to regret it! She was a hoot. I liked her immediately. (To watch a discussion between the author and Mimi, visit http://bit.ly/WIM_Goldschmidt.)

When I served on the CSWM with Mimi, I knew her as a researcher with contributions to oral microbiology who had played a pivotal role in the development of rapid tests, particularly for *Salmonella*, *Escherichia coli*, and fungal pathogens (4–9). I knew nothing about Mimi's Apollo work until I was at a reception at the ASM's annual meeting at which she was presented with the coveted Alice C. Evans Award, which recognizes contributions toward the full participation and advancement of women in microbiology. During her acceptance speech, a colleague leaned over and asked if I knew who Mimi's most famous student was. In my head I started listing the famous microbiologists who could have emerged out of the field of oral microbiology when he whispered, "Neil Armstrong," and winked. I asked Mimi and her response was, "Oh yes, and Buzz, Buzz Aldrin." What followed was the story of how a young female scientist battled against gender discrimination to gain a degree and become critically involved in one of the greatest technological accomplishments of the modern era.

As a young girl growing up in in Erie, PA, Mimi was an active Girl Scout. A scout counselor introduced her to the hidden biology that could be found when walking in the forest; Mimi became enthralled and at 10 years old knew she was going to be a biologist. She attended Case Western Reserve University and during a freshman biology class was introduced to microbiology. She was immediately repulsed—all the Latin terminology made her determined to avoid any class where the ability to spell such difficult names was required. Nonetheless, her junior year required an entire module in microbiology. Rather than focusing on terminology, her instructor helped her appreciate the role of microorganisms as pathogens, but also as important players in our health. She became enamored of the role of microorganisms as both adversaries and collaborators: "I fell in love with the idea that at the same time we can't live with them and we can't live without them." She also became acutely aware of the influence a mentor can have on a student's interest in science.

Mimi completed her bachelor's degree in 1947 and wanted to build a career within the field of microbiology as a scientist and educator. At the time, Case Western didn't have a microbiology department, so she spoke to her family about attending graduate school. He father forbad it: "No man is going to marry a woman with that much education." His primary concern was that if Mimi were unable to find a husband, she would remain dependent on her family. It was a supportive uncle who helped change her father's mind: "Even though she'll be a spinster, at least she'll be able to support herself." Although she was allowed to attend graduate school, she wasn't allowed to go any farther than Indiana, so she enrolled at Purdue University (Fig. 1). At the time, it was almost impossible for a woman to be accepted into the doctoral program without first demonstrating the ability to carry out research, so she obtained her master's degree before beginning her doctoral studies.

Initially Mimi began her doctoral work in the lab of the eminent Henry Koffler, but she was soon drawn to the research of Dorothy Powelson, who was studying the physiology of *Micrococcus pyogenes* (subsequently reidentified as a strain of *Staphylococcus aureus*). Working with Dr. Powelson, Mimi examined how culture media can alter the interpretation of diagnostic tests; it became the foundational work for a long career in the development of

FIGURE 1 Mimi Goldschmidt at Purdue University in 1948, holding a flask of *Penicillium chrysogenum* while working on her Master's degree. Photo courtesy of Millicent Goldschmidt.

diagnostics. Not insignificantly, her mentor was the only female faculty member in biological sciences at Purdue and was instrumental in helping show Mimi that women could be successful scientists. Coincidentally, while Mimi was earning her Ph.D., a chemistry undergraduate at Purdue was similarly being inspired by Dr. Powelson's lectures in bacteriology. That student was Rita Colwell, who would go on to be the president of the ASM and the first female director of the U.S. National Science Foundation (10).

During her graduate work, Mimi met and married fellow graduate student Eugene Goldschmidt. It caused quite a stir within the department, with male colleagues questioning her commitment to research and even whether her goal had really been to obtain an M.R.S. rather than a Ph.D. Even more troubling was the overt dismay over her pregnancy shortly before graduation. Faculty in the department felt that she had wasted a graduate slot that could have gone to a man, who would have had a productive career. Instead, they had just "trained a mother!" No one seemed struck by the contradiction of how acceptable it was for male scientists to get married and have families without anyone grumbling about the productivity of their research. Unsurprisingly, these same faculty also found it acceptable to pinch Mimi's rear when she got in the elevator.

After earning her doctorate in 1953, Mimi's career path followed Eugene's. When Eugene found a position at Fort Detrick, then the home of the U.S. Biological Warfare Program, a second pregnancy meant that Mimi was not allowed to work with pathogens. Instead, she worked on superficially a more mundane project—looking at the impact of chemical explosives on microorganisms. To do this, bacteria were placed in small, sterile pillows with little parachutes attached. Explosives were then placed in the ground and covered with thin metal sheets, and Mimi would place the microbial pillows on top. The heavily pregnant Mimi would then put on a hardhat, clamber over a pile of sandbags, and push the large handle of the detonator to set off the explosion. She would then clamber back over the sandbags and start chasing after the pillows as they slowly drifted back down to earth. One irony of this research was that it was used by Joshua Lederberg, the Nobel-winning chemist who shepherded in the science of astrobiology (known as exobiology at the time), as a demonstration of the militarization of U.S. science during the Cold War. He used such experiments to push NASA toward more meaningful "origin of life" studies rather than research that continued to serve the security state (11). Mimi was therefore already involved in the establishment of astrobiology as a field, long before beginning her work with NASA.

When the army asked Eugene to genetically engineer a weaponized superbug, he decided to leave Fort Detrick. After spending 2 years in Austin, the family settled in Houston and Eugene began his faculty appointment at the University of Houston. Mimi began working at Baylor College of Medicine for Robert "Bob" Williams, the interim chair of the Department of Microbiology. As Mimi was beginning her work at Baylor, big things were happening down the road at the Johnson Space Center.

The Apollo space program, the U.S. effort to establish a manned space program, began under President Dwight Eisenhower at the height of the Cold War. In 1961, in his famous "National Urgent Needs" speech to Congress, President John Kennedy set a timeline of 8 to 9 years for a manned mission to travel to the moon and return safely to Earth, dramatically enlarging and accelerating the program. While the original goal of the Apollo missions had been to demonstrate the strength and superiority of U.S. technology, the newly convened Space Science Board of the National Academy of Sciences quickly identified the potential research that could be carried out with lunar samples (11, 12). In order to legitimize the program as a scientific endeavor, the Apollo Program was therefore expanded to include the return of 50 kg of lunar material (Fig. 2).

FIGURE 2 A technician carries out mice challenge experiments by inoculating mice with lunar material from the Apollo 11 mission in the Lunar Receiving Laboratory (dated August 1969; photo courtesy of NASA image archive).

In addition to the engineering complexities of adding 50 kg to the payload of a returning lunar module, it also meant that a large amount of a poorly understood and potentially dangerous material was being brought back to Earth. No protocols for handling such material existed at the time, requiring that all the theory, protocols, and handling methods had to be developed in a few short years. As an example of just how little was known about the lunar samples, some researchers proposed that after millions of years in the vacuum of space and subjected to strong ionizing radiation, if the lunar samples were removed from a vacuum they would explode. More legitimately, procedures would be required to prevent terrestrial contamination of lunar samples, which would otherwise confound the interpretation of data critical to our understanding of Earth's evolution. Out of these problems grew the concept of the Lunar Receiving Laboratory (LRL), a specially designed facility that would allow the containment, initial examination, and eventual distribution of the lunar materials. Construction began in 1966 (13).

While the idea of life on other planets had existed for centuries, the potential dangers of infectious material on the moon brought into sharp focus the question of whether life actually occurred on another planetary body. In early 1965 a report by the Space Science Board, backed by other federal agencies, including the Public Health Service, the Department of Agriculture, and the U.S. Department of the Interior, called for the development of procedures to contain and analyze samples from the moon for potentially infectious material. Although most scientists thought the possibility of life on the moon was remote, the potential damage to the biosphere was too great to risk. The Apollo Program therefore developed the first practical methods to search for life on another planetary body, launching the practical science of astrobiology. To accommodate these additional biohazard precautions, the LRL had to be enlarged to include quarantine for the lunar command module, astronauts and samples, and extensive testing facilities to determine whether the lunar samples contained microbial life (14). Baylor Medical College scientists, including Mimi, were contracted to develop the biological assessment of lunar material, which became known as the Baylor Protocol (15, 16).

Mimi's primary role was the design of the LRL to allow the examination of lunar samples for the presence of viruses, bacteria, and fungi. The protocols had to be time-sensitive, as the samples had to be quickly distributed to other researchers to allow testing for the presence of chemically labile material. Although the scientists at the LRL had access to instrumentation that would allow instantaneous analysis of hazards such as radioactivity, such rapid analyses did not exist for microorganisms. Instead, time-consuming *in vitro* cultivation techniques and *in vivo* animal challenge experiments were used. Despite the crude methodology, many of the developed techniques are still relevant today, including the quarantine and sterilization methods, the chemically derived cultivation conditions, the use of germfree animals, and differentiating between near negligible and detectable limits for sample analysis (3). Nonetheless, the experiments were slow, laborious, and imperfect. Mimi recognized such limitations as a "big hole" in the microbial research methods.

As the Baylor Protocol continued to evolve, it became obvious that they needed a way to prevent the astronauts from contaminating the lunar samples during collection. The Apollo A7L extravehicular spacesuits were cumbersome, weighing almost 35 kg, which made even the simplest tasks difficult. This difficulty was compounded by the fact that the astronauts were not scientists and, as Mimi puts it, were more interested in "playing golf and planting a flag." So with the help of a strong wrestler (also a microbiologist) who could move around in the spacesuit, Mimi figured out the sampling procedure. She then gave Neil and Buzz a crash course in microbiology and the aseptic techniques that they would use on the moon (Fig. 3). It's not hard to imagine Mimi standing on the sidelines, barking orders as these alpha-male test pilots lumbered around practicing microbiology in their iconic spacesuits.

As the time for the Apollo 11 mission rapidly approached, meticulous planning and deliberation quickly gave way to seat-of-the-pants solutions. One problem was how to quarantine the astronauts as they moved from the lunar command module (after splashdown in the Pacific Ocean) to the LRL facility in Houston. Mimi came up with the idea of biological isolation suits, made of a tight-weave fabric and full-face respirator. The astronauts donned these suits before leaving the command module, preventing any of the microorganisms, either on or infecting the astronauts, from escaping to Earth's surface (Fig. 4).

By Apollo 14 it was clear that the lunar surface was an inhospitable environment for life and that lunar samples were sterile. The LRL facility ceased to serve as a site for biological quarantine, but it continued to function in preventing terrestrial contamination. As the Baylor program was closed down, Mimi was offered a permanent position at NASA. As exciting as the research opportunities were, Mimi wanted a position that included teaching and allowed her to be closer to her family, so she returned to Baylor. Upon her return, Bob Williams, who had suffered from polio during his tenure at Baylor Medical College and was confined to a wheelchair, no longer served

FIGURE 3 Neil Armstrong (right; holding the bag) and Buzz Aldrin (left; using the scoop) practice aseptic techniques in their A7L spacesuits. Such methods were developed to allow the collection of lunar material without contamination from terrestrial microorganisms (dated April 22, 1969; photo courtesy of NASA image archive).

as interim chair of the Department of Microbiology (at the time, the administration felt that someone in a wheelchair should not be the face of the department). Mimi struggled with the new chair, who required her to pay her own salary from grant funds if she wanted to stay in the department. However, when Mimi wrote a research proposal, the new chair refused to sign it. He told her that while the science was excellent, her proposed salary was too high for a married woman—she didn't need to support a family, so she didn't need such a large salary. Mimi barked back, "What does my worth as a scientist matter if I'm married or single?" Mimi realized that she had no scientific future in that environment and accepted a position at MD Anderson Medical Center, also on the Texas Medical Center (TMC) campus, where she made significantly more than even NASA had been paying. Eventually she made the move to another TMC institution, the University of Texas Dental Branch, where she went on to a successful and productive career.

Mimi continued to work as an active faculty member and research scientist until the age of 85. In the laboratory, her students remember Mimi as a great mentor and meticulous scientist. She was passionate about science and

Figure 4 The three Apollo 11 astronauts (Neil Armstrong, Michael Collins, and Buzz Aldrin) wait in the life raft for pick-up after splashdown wearing their quarantine suits. The astronauts would continue to wear these suits to the quarantine facilities at the Lunar Receiving Laboratory (LRL). Lieutenant Clancy Hatleberg closes the hatch on the command module, before it was also moved to the LRL for quarantine and chemical disinfection (dated July 24, 1969; photo courtesy of NASA image archive).

passing on knowledge, as well as compassionate toward her students. Her past students expressed to me that they felt valued and that their time was as valuable to Mimi as anyone in the university. Mimi also made a point of listening to them, whether it was about science, careers, or their personal lives. They were similarly inspired by her stories, allowing Mimi to serve as the living link between our more egalitarian present and the chauvinistic, rear-pinching, *Mad Men* world of the 1950s and 1960s.

As funny as Mimi's stories can be, it is hard to imagine how she survived the humiliations and sense of injustice of her early career without becoming embittered and resentful. I've run into other women scientists who persevered through similar experiences and were hardened by the experience or feel the need to hide their femininity in an attempt to be judged as equals. Mimi is none of those things; however, as a trailblazer, she does still carry some scars. As she likes to say, "A pioneer searches ahead to find the best possible route [for everyone else] and returns with her rear end full of arrows." But she survived by learning to be tough and throwing those arrows right back. Her experiences have made her a strong advocate for women scientists. While at UT Houston, Mimi's advocacy led to the first sexual harassment training at the institution. Within ASM she oversees the Career Development Grant for Post-doctoral Women, which she initiated with her own funds, and still provides travel funds in support of projects that will advance the career goals of its recipient. She also personally supports two graduate scholarships through the Texas ASM Branch. These significant awards are given annually to female graduate students who show promise in the field of microbiology and can be used in any way that the student feels will benefit her career. Mimi makes a special effort to contact each of the awardees and meet with them at one of the Texas ASM Branch meetings where they present their research. One such recipient said that the award was transformative in her career. As a high school dropout, this student fell in love with science and worked hard to complete her education at community college, be accepted into a university program, and ultimately be accepted into a graduate program in microbiology. Even now, as a fifth-year graduate student, the shadow of her high school performance made her feel like she didn't belong. Mimi's graduate fellowship showed that other scientists valued her work, and the award was a huge boost to her self-confidence as a scientist. The student is going to use the money to take a scientific writing course, along with presenting her research at an international conference in Germany, where she hopes the networking opportunities will help her find a postdoctoral position. She said that talking to Mimi and hearing stories about how she had to fight for acceptance have been inspirational.

Although Mimi may have a lower profile than some of the other women featured in this volume, her research had a profound impact on the U.S. Space Program and contributed greatly to the field of molecular diagnostics, and her stories have proved inspirational to generations of female scientists. Heidi Kaplan, Associate Professor at McGovern Medical School at UTHealth in Houston, remembers a workshop that Mimi gave at an Association of Women in Science branch meeting in Houston: "Mimi was the invited speaker. She showed up dressed in an enormous hat, outrageously colorful dress, and an oversized bead necklace…looking like a cross between a hippy and a bag lady. She then launched into a lecture about how important it is to dress appropriately for the professional environment. While she was doing this, she slowly started to strip! By the time she was finished talking, she revealed the smart business suit she was wearing under her outrageous costume. She challenged the audience about their assumptions of her as a person at the beginning of her lecture, and those assumptions at the end. It was hilarious and remarkably effective for the audience of female graduate students and postdocs. No one forgot the important message of dressing to be taken seriously as a scientist." Occasionally there are anecdotes that scrape along the edge of the politically correct spectrum, but Mimi grew up in a different age.

Mimi's stories serve as a reminder that we live in a different era, built upon the efforts of pioneering scientists, both men and women, who worked to create a merit-based academic environment in which gender is irrelevant. Mimi is also the embodiment of what so many of us strive for in our professional careers—the ideal work-life balance. Mimi became a scientist when women were not viewed as scientists, had a family when it was considered unprofessional, and remained a working scientist when mothers were expected to be in the home. Mimi has suggested that she may have been a higher-profile scientist if she hadn't had a family, but she measures her success through her children, who have both gone on to highly successful careers; her daughter is an accountant and a lawyer, and her son has a Ph.D. in neurobiology. And she continues to work tirelessly toward ensuring that men and women are viewed equally in terms of capability and advancement. At 90 she's still a firecracker, zooming around the ASM General Meeting on a scooter (last time we met I was in heels and had to jog to keep up), making grand entrances, raising hell, and telling her stories.

ACKNOWLEDGMENTS

Special thanks to Mimi, along with Heidi Kaplan, Naomi Bier, and Douglas Litwin for interviews and valuable feedback on this chapter. I also thank the ASM Communication staff for making it possible for me to record my interview and Philip Montgomery and Ruth SoRelle for permission to use material from the Texas Medical Center Women's History Project.

CITATION

Barton HA. 2018. Millicent C. Goldschmidt: scarred pioneer and protector of the biosphere, p 125–136. *In* Whitaker RJ, Barton HA (ed), *Women in Microbiology*. American Society for Microbiology, Washington, DC.

References

1. Lederberg J. 1960. Exobiology: approaches to life beyond the earth. *Science* 132:393–400.
2. Des Marais DJ, Nuth JA III, Allamandola LJ, Boss AP, Farmer JD, Hoehler TM, Jakosky BM, Meadows VS, Pohorille A, Runnegar B, Spormann AM. 2008. The NASA astrobiology roadmap. *Astrobiology* 8:715–730.

3. Summons RE, Sessions AL, Allwood AC, Barton HA, Beaty DW, Blakkolb B, Canham J, Clark BC, Dworkin JP, Lin Y, Mathies R, Milkovich SM, Steele A. 2014. Planning considerations related to the organic contamination of Martian samples and implications for the Mars 2020 Rover. *Astrobiology* 14:969–1027.

4. Goldschmidt MC, Fung DY, Grant R, White J, Brown T. 1991. New aniline blue dye medium for rapid identification and isolation of *Candida albicans*. *J Clin Microbiol* 29:1095–1099.

5. Goldschmidt MC, Lockhart BM. 1971. Simplified rapid procedure for determination of agmatine and other guanidino-containing compounds. *Anal Chem* 43:1475–1479.

6. Goldschmidt MC, Lockhart BM. 1971. Rapid methods for determining decarboxylase activity: arginine decarboxylase. *Appl Microbiol* 22:350–357.

7. Goldschmidt MC, Lockhart BM, Perry K. 1971. Rapid methods for determining decarboxylase activity: ornithine and lysine decarboxylases. *Appl Microbiol* 22:344–349.

8. Goldschmidt MC, Williams RP. 1968. Thiamine-induced formation of the mono-pyrrole moiety of prodigiosin. *J Bacteriol* 96:609–616.

9. Wheeler TG, Goldschmidt MC. 1975. Determination of bacterial cell concentrations by electrical measurements. *J Clin Microbiol* 1:25–29.

10. Hyde B. 1998. Rita Colwell: sea change at the NSF. *Curr Biol* 8:R404.

11. Wolfe AJ. 2002. Germs in space. Joshua Lederberg, exobiology, and the public imagination, 1958–1964. *Isis* 93:183–205.

12. Space Science Board. 1961. *Policy Positions on Man's Role in the National Space Program and Support of Basic Research for Space Science*. National Academy of Sciences, Washington, DC.

13. Mangus S, Larsen W. 2004. *Lunar Receiving Laboratory Project History*. Langley Research Center, Hampton, VA.

14. NASA. 1965. *NASA 1965 Summer Conference on Lunar Exploration and Science*. NASA, Falmouth, MA.

15. McLane JC Jr, King EA Jr, Flory DA, Richardson KA, Dawson JP, Kemmerer WW, Wooley BC. 1967. Lunar receiving laboratory. *Science* 155:525–529.

16. Campbell BO, Goldschmidt MC, Lipscomb HS, Knight JV, Melnick JL. 1967. *Comprehensive Biological Protocol for the Lunar Sample Receiving Laboratory Manned Spacecraft Center*. Baylor Medical College, Houston, TX. https://ntrs.nasa.gov/archive/nasa/casi.ntrs.nasa.gov/19680021536.pdf

Women in Microbiology
Edited by Rachel J. Whitaker and Hazel A. Barton
© 2018 American Society for Microbiology. All rights reserved.
doi:10.1128/9781555819545.ch16

Susan Gottesman: An Exceptional Scientist and Mentor

16

Carin K. Vanderpool[1]

Susan Gottesman (née Kemelhor) was born in New York, the older of two sisters, to parents who were "not scientists, but supportive." Of what spurred her interest in science at a young age, she said, "My father used to give me books of various kinds, and in fifth grade or so, he gave me *Microbe Hunters*. Other books he gave me were *The Count of Monte Cristo* and all kinds of other things. But *Microbe Hunters* captured my imagination." She said *Microbe Hunters* described the logic of how to do experiments, but was also "written in a way that was exciting. So, about that point, that's what I decided I wanted to do. And I didn't go very far away from it." She stayed interested in microbiology throughout her years in primary and secondary school, "even though I still didn't know very much about it." In high school, she had the opportunity to conduct research in a summer program at Waldemar, on Long Island. Gottesman said she and other students in the late 1950s benefitted from a post-Sputnik push by the U.S. government to increase science education for Americans. In this program, she had her introduction to DNA as genetic material, relatively new information that hadn't yet made it into her high school curriculum. She worked on one research project involving "magnets and cancer" and another where the goal was to develop a defined medium on which to grow bacteria. She said even though she was only a high school student, she knew that the scientific quality of the research in the program wasn't high, "but it was an experience with science and with doing things in the lab, and I knew I wanted more of that when I went off to college."

---

[1]Department of Microbiology, University of Illinois Urbana-Champaign, Urbana, IL 61801

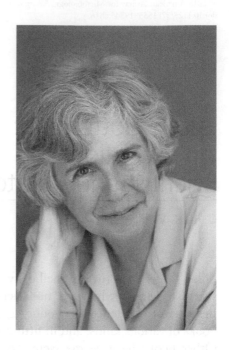

FIGURE 1 Photo of Dr. Susan Gottesman.

Gottesman studied biochemical sciences at Radcliffe College. As a freshman, she signed up for a freshman seminar on bacterial genetics. "The president of Radcliffe at that time was a woman named Mary Bunting, who was a microbiologist. She wasn't actually there most of the year because she was on the Atomic Energy Commission, but she had set up a lab in one of the Radcliffe buildings. I think there were four of us in this freshman seminar and some upperclassmen who were supposed to help us. We could hang around and play," doing experiments with pigmented *Serratia* strains.

Gottesman sought out more opportunities for lab work as an undergraduate. She said, "My junior year in college, I missed being in the lab, and I saw there was an ad for a tech in the [James] Watson lab, so I applied. You know, he hired a lot of Cliffies; he had a whole team of Radcliffe undergraduate women who worked in the lab. I worked with Gary Gussin; he was a graduate student at the time. Two of us came in and tried to help him finish his graduate project. I'm not sure I knew exactly what I was doing... I mean, I was doing TCA [trichloroacetic acid] precipitations and [Gary] was looking at tRNA charging and things. This was early molecular biology. Even though I didn't understand entirely what I was doing, it didn't matter. I liked being around the lab. Mario Capecchi [who would later earn the Nobel Prize for his contributions to development of gene targeting methods using embryonic stem cells] worked in the same room with Gary and me. Joan [Argetsinger] Steitz [a scientist renowned for her work in RNA biology] worked nearby, and they had tea every day at four. That was the English way and Jim Watson was big on being like an English lab. So, you got to meet everybody."

Of another research experience during her college years, Gottesman said, "I was a biochemical sciences major and we had tutors. My tutor, who was assigned to me by chance, was Boris Magasanik [a microbiologist]. So, that was one of the things that was keeping me in the bacterial world. Boris had come and talked to us in the freshman seminar about phase variation. Senior year I did my senior thesis with [Magasanik]. I went over and did a project in his lab—he was at MIT. That was the first time I had a chance to do my own project. I was working with a graduate student who happened to be a woman, Bonnie Tyler. I was looking at kinetics of β-Gal (β-galactosidase) induction. I learned to assay β-Gal, which was fun and was really my first independent research experience. In fact, β-Gal may still be the only thing I know how to assay. That experience has served me well."

Gottesman found that the combination of lab experiences she had at Radcliffe kept her interested in pursuing a scientific career. Her time at Radcliffe also had a big impact on her personal life. Her freshman year, she met her future husband, Michael Gottesman, in physics lab. She said, "He was a year ahead of me. I was wandering around the lab being friendly because the guy that was my lab partner... it was sort of like a date, but it wasn't too successful, so I was probably being more outgoing than I would have been otherwise. Michael was at another table doing ballistic pendulum measurements." Susan and Michael married her junior year, and Michael graduated the following spring. He was a medical student at Harvard when she graduated from Radcliffe. She said, "I was applying to graduate school in the Boston area. I asked Boris [Magasanik] who I should be interested in working with. He said, 'Well, there are these two guys who are starting their labs and they're both very good. Ethan Singer and Jon Beckwith.' They had worked together in Paris and had both come back. Ethan was at MIT, and Jon was at Harvard Medical School. I applied to Harvard with the idea of working with Jon. I was part of maybe his second wave of students. There were a few people finishing up as I came in, and the lab was expanding."

Jon Beckwith is one of the pioneers of modern molecular biology and a titan in the field of bacterial genetics. Gottesman remembered, "When I joined Jon's lab they were working on transcriptional regulation. They worked on cyclic AMP (cAMP)-CRP; they called it CAP. They isolated the first mutants that affected it. They had some *lac* fusions and they found a cAMP-independent mutant, the *lac*UV5 promoter. They were just doing the first experiments to really understand *lac* regulation. My project was to isolate an arabinose transducing phage so we could study arabinose regulation. [The response to] arabinose was supposed to be positive regulation, but nobody really knew if positive regulation existed. All we knew about [regulation] at that point was Lac repressor and lambda repressor and that's it." Gottesman was in Beckwith's lab at the time they made international headlines for isolating the first gene from a bacterial chromosome. She said, "So they went public with that and there was a lot of publicity about the first genetic engineering. This was the first whiffs of politics, and Jon is pretty political. So that was entertaining. There were only a couple of us who were not directly involved. It was an interesting place to be. I didn't feel like a student exactly. I was just part of the group."

Gottesman's family life changed with the birth of her first child, Daniel, while she was a graduate student in the Beckwith lab. After she finished her thesis, she and Michael moved with Daniel to Bethesda, MD, where she started a postdoc position at the NIH. She noted, "My early career was all affected by Michael's schedule to some extent. This was during a time when all M.D.'s were being drafted. So, the alternative to being drafted was to join the Public Health Service [and serve] at NIH." She joined the laboratory of Max Gottesman (no relation). She remembered, "They [Max Gottesman's group] were doing things that made most of what I had done for my thesis obsolete. They had come up with better ways to get lambda to hop around and pick up genes." There was a strong cohort of microbiologists at NIH at that time, and they were taking full advantage of the use of lambda phage as a powerful tool for bacterial genetics and molecular biology. Recalling her first exposure to a (still going strong) seminar series at NIH, "Lambda Lunch," she described, "They spoke in a language—lambda jargon—that took a long time for me to understand." Nonetheless, Gottesman embarked on a lambda project where she would begin experiments that would form the foundation for some of her most famous work: on the role of proteolysis in regulating bacterial physiology and stress responses. She said, "The postdoc project was looking at lambda excision. I started out with a lot of *in vivo* experiments, but the aim was to purify Xis [protease] and make it work *in vitro*. Because lambda Xis was functionally unstable, I got interested in mutations in proteases and what they did for the cell." During these post-doctoral years at NIH, the Gottesman family grew by one more with the birth of daughter Rebecca. Gottesman remembers that she shared a module in a narrow lab with Don Court, a tall, broad-shouldered man, and humorously remarked that it got to be "tight quarters" when she got to the point of being "very pregnant."

Asked about who were important or influential mentors for her, Gottesman reflected, "Boris is more of a bacterial physiologist but he was interested in regulation, too, so he was certainly an influence. And his graduate student Bonnie was really the only woman scientist I had any encounter with [up to that point], and she had a child when she was a graduate student. From my perspective, she made it look easy. So that had a bit of an influence on my decision to have a child while I was in graduate school. In terms of science and thinking genetically, I have to say it was Jon and that lab. The way they think about things and approach problems had the biggest long-term effect [on my career]. The community here [at NIH] of lambda people, Max, Bob Weisberg and others, they were also influential. Again, in terms of how you think about science, what kinds of things are exciting, what kinds of tools you use. I think I already knew when I was in Jon's lab—when he was on sabbatical for a year, we did some work with phage, making different types of tools, and it became clear to me that phage were a tool that it would be useful for me to understand. Coming to NIH to that phage group reinforced that. Phage [biology] was a field in which we had a lot of information about networks and how things were regulated, and that really appealed to me."

When her husband, Michael, finished his stint in the Public Health Service, the family moved back to Boston, where Michael completed a residency and Gottesman herself worked as a research associate in David Botstein's lab at MIT. After that, it was back to Bethesda, where she and Michael were hired into permanent positions at NIH. Gottesman recalled, "Ira Pastan is the one who hired us when we came back, both Michael and me—and he got out of the way and let us pursue what we wanted. Ira had been asked to establish the LMB [Laboratory of Molecular Biology] at NCI [National Cancer Institute]. Max [Gottesman] moved over with him, and Sankar [Adhya] was with him at that time. Don Court was here, too. Max had let me take the lambda excision project with me [to Botstein's lab], and I think he thought it would be nice to have it back [at NIH]. Michael [Gottesman] had moved into working in eukaryotic systems and cells. Ira, as lab chief, just went to the NCI scientific director and said, 'They're both good; I want to hire them.'"

Gottesman has remained a tenured senior investigator in the LMB since returning in 1976 and is currently co-chief of the LMB and head of the Biochemical Genetics Section.

Perhaps not surprisingly, given her training with some of the most talented bacterial geneticists in the history of microbiology, Gottesman has favored genetic approaches to solve problems. She said, "I think about [science] in terms of genetic circuits and what I can do to perturb them in order to ask the interesting questions. It's just the way I think about things. I'm not good at thinking about proteins in 3D, I'm not good at purifying proteins and making them do what I want, so I tend to think genetically about problems and about how to solve them. So, having an organism [*Escherichia coli*] in which you can think about doing almost anything genetically… I haven't been tempted to move very far from it. As long as we keep having interesting questions to ask, I want to ask them genetically, and as long as we can do that in *E. coli*, well, it keeps getting easier. We can ask more sophisticated questions and mutagenize in different ways." She reflected on how much has changed in the world of bacterial genetics over the course of her career. "I was just looking back at an old paper where we spent half the time trying to map the mutation and trying to figure out if things were linked, and we got a rough idea [of the location of the mutation] but we didn't know more than that. Now you can mutate and just sequence and know what you have. You can make exactly what you want in the chromosome and ask what it does. So, now you have to think harder… everything accelerates; we can do things faster and faster. So, you have to think a little bit harder about what are the questions you want to ask and not just do [an experiment] because you can do it. I'm not sure I'm quite doing as much of the thinking about some of our projects as I really want to do. But, I think it's still possible to ask interesting questions genetically in ways that you can't otherwise."

Asked which of her scientific accomplishments she thought represented her most exciting or creative work, Gottesman demurred. "Thinking about proteolysis as a regulatory mechanism and how small RNAs could regulate were the big things in terms of what I've done, but I tend to think about what was fun, mostly. Early in the proteolysis story, thinking that there should be [protein] degradation and what those [substrates] should look like, and then figuring out that network. That became possible because we had a clone of *sulA* [encoding a protein whose stability is regulated by proteolysis, with consequences for the DNA damage response and cell division] and we could show that it was degraded. [Another fun project was] looking at RyhB [one of the first small RNAs in bacteria for which Gottesman's group uncovered an important and previously unknown role in the iron starvation response], seeing what it was regulating, and then finding in the literature unexplained examples of positive regulation [by a repressor protein, which Gottesman's group showed were actually mediated by the small RNA RyhB]. It's when your predictions bear out that it's fun."

Gottesman's research career has been long and eventful. She's been recognized for her outstanding research contributions with numerous awards. She was elected to the American Academy of Microbiology and the National Academy of Sciences. She has won a Lifetime Achievement Award from the American Society for Microbiology, the Herbert Tabor Research Award from the American Society for Biochemistry and Molecular Biology, and the Selman A. Waksman Award in Microbiology. She humbly credits her success in large part to the good fortune of being at NIH, which she said has allowed her a degree of freedom she might not have had at another institution. "Because we're not applying for grants, we don't always have to know where we're going or why we're doing it ahead of time in quite the same fashion [as an investigator at an academic institution reliant on external grant support]. That meant I could sort of do things that took a while to develop—certainly the small RNA business bounced around for a while before it became clear that something was going on that we should spend some time on. The other major thing is that because we're mostly postdocs and postdocs take things with them [take their projects to independent positions], we also sort of have to keep moving into different things. So, projects that do that [move flexibly into new areas] easily become a little more attractive. But that [feature] does make it [our research] a little random. Different people [postdocs who come to the lab] are interested in doing different things. Sometimes I feel like I'm being dragged in some directions by someone's interest into things that maybe I shouldn't be doing, when instead [I should be] thinking more about what *I* really want to do and find people to do it. It was clear when we got that first list of small RNAs that those were going to be gold and we should go after them, and it was pretty straightforward. And we've done that. Now it's a little harder, maybe a little more random."

Asked who was most influential in terms of the development of her own mentoring style, Gottesman was very self-deprecating. She said, "I probably made it up as I went along and didn't know what I was doing most of the time.... Also, I don't see a lot of graduate students, which is probably the group that needs more guidance and to be set on the right track. I get people [postdocs] who have already partially developed their own style and see where they're going, and so maybe I can get away with doing less mentoring. Sometimes it's just coming up with projects that might work and shoving them one direction or another." When pushed further, and asked what she believes is the most important advice she gives to her trainees, Gottesman replied, "Sometimes it's to ignore what I say. I mean, day to day, don't take everything that people tell you too seriously until you sort it out. Sometimes [my advice is meant to] push the really good people to do what they can do and not be afraid to do it. It really varies with who it is and what they want to be doing, and what they should be doing." Regarding advice she thinks is particularly relevant for women trainees, Gottesman stated, "I think sometimes they [women] doubt themselves more than men do, or at least they're more obvious about it. [Women doubt] what they can do, how good they are, what direction they should go in.... So, [women] sometimes have to be told that they're really doing good stuff, that they're going in the right direction. But, you know, not everybody should be doing this [career] anyhow. In the first place, I think unless you really love the kind of things we do, [an independent research career is] too much trouble. So, that's question 1. Do you really love it? If you do, then [you can ask yourself, are you] good enough to do it?"

As a fourth-year Ph.D. student at the University of Minnesota, I was beginning to think about life after graduate school. Though unsure where I wanted my career to go, I had resolved to do postdoctoral training and was overwhelmed with the possibilities. At this critical juncture, a paper from Susan Gottesman's lab (1) was presented in my journal club, and my path started to take shape. That influential paper, by Wassarman et al., represented a collaboration between Gottesman and her colleague at NIH, Gisela Storz, and described one of the first efforts to define the extent of RNA-mediated regulation in bacteria on a genome-wide scale. I was stunned. I found the concept that a large class of regulators—small RNAs—had been "hiding" in bacterial genomes, virtually unknown and uncharacterized, to be highly intriguing. About a year later, I started my postdoc work in Gottesman's lab.

I was a postdoc in the Gottesman lab for about 3 years before moving to start my own lab in the Department of Microbiology at the University of Illinois. Those 3 years were among the most scientifically exciting and fulfilling of my career thus far. We were still in the early days of small RNA discovery. When I joined the lab, Susan handed me a list of genetic loci encoding small RNAs and told me to pick a few that looked interesting to work on. We knew so little about the depth and breadth of small RNA functions that everything we learned was new and exciting. Susan's enthusiasm for science and pursuing unexpected interesting results provided a fertile environment for exploration. She gave her trainees a supportive and open environment and near total freedom to pursue projects that interested us. Her excitement to hear new results amplified my own and had me knocking on her door nearly every day to show her a plate or a Northern blot or to discuss a new idea. If she was in her office, her door was always open, and she would get out a notebook so that she could take notes and draw pictures to help work through the logic of a problem or analyze a new result. My labmates and I took pride in bringing her a result that intrigued her enough to make her chuckle and say, "Well, that's entertaining."

After about a year in her lab, my work was going well and Susan asked if I'd ever thought about what I wanted to do next. Still not confident I could pull off an independent research career in academia, I believe I responded vaguely and asked her how she decided on her own career path. She told me that it just didn't occur to her that she should try to do anything else, that she found research to be fun and there was really nothing she could imagine doing instead. She assured me that I certainly had "what it takes," for a successful career as an independent scientist, so the real question was whether I found that route or another to be more appealing? This almost simplistic way of framing the issue felt a bit revolutionary to me and freed me of some of the self-doubt that clouded my ability to decide on a path to pursue. Susan's support and career advice paved the way for my successful transition to a tenure-track faculty position, and her ongoing mentoring has guided me through numerous career milestones. I feel incredibly fortunate to have been mentored by such an outstanding scientist, who I continue to admire and try to emulate. For this profile, I interviewed Susan, asking her questions about her early life and career in science, as well as questions about her current scientific interests and point of view. I hope I have captured her accurately and in a way that conveys her exceptionality as a scientist, mentor, and human.

CITATION

Vanderpool CK. 2018. Susan Gottesman: an exceptional scientist and mentor, p 137–145. *In* Whitaker RJ, Barton HA (ed), *Women in Microbiology*. American Society for Microbiology, Washington, DC.

References

1. Wassarman KM, Repoila F, Rosenow C, Storz G, Gottesman S. 2001. Identification of novel small RNAs using comparative genomics and microarrays. *Genes Dev* 15:1637–1651.

Women in Microbiology
Edited by Rachel J. Whitaker and Hazel A. Barton
© 2018 American Society for Microbiology. All rights reserved.
doi:10.1128/9781555819545.ch17

Carlyn Halde: Free Spirit   17

Wendy J. Wilson[1] and Shirley Lowe[2]

Carlyn Halde was born in Southern California in 1924 and grew up in an unconventional family: her parents were artists and moved to follow their work. Her father created cornices inside ornate buildings with Art Deco designs that represented the luxury and exuberance of the 1930s and 1940s. Her mother was a photographer with a small business of her own. Originally from the Midwest, her parents came to Southern California during an exciting time of growth, but they often went back to the Midwest to work on projects. The family settled in Alhambra (a suburb of Los Angeles) when Carlyn entered the 4th grade. California was experiencing exciting advances: women enjoyed rights that eluded women in most other states, and progressive movements gestated for labor unions, pensions, and social welfare programs (1). These utopian ideals didn't quite materialize at the time, but the forward-thinking values spawned activists who continued to nudge society forward. It was in California where Carlyn truly felt at home. She lived an idyllic childhood filled with love and creativity. Years later, when she was plotting her future career, she ensured that California was where she could settle and build a life. Carlyn loved the California ethos. No doubt it factored into her life of volunteering and support of human rights and health equality.

Through her parents, Carlyn embraced creative pursuits at an early age. Instead of the arts, however, she gravitated toward nature. As a child she contracted rheumatic fever, and her doctor's orders were no physical activity. Books became Carlyn's best friends, and she nurtured her love of nature by

[1]MSEPS, Las Positas College, San Leandro, CA 94577
[2](Retired) Department of Microbiology and Immunology, University of California, San Francisco, San Francisco, CA 94143

reading many books in the local library about nature, animals, and adventures in exotic locales. She looked longingly up at the mountains that surrounded Los Angeles (they could be seen clearly then) and dreamed of the day she could explore them. It was good medicine; she remained an avid reader her whole life (2).

Carlyn was naturally introverted, and getting sick did not help her to develop friendships as a child. But she and her brother spent many magical times together growing up. She was a bright student and seemed to have a knack for finding mentors that fed her dreams. In elementary school, she was inspired by her science teacher Florence Bevington. At Alhambra High School her biology teacher, Elizabeth Hager, inspired girls to pursue higher education. Carlyn followed Elizabeth's advice and entered University of California, Los Angeles (UCLA), in 1941 as a zoology major with an aim to be a high school science teacher.

COLLEGE AND GRADUATE SCHOOL LIFE
It was at UCLA where Carlyn truly blossomed. Restored to health, she embraced the campus and immersed herself in all aspects of her favorite topic, biology. At UCLA Carlyn developed lifelong friendships as a member of the Lambda Sigma Society (https://lambdasigma.org/), a co-ed national honor society dedicated to fostering leadership, scholarship, fellowship, and the spirit of service among college students. Dr. Odra Plunkett, a microbiology professor, was the group's faculty advisor. Through this group, the books Carlyn read as a child jumped off the pages and into her life: she went camping and hiking and learned about plants and animals in the diverse habitats that surrounded Los Angeles, including those mountains that seemed so unobtainable as a child. She formed close friendships with like-minded students. She was in heaven! The group even went on several week-long camping trips to Mexico on which academic discussions of nature were common. Their camaraderie and support were treasured throughout her life.

As an undergrad, Carlyn got a job in the lab of Dr. Plunkett. He was a down-to-earth professor specializing in medical mycology research. He insisted that all his lab charges called him Pappy. He and his wife would have the entire crew over for dinner and foster their professional development. Dr. Plunkett sparked Carlyn's lifelong interest in medical mycology and helped cultivate it into a career. He didn't care that she was female: he saw a bright person and nurtured her talent. She liked it so much that she decided to stay after her B.S. in zoology and get a master's degree in microbiology with Pappy. She still wanted to be a high school science teacher like her mentor Elizabeth Hager. But we think she was gathering up courage to pursue a scientific career in a world dominated by men. Her master's thesis further immersed her in medical mycology research, and she also helped teach Dr. Plunkett's mycology and medical mycology laboratories. She loved mycology (Fig. 1)!

Upon graduation, Carlyn nevertheless went with her original plan of becoming a high school teacher, securing a job at San Fernando High School. She found it to be a dreadful experience: there were no supplies, and it was difficult to engage the students. Luckily, before the year was out, she was invited to teach medical mycology at the University of Hawaii, a grand adventure to which she said, "Absolutely!" It was a new subject at the time to everyone on the Hawaiian Islands: to the microbiology faculty, hospital medical technologists, and even an Army colonel pathologist who supervised the reference laboratory serving the entire Pacific Arena. She also worked at the lab of the Army Hospital (now Tripler Army Medical Center). On weekends she thoroughly enjoyed camping and hiking expeditions all around the islands with her University of Hawaii friends and hanging out at the famous Waikiki Beach (which at the time had only three hotels). She even considered learning to surf when she met Duke Kahanamoku, a legendary Hawaiian surfer.

Figure 1 Carlyn as a young scientist with her fungal cultures.

While in Hawaii, Carlyn had one mishap: she inoculated herself with *Cryptococcus neoformans*, isolated from a patient who died of the infection. Our modern understanding of *C. neoformans* is as an opportunistic pathogen, but at the time everyone thought that Carlyn might die, and she was extremely worried. But she didn't even get sick, and she later published an article about the incident so that scientists and health professionals could learn from her experience (3). The laboratory accident didn't deter her career plans.

After spending 2 years in Hawaii, she decided that she wanted to be a professional medical mycologist, and a Ph.D. would be necessary. She noted that those who got hired at UCLA earned their Ph.D.s outside of California, and since she wanted to settle in California long-term, she applied elsewhere and decided to study under one of the most respected mycologists of the time, Dr. Norman Conant at Duke University. With Dr. Plunkett's recommendation, Dr. Conant accepted her, undoubtedly impressed with all she had accomplished already. At the same time, all her friends were also applying to a new scholarship called the Fulbright, so Carlyn applied to study medically important fungi in Manila, Philippines. She didn't think she had much of a chance, but she received one of the first Fulbright Scholarships. Dr. Conant agreed that she should start her Ph.D. program a year later so that she could take advantage of the opportunity, and off she went to Manila on another adventure.

Through her Fulbright, Carlyn worked at the University of Philippines Institute of Hygiene with dermatology professor Dr. Guitarez; she accompanied him on his rounds and observed many skin infections. She took samples of the fungal infections and brought them to the lab for culture and examination. Obtaining samples for identification was new to the hospital, and Carlyn shared her results by teaching the hospital staff about medical mycology. She also worked with Dr. Socorro Simuangco at the University of Santo Tomas Medical School, also in Manila, in much the same way, collecting many specimens. During her time in the Philippines, Carlyn traveled into the remote mountain areas. It was an adventure just getting to the communities of indigenous people. She loved interacting with the people and seeing their customs, and of course, she was always prepared to help them with their skin infections (and take a sample).

As her Fulbright was winding down, Carlyn was not quite done with her adventure, so she arranged a trip around Southeast Asia before reporting to Duke University. A staff member at the World Health Organization told her that if she gave talks in the places she was going, he could coordinate her travel to meet with medical professionals there. She went to Borneo,

Surabaya, Bali, and Jakarta, then on to Columbo, Madras, Bombay, Delhi, and Calcutta, and, finally, to Rangoon Bangkok, Hong Kong, and Japan. At each place she met with medical school professionals who shared their patient cases, and she shared her medical mycology knowledge. After a visit with her family in her beloved California, she went off to start her Ph.D. at Duke University.

As a world traveler, Carlyn had seen many places and different cultures, but that did not prepare her for the racial discrimination she encountered in North Carolina. She was upset that she could not sit with her black colleagues on the bus to work and was upset that they had to deal with a segregated society. She didn't enjoy her experience there at all. She also experienced sexism firsthand. It bothered her the entire 2 years she was working on her Ph.D. thesis, "The Relation of Nutrition to the Growth and Morphology of *Trichophyton conentricum* Blanchard 1896." When she finished her doctorate in medical microbiology, she didn't even stick around for graduation, preferring to go back to California as soon as possible. Carlyn never forgot her experiences with segregation and sexism and became a strong supporter of movements and organizations like the ACLU to end discrimination based on race and gender (4).

WORKING LIFE

After completing her Ph.D. in 1954, Carlyn secured a job in Northern California, working for 2 years in the clinical labs at the Presbyterian Hospital and the Medical Center of San Francisco. She next moved for 3 years to UCLA, where she oversaw the Dermatology Research Laboratory and collaborated with Thomas Sternberg and Victor Newcomer in ground-breaking research demonstrating the efficacy of amphotericin B to treat animals and humans with coccidioidomycosis (5), caused by an environmental fungal pathogen in soil and endemic to parts of Central and South America and the Southwest United States (including California). This was, and continues to be, an important public health pathogen, with up to 5% of infected individuals developing an acute lung infection that can disseminate to other parts of the body, such as the brain, soft tissues, joints, and bone. Prior to the introduction of amphotericin B, disseminated infections were often fatal (6). For decades amphotericin B remained the only effective therapy for invasive fungal diseases, until the development of the azole antifungals in the early 1980s (7).

In 1958, Carlyn set off for another adventure: the University of Indonesia Medical School in Djakarta asked if she could help set up their mycology laboratory and teach. Her answer was, "Absolutely!" When she finished in 1961, Carlyn went back to her parents in Alhambra with a proposal: would they like to accompany her on a grand tour of Africa? Maybe most parents would decline, but Carlyn's parents said yes. They traveled for 6 months together throughout Africa, and while she went home to look for a job, she arranged for her parents to continue their journey into the Middle East and Asia.

In 1964, Carlyn accepted employment at the University of California, San Francisco (UCSF), Medical School of Pharmacy. She taught medical microbiology (including medical mycology) to thousands of medical, dental, and pharmacy students over her 31-year career at the institution. She established an important resource program for medical communities (pathologists, dermatologists, ophthalmologists, and dentists) and published on a variety of topics, including mycological infections (8), taxonomic questions (9, 10), animal models for fungal drug development (11), and interesting case studies (12). She wrote a book on mycology for clinicians (13) and wrote with Dr. Miriam Valesco of the California Department of Public Health on the need for mycosis reporting requirements (14).

Carlyn made her expertise available and easily accessible to anyone who needed her. She offered weekend and night courses for laboratory technicians and physicians, and supervised the training of dermatology and ophthalmology residents on how to recognize fungal infections. She was a consultant for the California Department of Public Health Lab. With her UCSF colleague Dr. Raza Aly of the Department of Dermatology, she gave countless mycology workshops at UCSF to the medical and public health communities in Northern California.

While she was a young professor at UCSF, Carlyn was tasked with teaching medical microbiology and went to the San Francisco Tuberculosis Association, gathering information on tuberculosis for her students. This grassroots organization, started in 1908, was the very first nonprofit health organization in San Francisco. Its mission was, and is, to fight lung disease, advocate for clean air, and advance public health for local communities. Carlyn was an advocate for health equality herself, and this sparked a relationship that continued throughout her whole life. Early in her career, she was a recipient of the Strobel Research Fund to support her work on tuberculosis. In 1971, she joined the Board of Directors, serving in that position for 17 years; she also served on several other committees (e.g., for lung disease, anti-smoking, and fund raising). The organization changed its name to the Breathe California Golden Gate Public Health Partnership, and she was there to help celebrate their 100-year anniversary (15).

While teaching the UCSF dental students in the 1970s, Carlyn was introduced to a fledgling nonprofit that became very dear to her heart, Project Concern International, whose mission is to provide health care to disadvantaged people. She assisted in creating and participated in a program that allowed dental students to spend a couple of weeks in the Casa de Todos clinic in Tijuana, providing care to people who had no access to dentists. Carlyn continued to support the efforts of Project Concern throughout her life. According to their website, "She could always be counted on to recruit her best and brightest students to serve with us on the ground, often in countries far afield like Vietnam and Hong Kong" (16, 17).

Carlyn was active in the American Society for Microbiology (ASM), her local Northern California ASM branch (http://www.asmbranches.org/brcano/), the Medical Mycological Society of the Americas (MMSA), and the International Society for Human and Animal Mycology (ISHAM) (18, 19). She attended meetings faithfully well into her 80s, and she generously contributed both time and funds to these organizations. She established a travel grant in collaboration with MMSA to support a medical microbiology student from Latin America, Puerto Rico, or the Caribbean to participate in ASM Microbe. As an example of her inclusiveness, she stipulated that student presenters who have a medical mycology-related abstract accepted to the ASM Microbe meeting would automatically be considered, with no application necessary. Another example of her generosity to mycology education was the Carlyn Halde Membership Fund, established long ago for ISHAM, to provide memberships for 1 to 2 years when circumstances made it difficult for applicants to afford the membership fee (20). And she fostered budding interest in medical mycology for local Northern California microbiology students by funding grants for medical mycology talks at their ASM Northern California Branch meetings (http://www.asmbranches.org/brcano) (21).

Carlyn's time at UCSF was not without personal struggle, as epilepsy plagued her in her 40s, when she was trying to establish her research career. Medications made it difficult to concentrate, and she did not have as much stamina. Her colleagues were less than supportive. She had an inner strength few people possessed, and not being particularly intimidated by others, she continued to find ways to teach and further the knowledge and importance of medically important fungi. Another point of contention was that because her career was mostly centered around teaching and service, she at times felt ostracized by the research faculty, composed primarily of men. They didn't want to teach, but they still didn't seem to value her contributions to the department. She felt unappreciated; that was the thing that bothered her most, and she didn't receive her full professorship until age 60. Now, thankfully, times have changed, and the department is very appreciative of their teaching faculty. Carlyn retired at age 65 because she unselfishly wanted to release her position to a young scientist; however, she remained an amazing resource to all the young faculty and continued to lecture on medical mycology and tuberculosis and teach in the student laboratories until she was 84. The only reason she stopped then was because she could no longer stand up for the duration of a lecture. Carlyn showed her enduring commitment to UCSF by becoming a Dean's Associate (those representing the core philanthropic leadership) by contributing $100,000 of her estate to the School of Pharmacy.

We know Carlyn liked adventure and was very interested in the natural world and its wonderful diversity of cultures. She was invited to join the prestigious Society of Woman Geographers (SWG), an international community of geographers supporting research and exploration and providing opportunities to share ideas and intellectual exchange. Their motto is "For women who know no boundaries." This society of influential women covers a great variety of member interests, "from anthropology to geography to zoology." She treasured these interactions and their shared values of lifelong learning and curiosity and looked forward to inspiring talks from its members. SWG has an oral history program to record interviews with members, and Carlyn's life growing up until the time she completed her Ph.D. at Duke University was recorded through this project. We thank them for this invaluable resource (2).

Throughout her life, Carlyn embodied the free-spirit lifestyle. She might not have identified herself as such, but she was independent minded: she did things because she wanted to do them. She valued her friends and colleagues unconditionally. She cared about what she believed in and supported those causes wholeheartedly with her time and money. She supported her life passions: medical mycology; microbiology; the environment; fairness, equality, and justice for all; her local community; and the world communities. She was generous with her time and money. She did not let relationships define her or people change her, and she didn't care what people thought of her.

Carlyn touched so many people with her enthusiasm, presence, and generosity, leaving at least 28 groups funding from her estate when she passed away in 2014. Carlyn gave us much to think about. In summary, we learned from Carlyn how one can become a successful scientist and yet live the life one imagined. The blueprint is the same as always: be true to yourself,

Figure 2 Carlyn enjoying the outdoors.

follow your natural curiosity, find mentors and positive people that take an interest in you, have a goal, and set it in motion. Remember to give back to your community. And have fun along the way (Fig. 2)!

We love Carlyn and will never forget her free spirit soul and the contributions she made to medical mycology, microbiology, public health equality, and the freedom to pursue one's dreams, whatever they are, without persecution.

ACKNOWLEDGMENTS

We thank the American Society for Microbiology for including Carlyn's story in this women's history book and for administering the Carlyn Halde Fund for mycology student international travel to ASM Microbe. We also thank the Society of Woman Geographers, who were prescient to include Carlyn in their oral history project, which helped us tremendously in writing about Carlyn's early life and explorations.

CITATION

Wilson WJ, Lowe S. 2018. Carlyn Halde: free spirit, p 147–157. *In* Whitaker RJ, Barton HA (ed), *Women in Microbiology*. American Society for Microbiology, Washington, DC.

References

1. Wikipedia contributors. 2017. *History of California 1900 to present, on Wikipedia, The Free Encyclopedia.* https://en.wikipedia.org/wiki/History_of_California_1900_to_present.
2. Cordes F. 2013. *Interview with Carlyn J. Halde, Ph.D. (from childhood to Ph.D.), Oral History Program.* The Society of Woman Geographers. http://www.iswg.org/resources/oral-histories.
3. Halde C. 1964. Percutaneous *Cryptococcus neoformans* inoculation without infection. *Arch Dermatol* **89**:545.
4. American Civil Liberties Union, Northern California. *Legacy donors. Carlyn Jean Halde.* https://www.aclunc.org/donate/legacy-donors.
5. Halde C, Newcomer VD, Wright ET, Sternberg TH. 1957. An evaluation of amphotericin B in vitro and in vivo in mice against *Coccidioides immitis* and *Candida albicans*, and preliminary observations concerning the administration of amphotericin B to man. *J Invest Dermatol* **28**:217–231.
6. Malo J, Luraschi-Monjagatta C, Wolk DM, Thompson R, Hage CA, Knox KS. 2014. Update on the diagnosis of pulmonary coccidioidomycosis. *Ann Am Thor Soc* **11**:243–253.
7. Maertens JA. 2004. History of the development of azole derivatives. *Clin Microbiol Infect* **10**:1–10.
8. Elliott ID, Halde C, Shapiro J. 1977. Keratitis and endophthalmitis caused by *Petriellidium boydii*. *Am J Ophthalmol* **83**:16–18.
9. McGinnis MR, Ajello L, Beneke ES, Drouhet E, Goodman NL, Halde CJ, Haley LD, Kane JL, Land GA, Padhye AA, Pincus DH, Rinaldi MG, Rogers AL, Salkin IF, Schell WA, Weitzman I. 1984. Taxonomic and nomenclatural evaluation of the genera *Candida* and *Torulopsis*. *J Clin Microbiol* **20**:813–814.
10. Mcginnis MR, Rinaldi MG, Halde C, Hilger AE. 1975. Mycotic flora of the interdigital spaces of the human foot: a preliminary investigation. *Mycopathologia* **55**:47–52.
11. Waldorf AR, Halde C, Vedros NA. 1982. Murine model of pulmonary mucormycosis in cortisone-treated mice. *Sabouraudia* **20**:217–224.
12. Halde C, Fraher MA. 1966. *Cryptococcus neoformans* in pigeon feces in San Francisco. *Calif Med* **104**:188–190.
13. Halde C. 1987. *Basic Mycology for the Clinician.* Thieme Medical Publishers, Inc, New York, NY.
14. Halde C, Valesco M, Flores M. 1992. The need for a mycoses reporting system. *Curr Top Med Mycol* **4**:259–265.
15. Breathe California Golden Gate. 30 July 2014. *In memory of Carlyn Halde. Breathe California donor profile.* https://www.youtube.com/watch?v=LfxCPRYI38E.
16. Project Concern International. *Honoring Dr. Carlyn Halde.* https://www.pciglobal.org/honoring-dr-carlyn-halde-gg/.
17. Medical Mycology Society of the Americas. *The Carlyn Halde Latin American Student Travel Award.* http://www.mycologicalsociety.org/carlyn_halde_latin_american_student_travel_award.
18. International Society of Human and Animal Mycoses. *Carlyn Halde Fund.* https://www.isham.org/membership/carlyn-halde-fund.

19. International Society of Human and Animal Mycoses. *Carlyn Halde*. https://www. isham.org/about-isham/mycological-heroes/carlyn-halde.
20. Shoupe M. 17 July 2009. *Wet*. https://www.youtube.com/watch?v=2qbkYHFd7Ww.
21. Northern California Branch, American Society for Microbiology. http://www. asmbranches.org/brcano/.

Women in Microbiology
Edited by Rachel J. Whitaker and Hazel A. Barton
© 2018 American Society for Microbiology, Washington, DC
doi:10.1128/9781555819545.ch18

Jo Handelsman: Adviser, Teacher, Role Model, Friend

18

Patrick D. Schloss[1]

Effective mentoring can be learned, but not taught. Good mentors discover their own objectives, methods, and style by mentoring. And mentoring. And mentoring some more.

> **Jo Handelsman and colleagues,** *Entering Mentoring:*
> *A Seminar To Train a New Generation of Scientists* **(1)**

During the time I spent as a postdoc in the lab of Dr. Jo Handelsman (2002 to 2006) at the University of Wisconsin, we frequently discussed issues encountered while mentoring undergraduate scientists. A common theme throughout those talks was the title of the National Academy Press book *Adviser, Teacher, Role Model, Friend: On Being a Mentor to Students in Science and Engineering* (2). We asked ourselves, is an effective mentor really able to be an adviser, teacher, role model, and a friend? Which attribute is most important? Are they all critical to a productive mentoring relationship? These intriguing conversations each illuminated different parts of the mentor-mentee relationship. Because Jo exemplifies all of these and more with each of her trainees, I had the fortune of learning by example. My time in her lab was paramount to my development as a scientist, and more than a decade later, I can honestly say that I am most grateful for what she taught me about mentoring (Fig. 1).

The morning of my thesis defense in November of 2001, I emailed Jo Handelsman to ask for a position as a postdoctoral trainee in her lab in the

---

[1]Department of Microbiology & Immunology, University of Michigan, Ann Arbor, MI 48109

Figure 1 Jo Handelsman mentoring former graduate student Zakee Sabree in her laboratory at the University of Wisconsin. Zakee is an assistant professor in the Department of Evolution, Ecology and Organismal Biology at the Ohio State University. Photograph credit: Jeff Miller, University of Wisconsin-Madison (http://www.hhmi.org/news/howard-hughes-medical-institutes-million-dollar-professors-0).

Department of Plant Pathology at the University of Wisconsin. As I look back on my nondescript email, I find it amazing that I got a response only an hour later wishing me good luck and telling me to give her a call once I had calmed down from the defense. A few months after "cold-emailing" her, I drove my young family to Madison to start my position in Jo's lab.

I was interested in Jo's lab because I was intrigued by her use of metagenomics—a word that she coined for an approach that her research group had been using to study the treasure trove of microbial natural products in soil (3, 4). Although today we tend to think of metagenomics as shotgun sequencing of DNA from an environmental sample, her initial work described how she and her team could assess the genomic and functional characteristics of uncultured bacteria. They were able to clone DNA from uncultured populations and coerce other commonly used species, usually *Escherichia coli*, to express the foreign genes so they could be studied in the context of a biological system. At the time, sequencing was quite expensive and we needed to be selective about what we sequenced. Jo's group used genetic selections and screens to identify the cloned DNA that they would sequence. Others, most notably Ed DeLong, sequenced DNA fragments cloned into fosmids that harbored rRNA genes, a process that led to the discovery of bacteriorhodopsins in the ocean (5). Later, Jill Banfield's lab pioneered shotgun sequencing using bulk samples of community DNA, thereby leading to what most think of as modern-day metagenomics (6). Jo's lab pioneered their functional approach by sequencing the cloned DNA after demonstrating that it conferred a desirable function on *E. coli*. Typically the gene products included enzymes like proteases or lipases, but Jo's passion was, and continues to be, the discovery of novel antibiotics and antibiotic resistance genes. It's funny how things worked out; despite joining Jo's lab to learn and apply her brand of metagenomics, I never built my own meta-genomic libraries or attempted to find any novel enzymes or small molecules.

Only after I joined her lab did I learn that Jo is also a person who thinks deeply about what it means to be a teacher and mentor. In recognition of this, shortly after I joined Jo's lab, she was named a Howard Hughes Medical Institute (HHMI) Professor and given a mandate to develop instructional materials related to scientific teaching and mentorship. One of the resources that Jo would use with trainees when discussing how to mentor under-graduates in a laboratory setting was a book published by the National Academy Press, *Adviser, Teacher, Role Model, Friend: On Being a Mentor to Students in Science and Engineering* (2). Now, after more than 15 years and as a professor myself, I mentor junior scientists and help them to plan their careers. I regularly think about these attributes in my own relationships with my trainees and hope that they can see them in our interactions. As I think back on Jo's role as my mentor and as a mentor to others, I can see that she has been an adviser, a teacher, a role model, and a friend.

ADVISER

In the mentoring workshop sessions that Jo facilitated, we would frequently discuss that mentoring can go in two directions. A mentor should provide good advice to their mentee on such topics that include the direction of their project, career development, and integration into their professional network. But sometimes, it is important for a mentee to know how to mentor their mentor, to nudge the mentor to provide feedback on a manuscript and remain engaged in the mentee's project, and perhaps to help focus the mentor's attention on what is truly important.

During my time in Jo's lab there was a point when I was struggling to make sense of metrics for measuring the diversity of microbial communities. We were interested in finding the total number of bacterial taxa in soil by sampling the lowest number of 16S rRNA sequences required to detect low-abundance taxa. The main problem with such an approach is that it is difficult to validate the solution, and thereby the methods, without already having a solution in hand. Not making much headway on the problem, I started hiding in fear that Jo would corner me and ask why I had not made more progress on the project.

Since I did not trust my programming abilities or myself, I started using the word usage distribution in books to effectively ask, "How far would I have to read to know the number of different words in a book?" I started with *Goodnight, Moon*, because I had been reading it multiple times each night to my daughter and had memorized the book. After about 3 weeks, Jo popped her head in my office and asked, "How are things going?" I felt embarrassed to admit the truth.

I told Jo about integrating *Goodnight, Moon* into my research, and she loved the idea on the condition that I also included her favorite book, *The Portrait of a Lady*. From that point forward, she actively encouraged me to push the "books model" as a way to think about bacterial community diversity. We eventually published the model, and many educators have since told me how they have used it to teach their students about microbial community diversity (7). Years later, Jo was awarded the D. C. White Research and Mentoring Award for excellence in mentoring at the 2011 ASM General Meeting. I was late to her talk. As I walked into the packed room, and without missing a beat, I heard Jo say, "Coming into the room right now is the author of this model." Jo was on a slide describing the books model. Where another adviser might have told me that the idea was silly, Jo embraced it and encouraged me to think of accessible models that people can use to understand abstract ideas.

Just as Jo was effective at mentoring her trainees, I frequently heard stories of Jo mentoring administrators who she would normally look to for mentoring. In one instance, she met with Harvard president Larry Summers after he made disparaging comments about the ability of women to do math. By all accounts, she was able to have a frank but constructive conversation on how to meaningfully empower female faculty and students on his campus. Afterwards, she was able to take this experience and turn it into a published policy statement (8).

What stands out to me the most about her role as an adviser, however, was that between 2014 and 2017, Jo served the Obama administration as the Associate Director for Science in the Office of Science and Technology Policy (OSTP). In an interview at the 2017 Microbe meeting (Fig. 2), Jo commented that a highlight of this position was the opportunities she had to engage President Obama in discussions about science (9). (Watch the interview with Jo here: http://bit.ly/WIM_Handelsman.) She recounted

Figure 2 Jo Handelsman talks with her former trainees Patrick Schloss, Nichole Broderick, and Courtney Robinson at the 2017 Microbe Meeting in New Orleans. The three are currently tenure-track faculty at the University of Michigan, the University of Connecticut, and Howard University.

that when the administration was coming to a close and she went to say good-bye to the President, he thanked her for teaching him about science and for "his microbes." Whenever Jo would go to brief the President on some policy, she would take a stuffed microbe for his collection. Whether it was to discuss biodefense (e.g., Black Death) or climate change (e.g., cyanobacteria), she would give him a memento to remember their conversation. In these discussions, the President engaged Jo in a conversation linking back to the other microbes she had brought him. The effectiveness of this exchange would be on display when the President then used the stuffed microbes to tell others about the science. At OSTP, Jo was able to "mentor up" for many of the policies in science that were important to her, including preserving the integrity of soil and demonstrating that many of our supposed science-driven forensic techniques were biased and anything but scientific. Just as Jo encouraged me to break down complex problems into simple concepts that may seem silly to others, she had used the same approach to advise and teach President Obama.

TEACHER

The typical postdoctoral trainee receives minimal practice honing his or her teaching skills. Again, because of her HHMI Professorship, I was able to benefit from the concept of "scientific teaching" that Jo developed at the University of Wisconsin and later at Yale University (10). As she and her colleagues would state in a Policy Forum article in *Science*, scientific teaching is "teaching that mirrors science at its best—experimental, rigorous, and based on evidence" (11). Talking with Jo about teaching, it becomes clear that her command of the educational literature is on par with her command of the microbial ecology literature. She would frequently question why we worked so tirelessly to follow the scientific method in our research, but as soon as we stepped foot in the classroom, we would shun the same tools—the literature and experimentation—that make us successful in the laboratory. She practiced what she taught: integrating an educational evaluator into all of her work and publishing the results of her "Entering Mentoring" and "Scientific Teaching" workshops (11, 12). The concept that scientists should scientifically teach science left an indelible mark on me as an educator. A curriculum becomes a series of experiments rather than lecture notes that are etched in stone. Seeing teaching as a series of experiments also takes a considerable burden off the instructor; we should expect some of the experiments to fail and some to succeed. Over time, the instructional materials improve because we start asking the right questions.

As Jo and her colleagues lay out, effective instructional materials should also allow students to see the scientific method at work (10). Laboratory exercises should not be a series of "cookbook" lessons where the instructor knows the answer ahead of time because in a research environment, scientists rarely know the answer ahead of time. I remember the first time I saw Jo break down an effective lesson plan on the topic of bacterial ice nucleation. The first step was to identify the preconceived misconceptions that students had. As a 20-something with a Ph.D., I just *knew* that water froze at 0°C. Jo then had us measure the temperature of the liquid water in test tubes sitting in our ice buckets. The temperatures were just below 0°C. Then we dropped a culture of *Pseudomonas syringae* into the tubes—the water froze solid. Even today, watching YouTube videos of others dropping *P. syringae* into water and seeing it flash freeze makes me excited (13). She then challenged us to take this lesson, which she intentionally taught us in a cookbook manner, and convert it into a student-driven laboratory exercise using other things we found around the room. From this seed of a lesson plan we could see the value of educational concepts such as constructivism, active learning, and, most importantly, scientific teaching.

I was quite taken by the concept of scientific teaching. When I started to plan the research talk I would give while interviewing for faculty positions,

I realized that a research talk is really an opportunity to teach about what I have learned. With Jo's encouragement, my presentation involved handing out brown paper bags of M&Ms to people at the seminar. I would have them break into groups and define what would make up a species of candy. Then they had to determine how far to sample before they would know the number of candy species in the bag. Having seen the difficulty of the exercise, they were eager to hear how I approached the problem based on 16S rRNA gene sequences extracted from soil, feces, or any other environment. More than a decade later, people still come to me and tell me they remember this activity. The concept of scientific teaching is powerful and continues to have a significant impact on my teaching. It is also evident in the growing number of experiential laboratory activities, including the Small World Initiative that Jo started at Yale to have classrooms from around the world isolate bacteria from various environments to identify and characterize potential antibiotics (http://www.smallworldinitiative.org).

ROLE MODEL

While I was in Jo's research group, one had to walk through a bank of computers to get to her office. By sitting at these computers analyzing data, I benefited from overhearing Jo's conversations and having numerous informal opportunities to chat with Jo. Learning from Jo in this environment was unique and unscripted. She would ask people sitting in the computer room for their thoughts on writing, conversations from seminars would spill over, and there were numerous opportunities to expand our professional network by having Jo introduce us to other scientists who were visiting. This was one place where I would learn how important it was to Jo that her lab be diverse in terms of gender, racial, and cultural composition but also in terms of scientific background—not only because it was just but also because it allowed the group to answer questions in a way different from anyone else. I learned a lot about being a professor from sitting at those computers and watching Jo.

Aside from being the gateway to Jo's office, this room had another architectural component that was unique. Lining the walls of the room was every issue of the *Journal of Bacteriology* and *Applied and Environmental Microbiology*, two journals published by the American Society for Microbiology (ASM). ASM has taken on a special significance to me because of the importance Jo gave it both through the display of these journals and also because of her dedication as a reviewer and editor for *Applied and Environmental Microbiology* and then as President of ASM. She instilled in me the belief that if I wanted the bioinformatics tools we developed to have the greatest possible audience, we should publish them in an ASM journal. With this example, I see serving the society as a reviewer, editor, and now Chair of the Journals Board as a great honor and privilege.

Another venue where Jo served as a role model to the members of her research group was through her tireless efforts to address gender and racial discrimination in science as a whole. Jo worked with a colleague from the Engineering school, Molly Carnes, to cofound and direct the Women in Science and Engineering Leadership Institute at the University of Wisconsin. Similar to the methods she extolled in scientific teaching, Jo used data-guided approaches to quantify bias and make suggestions to overcome it.

While I was a postdoc, Jo and her colleagues set out to demonstrate that the process used to nominate and select NIH Pioneer Award winners was plagued with biases against women (14). I still recall Jo sharing the work with us to demonstrate the power of our implicit biases. Later she would go on to show that male and female faculty express biases against female students in how they review identical application materials that were submitted with either a male or female name (15). She also partnered with the ASM to quantify the representation of female presenters at its meetings and develop a strategy for overcoming bias against female presenters. In the end, she was able to demonstrate that with very simple interventions it was possible to have a more equitable gender makeup of speakers (16). Issues of gender and racial bias raise important questions on college campuses. Through her passion for the issues involved and scientific approach, she has been a role model to many for how to engage these topics in a scientific and data-informed manner.

FRIEND

In our discussions of mentoring, "friend" was the most frequent role to raise eyebrows. Does someone really need to be your *friend* to be an effective mentor? Of the four, it is probably the least critical that your mentor is your friend. Needless to say, a mentoring relationship is considerably different when someone shares that bond of friendship or personal understanding with their mentor. During my time in Jo's lab, I had a series of personal ups and downs: my wife gave birth to two children, my mom passed away, and we learned that one of my children was profoundly deaf. None of these should have necessarily impacted my research or career development. Regardless, Jo exhibited great empathy to share in my excitement and frustrations when other friends would feel awkward bringing up the difficulties. Although she would regularly say, "See you in the morning!" on a Friday afternoon, she knew and appreciated the value of family and the importance of maintaining a work-life balance. Jo's father lived in Madison and would regularly come into the lab to talk to Jo and members of the lab. His presence with Jo and the time Jo spent with him made clear the value that Jo placed on her family. Though it has been years since I left the lab, Jo still asks about my wife and children and can remember interacting with them while I was in her lab.

Friendships can withstand tension and grow stronger. During her time in the OSTP, Jo took on the herculean effort of bringing together government agencies, nonprofit organizations, and companies to launch the National Microbiome Initiative (17). No doubt to her annoyance, I took to Twitter to indicate that I thought that this was a bad idea. I nearly instantly got a message back saying, "What would you do differently?" and an opportunity to talk through my concerns on the phone. At the end of the conversation, I do not think we were any closer to agreeing on the benefits of big top-down team-based science versus a small bottom-up investigator-driven approach. Regardless, she encouraged me to write a letter to the editor of *Science* to describe what my concerns were with the initiative (18). Had there been no personal bond, Jo could have ignored me, but instead, our relationship was important enough to her to reach out to me and hear my thoughts. At the same time, as a former trainee, I learned that I could disagree with my mentor and not be punished. This is possible only if the mentor and mentee have a strong foundation of friendship and, with it, respect underlying their relationship.

MOVING ON

After leaving the OSTP, Jo decided to move her research group from Yale University back to the University of Wisconsin, where she is now the Director of the Wisconsin Institute for Discovery. Jo continues to train exceptional scientists who are benefiting enormously from her mentorship. Among the people in the research group while I was there, my colleagues and I are established as tenure-track faculty at the University of Michigan, Oklahoma University, Howard University, the Ohio State University, the University of Connecticut, the University of Kentucky, and Colorado State University. In addition, other trainees are researchers in labs across government agencies, industry, and academia. We all take with us the mentoring skills that we learned from being part of Jo's group.

Mentoring is clearly important to Jo. If one were to take nothing else away from their interaction with Jo, they should appreciate the priority that she placed on trainees. Several years ago, I was invited by students to speak at their department's seminar series. It had been a busy year and I didn't really feel like going. I had been working very hard to limit the amount of travel I do to help maintain a work-life balance. When I told my wife about the invitation, she responded, "Well you have to go, right? Isn't that Jo's rule—if students invite you, then you go. Right?" And so, I went reminded of my mentor's training and my obligation to help the next generation of young scientists.

While Jo was working in the OSTP, she invited me to have lunch with her at the White House the next time I was in Washington, DC. I hadn't seen Jo in a number of years, and so I was little sheepish to go see her and was a bit overwhelmed at the prospect of being in the White House. As her previous meeting finished and I sat in a chair outside her office, those memories of sitting in the computer room outside her office all came back to me. As she greeted me, I felt obligated to show her some piece of data from the previous week or to update her on the latest draft of a manuscript. That bond of mentorship still remains long after I left her research group. Many have benefited enormously from Jo's mentorship, and it is our obligation to carry these lessons on.

CITATION

Schloss PD. 2018. Jo Handelsman: adviser, teacher, role model, friend, p 159–169. *In* Whitaker RJ, Barton HA (ed), *Women in Microbiology*. American Society for Microbiology, Washington, DC.

References

1. Handelsman J, Pfund C, Lauffer SM, Pribbenow CM. 2005. *Entering Mentoring: A Seminar To Train a New Generation of Scientists*. Board of Regents of the University of Wisconsin System, Madison, WI.
2. National Academy of Sciences, National Academy of Engineering, Institute of Medicine. 1997. *Adviser, Teacher, Role Model, Friend: On Being a Mentor to Students in Science and Engineering*. National Academy Press, Washington, DC.
3. Handelsman J, Rondon MR, Brady SF, Clardy J, Goodman RM. 1998. Molecular biological access to the chemistry of unknown soil microbes: a new frontier for natural products. *Chem Biol* 5:R245–R249.
4. Rondon MR, Goodman RM, Handelsman J. 1999. The Earth's bounty: assessing and accessing soil microbial diversity. *Trends Biotechnol* 17:403–409.
5. Béjà O, Aravind L, Koonin EV, Suzuki MT, Hadd A, Nguyen LP, Jovanovich SB, Gates CM, Feldman RA, Spudich JL, Spudich EN, DeLong EF. 2000. Bacterial rhodopsin: evidence for a new type of phototrophy in the sea. *Science* 289:1902–1906.

6. Tyson GW, Chapman J, Hugenholtz P, Allen EE, Ram RJ, Richardson PM, Solovyev VV, Rubin EM, Rokhsar DS, Banfield JF. 2004. Community structure and metabolism through reconstruction of microbial genomes from the environment. *Nature* 428:37–43.

7. Schloss PD, Handelsman J. 2007. The last word: books as a statistical metaphor for microbial communities. *Annu Rev Microbiol* 61:23–34.

8. Handelsman J, Cantor N, Carnes M, Denton D, Fine E, Grosz B, Hinshaw V, Marrett C, Rosser S, Shalala D, Sheridan J. 2005. Careers in science. More women in science. *Science* 309:1190–1191.

9. American Society for Microbiology. 2017. Women in microbiology—Jo Handelsman. https://www.youtube.com/watch?v=vNNGR0ERX_M. Accessed 25 September 2017.

10. Handelsman J, Miller S, Pfund C. 2007. *Scientific Teaching*. WH Freeman & Company, New York, NY.

11. Handelsman J, Ebert-May D, Beichner R, Bruns P, Chang A, DeHaan R, Gentile J, Lauffer S, Stewart J, Tilghman SM, Wood WB. 2004. Scientific teaching. *Science* 304:521–522.

12. Pfund C, Maidl Pribbenow C, Branchaw J, Miller Lauffer S, Handelsman J. 2006. The merits of training mentors. *Science* 311:473–474.

13. Martin MO. 2011. Bacterial ice nucleation protein. https://www.youtube.com/watch?v=pH-afIrfUbQ. Accessed 26 September 2017.

14. Carnes M, Geller S, Fine E, Sheridan J, Handelsman J. 2005. NIH Director's Pioneer Awards: could the selection process be biased against women? *J Womens Health (Larchmt)* 14:684–691.

15. Moss-Racusin CA, Dovidio JF, Brescoll VL, Graham MJ, Handelsman J. 2012. Science faculty's subtle gender biases favor male students. *Proc Natl Acad Sci USA* 109:16474–16479.

16. Casadevall A, Handelsman J. 2014. The presence of female conveners correlates with a higher proportion of female speakers at scientific symposia. *mBio* 5:e00846-13.

17. Alivisatos AP, Blaser MJ, Brodie EL, Chun M, Dangl JL, Donohue TJ, Dorrestein PC, Gilbert JA, Green JL, Jansson JK, Knight R, Maxon ME, McFall-Ngai MJ, Miller JF, Pollard KS, Ruby EG, Taha SA, Unified Microbiome Initiative Consortium. 2015. A unified initiative to harness Earth's microbiomes. *Science* 350:507–508.

18. Schloss P. 2015. Nurturing the microbiome field. *Science* 350:1044.

Women in Microbiology
Edited by Rachel J. Whitaker and Hazel A. Barton
© 2018 American Society for Microbiology. All rights reserved.
doi:10.1128/9781555819545.ch19

Caroline Harwood: With Grace, Enthusiasm, and True Grit

19

Rebecca E. Parales[1] and Margaret McFall-Ngai[2]

We have the honor of writing this short piece about a remarkable scientist who happens to be a woman, Caroline (Carrie) Harwood. One of us, R.E.P., was one of Carrie's first graduate students and the other, M.M.-N., is a long-time personal friend of Carrie and her family. We (the authors) met over the traditional New Year's dinner feasts at Carrie's house, hosted by Carrie and her husband and collaborator, E. Peter (Pete) Greenberg. These delightful, calorific experiences took place for the several years that Pete and Carrie were on the faculty at Cornell University and then at the University of Iowa. Our better halves were also critical participants: Juan Parales (R.E.P.) was a technician in Pete's lab, and Ned Ruby (M.M.-N.) was a postdoc with Pete at Harvard and was Pete's best man at Pete and Carrie's wedding. It was at that wedding in the spring of 1984 that M.M.-N. first met Carrie. R.E.P. had met Carrie two years earlier, when Carrie was a postdoc at Yale in Nick Ornston's lab and R.E.P. joined Nick's lab as a technician. But before digressing too far into personal impressions, we will first describe the professional Carrie.

To love what you do and feel that it matters; how could anything be more fun?

Katherine Graham, publisher/Chair of the Board,
Washington Post, **1969–1991; first female Fortune 500 CEO**

One of Carrie's defining features is great enthusiasm for her work, an infectious attitude that seems to have begun early in her intellectual journey

[1]Department of Microbiology and Molecular Genetics, University of California, Davis, Davis, CA 95616
[2]Pacific Biosciences Research Center, University of Hawaii at Manoa, Honolulu, HI 96822

Figure 1 Carrie Harwood, with her signature enthusiastic smile.

(Fig. 1). As is the case for most academics, Carrie's educational experiences shaped her career. She did all of her training in New England. These adventures started at Concord Academy, a girls' high school in Concord, MA. The website recounting the history of the school mentions that "[from 1963 to 1971] Concord Academy was regarded as perhaps the finest independent secondary school for girls in the country." This distinction may well have been because Carrie was a student there for 4 of those 8 years; she has often laughed that Queen Noor of Jordan (then Elizabeth Halaby) was her classmate, although these two students certainly had vastly different career trajectories.

Remaining in a relatively rarified academic environment, Carrie attended Colby College in Maine, recognized as one of the finest liberal arts colleges in the world. At Colby and subsequently for a master's degree at Boston University, Carrie studied biology. In her Ph.D. dissertation research and thereafter, Carrie's focus has been on the world of microbes. She studied at the University of Massachusetts Amherst under Ercole Canale-Parola, a world-famous authority on the structure, physiology, and biochemistry of spirochetes, a group of bacteria with both free-living and host-associated species (1–5). Carrie studied free-living spirochetes from a wide array of marine environments, with an emphasis on deciphering the biochemistry associated with fermentation pathways in members of this group of microbes. The time Carrie spent at the Marine Biological Laboratories in Woods Hole, MA, first attending the summer Microbial Diversity course as a graduate student and ultimately directing the course from 2000 to 2003, was also influential on her career and those of many of her students, who she encouraged to attend the course and in some cases serve as teaching assistants. Following graduate school, Carrie jumped in the car and headed south for a couple of hours to wind up at Yale University, for her postdoc training with Nick Ornston. There Carrie began to study the microbes involved in the degradation of aromatic compounds, particularly *Pseudomonas* spp. (6, 7), a topic that has remained a principal focus for her entire research career. Following her postdoc work, Carrie began to move west. She started her faculty career at Cornell University in 1984. After ~4 years, Carrie and Pete moved farther west to the University of Iowa, where they held positions for the next 16+ years. In 2005, Carrie and Pete moved their family and labs just about as far west as one could manage, to the University of Washington (UW) in Seattle. Through these various positions, Carrie fashioned a career that put her at the forefront of microbiology, with seminal contributions in the fields of biodegradation, signal transduction, and biofuel research. She became internationally recognized for her research contributions on the mechanisms by which microorganisms degrade environmental pollutants. In addition, within and outside of this context, she has also focused on how microbes sense and respond to environmental signals in a wide array of contexts, from plant and animal symbionts (both beneficial and pathogenic) to free-living bacterial phylotypes.

Carrie's work has been all about deciphering how things work at the biochemical and molecular levels. In work that expanded on studies she began during her postdoc with Nick Ornston, Carrie established that chemotaxis is a critical component in the aerobic biodegradation of a wide range of environmental pollutants, including benzene, toluene, naphthalene, chlorobenzoates, trichloroethylene, and the herbicide 2,4-dichlorophenoxyacetic acid (8–13). These and other studies of *Pseudomonas* chemotaxis (14–18) led Carrie to explore the function of the *wsp* signal transduction system in the opportunistic pathogen *Pseudomonas aeruginosa* (19–28). She discovered that this system, which is comprised of homologs of the standard bacterial chemotaxis system, regulates the sensing of surfaces and the subsequent biofilm formation by populations of this microbe; she found that *wsp*-associated signaling modulates intracellular levels of cyclic diguanylate (c-di-GMP). These findings revealed c-di-GMP as an attractive target for developing new drugs to combat chronic infections in cystic fibrosis (CF) patients. During her postdoctoral studies of bacteria that degrade aromatic compounds aerobically, Carrie began to wonder how such chemicals were turned over in anoxic environments. Surprisingly, literature searches revealed that almost nothing had been published on the topic. Upon arriving at Cornell, she teamed up with Professor Jane Gibson, a noted biochemist who studied phototrophic bacteria, and they began to dissect anaerobic aromatic acid degradation by the anoxygenic phototroph *Rhodopseudomonas palustris*. Their analyses demonstrated that *R. palustris* initiates aromatic- and alicyclic-acid degradation by forming coenzyme A thioesters (29–31). She then continued to dig deeper, identifying and characterizing the enzymes and determining the sequence of reactions and regulatory controls associated with anaerobic pathways for the biodegradation of these compounds (32–44). During these studies Carrie gained an appreciation for the diversity of metabolic options available to *R. palustris*, which can grow aerobically as a chemoheterotroph or a chemoautotroph and anaerobically as a photoautotroph or a photoheterotroph (with aromatic acids as a favorite substrate); it can also fix nitrogen (45). She led an interdisciplinary team of scientists to determine the genome sequences of multiple isolates of *R. palustris* (45, 46), and the genome data revealed further metabolic diversity: three different nitrogenases appeared to be encoded (45). Of course, this led Carrie to want to understand why, and she began to analyze nitrogen fixation in *R. palustris* and the concomitant production of hydrogen (47, 48). As a by-product of this research, she pioneered the concept of a light-driven process for hydrogen production as a nonpolluting energy source and developed a unique metabolic selection process that utilizes light energy to channel reducing equivalents from organic compounds to significant quantities of hydrogen (49–55).

In collaborations with Pete, Carrie has also made significant contributions to the field of quorum sensing, identifying new and unusual homoserine lactone autoinducer molecules in the alphaproteobacteria *R. palustris* and *Bradyrhizobium*; their team is also starting to unravel the role of quorum sensing in the interaction between endophytic bacteria and their poplar tree host (56–62). These studies, which are still at the early stages, hold promise to reveal new quorum-sensing targets and a mechanism of interkingdom communication. Carrie has made and continues to make groundbreaking, benchmark contributions to nearly every corner of the field of microbial physiology. For all of these contributions to science, Carrie has received much-deserved recognition, including election to membership in the National Academy of Sciences and Washington State Academy of Sciences, both in 2009, election as a fellow of the American Association for the Advancement of Science in 2008 and fellow of the American Academy of Microbiology in 2000, and the Procter and Gamble Award in Applied and Environmental Microbiology in 2010.

Carrie's contributions as a faculty member were always within the context of the laboratory environment that she created. For the first few years, Carrie's lab was populated entirely by women; this was not intentional, but at that time she was the only female faculty member in the microbiology departments at both Cornell University and the University of Iowa, and she likely attracted women in search of a role model. A favorite gift given to many of her female students and friends was a mug with the classic picture of "Rosie the Riveter" showing off her muscles and the quote, "We Can Do It!" This image aptly represents Carrie as a mentor: she is extremely supportive and her office door is always open, no matter how many items she has on her daily list of things to do. Carrie is an excellent writer and as part of her mentoring activities, she edits and reedits every abstract and manuscript that is generated in her lab. Every student who goes through her lab sees very few words survive of that first abstract for the annual meeting of the American Society for Microbiology (ASM). One former graduate student (Jayna Ditty) remembers turning in a draft of her first manuscript as a hard copy, and after making so many red marks on the first page, Carrie told Jayna to "just give me a copy on a disk." But everyone who works in Carrie's lab becomes a better writer as a result of her mentoring efforts. And Carrie's overall enthusiasm for science fosters a lab atmosphere that is collaborative rather than competitive or hierarchical. Everyone is treated equally—undergraduates, technicians, graduate students, and postdocs—and everyone is expected to work hard. Each student has his or her own research project, but everyone helps one another when needed, offering advice, support, or even an extra pair of hands for a complex experiment. Even before Carrie and Pete began collaborating scientifically, their two lab groups interacted often, starting with joint happy hours on Friday afternoons at Iowa. Carrie and Pete's lab groups have always been an extension of their family; every year everyone (including their significant others, kids, and even visiting relatives and friends) was invited for Thanksgiving dinner, a Christmas/holiday party, and various barbeques in the backyard. Although these events were nominally potlucks, Carrie and Pete cooked all the main parts of these feasts and provided an extensive selection of beer and wine. And every year any attending current and former lab members continue to meet for "family reunions" at the ASM meeting to catch up. The affection of her students for their experiences in her lab is reflected in the quotes below (*see* Boxes 1–4).

I never realized until lately that women are supposed to be the inferior sex.

Katherine Hepburn, American Film Institute,
greatest female star of classic Hollywood cinema

Box 1 Federico Rey, assistant professor, University of Wisconsin-Madison

I started graduate school with the idea of doing research in vascular biology until I heard Carrie give a talk on a bacterium that could adapt to different environments, degrade pollutants, and even make H_2 biofuel! Her enthusiasm for this cool organism was so contagious that I ended up wanting to work with it...finding a new interest in microbiology and joining her lab. Working with Carrie was fun and inspiring. She creates a fertile environment for students to thrive that includes a balanced dose of creative freedom and sensible guidance.

Carrie taught me the importance of communicating science. Before joining her lab I used to spend all my time thinking about hypotheses and experiments, but very little (or none) thinking about how to tell these concepts to my colleagues or the rest of the world, and as a nonnative [English] speaker I presented some additional challenges that she embraced. I remember giving a terrible talk at a small meeting early in my Ph.D., after which she said: "too many words too many times!" After this, she worked with me before almost every other talk I gave for the following 2 years. I cannot overstate how much this has helped me through my career.

Other happy memories nurtured both science and life in general, and showed Carrie's genuine interest in both. The lab moved from Iowa City to Seattle in the 3rd year of my Ph.D. The initial temporary space assigned to the lab was limited and I was having a hard time finding the right place to write my papers and thesis. When Carrie realized this, she offered to share her office with me...I spent the last year and a half of my Ph.D. working there. I really treasure this time and I still miss it very much. Our morning chats over coffee, many times about science but also about family and life, are among the best memories I have of my time in her lab.

In addition to an exceptional research career, Carrie has been unflagging in her professional service to her department, her university, professional societies, the nation, and beyond to the international arena of microbiology. She is widely appreciated for these efforts, as she is generous, hardworking, focused, and, most of all, very responsible. Carrie can always be counted on to do the highest quality job possible. She has been known to say of herself, "I have grit," which an online dictionary defines as "[tending] to stick to their goals despite numerous issues, problems, setbacks and failures. The person has firmness of mind and unyielding courage." One example of Carrie's grit

Box 2 Jake McKinlay, assistant professor, Indiana University

I couldn't have asked for a better mentor in Carrie. No matter what was going on in her life or where she was in the world, she was always somehow able prioritize my needs. Her guidance in science and professional development, refreshingly frank (aka brutally blunt) feedback, respect for work-life balance, and ability to lead by example enabled me to take my career to the next level. While most of us in science endlessly complain about the demands on our time, I never heard this talk from Carrie. In contrast, I only remember her telling me that she had the "best job in the world." I share this perspective now and strive to be the leader that Carrie was for me, though I will probably always enjoy complaining.

Box 3 Kathryn Fixen, research assistant professor, University of Washington

Our meetings in Carrie's office were always presided over by a picture of Annie Oakley, who perfectly exemplifies how I see Carrie. She is a strong female role model, not afraid to speak her mind in a field dominated by men. She is at her most brilliant when faced with a challenge in which she is not expected to succeed, and her integrity and honesty are what make her one of the most respected people in our field. Above all, she truly cares about the people around her and does her best to help them succeed. I know I am going to have a picture of her hanging on my office wall as a reminder of the type of person I strive to be.

Carrie always made me feel like she valued my ideas, which was incredibly empowering. One of the best memories I have of Carrie occurred just after I had given birth to my son. Between the sleep deprivation, feeding, and diapering, I was feeling the crisis in confidence that many new parents feel. Carrie came to visit me, and we talked for hours about science and just life in general. Her visit was an unspoken reminder of her support and that she still saw me as a valuable member of her team. It was exactly what I needed at the time.

comes to mind. Shortly after setting up her lab in Iowa, Carrie, who was 9 months pregnant, was attending a lunch for faculty and new graduate students at the Iowa Memorial Union when she went into labor. She calmly finished lunch, walked across the river to the medical school campus, where her lab was located, on the way telling Pete (who was also at the lunch) that she needed to stop at her office and she would meet him to walk to the hospital (which was two buildings away). She gave birth to Ted within the hour and was back in lab in about a week. Now, in addition to being the Gerald and Lyn Grinstein Endowed Professor in Microbiology at UW, Carrie is the Associate Vice Provost for Research at UW, on the Howard Hughes Medical Institute Review Board, on the Editorial Board of *Proceedings of the National Academy of Sciences of the United States of America*, and a member-at-large of the American Association for the Advancement of Science. She also served as an editor for *Applied and Environmental Microbiology* and *Annual Reviews in Microbiology*. This pattern of deep commitment to service characterizes Carrie's entire career.

Box 4 Nancy Nichols, research microbiologist, USDA

Carrie is contagiously enthusiastic about her research. A favorite memory from my time in Carrie's lab is from a day when she and a couple of students had been out to collect environmental samples. They were back in the lab and using a microscope to check their samples. Carrie was right there with them, and there was a chorus of "oohs" and "ahs" as we all looked at the samples. It was one of those days when science was just plain fun.

Carrie is excited by possibilities. In the lab, she wanted to hear about results while they were still in the "maybe I've discovered something with this experiment, or maybe not yet" stage. "Potentially exciting" was how she termed it, and she always had an enthusiastic smile when we had something to tell her.

A woman is like a tea bag—you never know how strong she is until she gets in hot water.

Eleanor Roosevelt, American activist, politician, and diplomat; first chair of the Presidential Commission on the Status of Women

Carrie shares several characteristics with the three other famous women quoted in this piece, Katherine Graham, Katherine Hepburn, and Eleanor Roosevelt. They were raised within about 250 miles of one another; they went to all-girls high schools (respectively: Concord Academy, Concord MA; The Madeira School, McLean, VA; Oxford School for Girls, Hartford, CT; and Allenswood Girls Academy, London); they are all patrician. Carrie is the eldest of six children, five girls (Carrie, Wheezy, Betty, Jane, and Kit) and one boy (Charlie). They were born to Charles and Barbara Harwood (aka Chuck and Beazy), who themselves came from old and well-established New England families. Carrie and Pete, with their children, Barbara and Ted, have always had a stunning, yet comfortable, home, but... there is a Gainsborough in their dining room. All four of these women also faced significant challenges in their lives. Carrie has been phenomenally strong in the face of very difficult times, including the diagnosis of her first child (daughter Barbara) with CF and the constant medical vigilance associated with it. She has handled sustained times of trial with love, grace, and exceptional fortitude and has been, and remains, unflaggingly positive. As a result of the diagnosis, both Carrie and Pete have directed a significant part of their research efforts toward understanding the molecular basis of the pathogenesis of *Pseudomonas aeruginosa* in CF patients (35–44, 63, 64), another connection between work and family.

We appreciate this opportunity to put in writing some thoughts about a precious friend and colleague.

CITATION
Parales RE, McFall-Ngai M. 2018. Caroline Harwood: with grace, enthusiasm, and true grit, p 171–182. *In* Whitaker RJ, Barton HA (ed), *Women in Microbiology*. American Society for Microbiology, Washington, DC.

References
1. Harwood CS, Canale-Parola E. 1981. Branched-chain amino acid fermentation by a marine spirochete: strategy for starvation survival. *J Bacteriol* **148:**109–116.
2. Harwood CS, Canale-Parola E. 1981. Adenosine 5'-triphosphate-yielding pathways of branched-chain amino acid fermentation by a marine spirochete. *J Bacteriol* **148:**117–123.

3. Harwood CS, Canale-Parola E. 1982. Properties of acetate kinase isozymes and a branched-chain fatty acid kinase from a spirochete. *J Bacteriol* 152:246–254.

4. Harwood CS, Jannasch HW, Canale-Parola E. 1982. Anaerobic spirochete from a deep-sea hydrothermal vent. *Appl Environ Microbiol* 44:234–237.

5. Harwood CS, Canale-Parola E. 1983. *Spirochaeta isovalercia* sp. nov., a marine anaerobe that forms branched-chain fatty acids as fermentation products. *Int J Syst Bacteriol* 33:573–579.

6. Harwood CS, Rivelli M, Ornston LN. 1984. Aromatic acids are chemoattractants for *Pseudomonas putida*. *J Bacteriol* 160:622–628.

7. Harwood CS, Ornston LN. 1984. TOL plasmid can prevent induction of chemotactic responses to aromatic acids. *J Bacteriol* 160:797–800.

8. Harwood CS, Parales RE, Dispensa M. 1990. Chemotaxis of *Pseudomonas putida* toward chlorinated benzoates. *Appl Environ Microbiol* 56:1501–1503.

9. Harwood CS, Nichols NN, Kim M-K, Ditty JL, Parales RE. 1994. Identification of the *pcaRKF* gene cluster from *Pseudomonas putida*: involvement in chemotaxis, biodegradation, and transport of 4-hydroxybenzoate. *J Bacteriol* 176:6479–6488.

10. Grimm AC, Harwood CS. 1997. Chemotaxis of *Pseudomonas* spp. to the polyaromatic hydrocarbon naphthalene. *Appl Environ Microbiol* 63:4111–4115.

11. Grimm AC, Harwood CS. 1999. NahY, a catabolic plasmid-encoded receptor required for chemotaxis of *Pseudomonas putida* to the aromatic hydrocarbon naphthalene. *J Bacteriol* 181:3310–3316.

12. Parales RE, Ditty JL, Harwood CS. 2000. Toluene-degrading bacteria are chemotactic towards the environmental pollutants benzene, toluene, and trichloroethylene. *Appl Environ Microbiol* 66:4098–4104.

13. Hawkins AC, Harwood CS. 2002. Chemotaxis of *Ralstonia eutropha* JMP134(pJP4) to the herbicide 2,4-dichlorophenoxyacetate. *Appl Environ Microbiol* 68:968–972.

14. Ditty JL, Grimm AC, Harwood CS. 1998. Identification of a chemotaxis gene region from *Pseudomonas putida*. *FEMS Microbiol Lett* 159:267–273.

15. Ditty JL, Harwood CS. 1999. Conserved cytoplasmic loops are important for both the transport and chemotaxis functions of PcaK, a protein from *Pseudomonas putida* with 12 membrane-spanning regions. *J Bacteriol* 181:5068–5074.

16. Nichols NN, Harwood CS. 2000. An aerotaxis transducer gene from *Pseudomonas putida*. *FEMS Microbiol Lett* 182:177–183.

17. Ferrández A, Hawkins AC, Summerfield DT, Harwood CS. 2002. Cluster II *che* genes from *Pseudomonas aeruginosa* are required for an optimal chemotactic response. *J Bacteriol* 184:4374–4383.

18. Ditty JL, Harwood CS. 2002. Charged amino acids conserved in the aromatic acid/H$^+$ symporter family of permeases are required for 4-hydroxybenzoate transport by PcaK from *Pseudomonas putida*. *J Bacteriol* 184:1444–1448.

19. Hickman JW, Tifrea DF, Harwood CS. 2005. A chemosensory system that regulates biofilm formation through modulation of cyclic diguanylate levels. *Proc Natl Acad Sci U S A* 102:14422–14427.

20. Güvener ZT, Harwood CS. 2007. Subcellular location characteristics of the *Pseudomonas aeruginosa* GGDEF protein, WspR, indicate that it produces cyclic-di-GMP in response to growth on surfaces. *Mol Microbiol* 66:1459–1473.

21. Hickman JW, Harwood CS. 2008. Identification of FleQ from *Pseudomonas aeruginosa* as a c-di-GMP-responsive transcription factor. *Mol Microbiol* 69:376–389.

22. Baraquet C, Murakami K, Parsek MR, Harwood CS. 2012. The FleQ protein from *Pseudomonas aeruginosa* functions as both a repressor and an activator to control gene expression from the *pel* operon promoter in response to c-di-GMP. *Nucleic Acids Res* 40:7207–7218.

23. O'Connor JR, Kuwada NJ, Huangyutitham V, Wiggins PA, Harwood CS. 2012. Surface sensing and lateral subcellular localization of WspA, the receptor in a chemosensory-like system leading to c-di-GMP production. *Mol Microbiol* 86:720–729.

24. Huangyutitham V, Güvener ZT, Harwood CS. 2013. Subcellular clustering of the phosphorylated WspR response regulator protein stimulates its diguanylate cyclase activity. *mBio* 4:e00242-13.

25. Irie Y, Borlee BR, O'Connor JR, Hill PJ, Harwood CS, Wozniak DJ, Parsek MR. 2012. Self-produced exopolysaccharide is a signal that stimulates biofilm formation in *Pseudomonas aeruginosa*. *Proc Natl Acad Sci U S A* 109:20632–20636.

26. Baraquet C, Harwood CS. 2013. C-di-GMP represses bacterial flagella synthesis by interacting with the Walker A motif of the enhancer binding protein FleQ. *Proc Natl Acad Sci U S A* 110:18478–18483.

27. Baraquet C, Harwood CS. 2015. A FleQ DNA binding consensus sequence revealed by studies of FleQ-dependent regulation of biofilm gene expression in *Pseudomonas aeruginosa*. *J Bacteriol* 198:178–186.

28. Matsuyama BY, Krasteva PV, Baraquet C, Harwood CS, Sondermann H, Navarro MV. 2016. Mechanistic insights into c-di-GMP-dependent control of the biofilm regulator FleQ from *Pseudomonas aeruginosa*. *Proc Natl Acad Sci U S A* 113:E209–E218.

29. Geissler JF, Harwood CS, Gibson J. 1988. Purification and properties of benzoate-coenzyme A ligase, a *Rhodopseudomonas palustris* enzyme involved in the anaerobic degradation of benzoate. *J Bacteriol* 170:1709–1714.

30. Merkel SM, Eberhard AE, Gibson J, Harwood CS. 1989. Involvement of coenzyme A thioesters in anaerobic metabolism of 4-hydroxybenzoate by *Rhodopseudomonas palustris*. *J Bacteriol* 171:1–7.

31. Gibson J, Dispensa M, Fogg GC, Evans DT, Harwood CS. 1994. 4-Hydroxybenzoate-coenzyme A ligase from *Rhodopseudomonas palustris*: purification, gene sequence, and role in anaerobic degradation. *J Bacteriol* 176:634–641.

32. Kim M-K, Harwood CS. 1991. Regulation of benzoate-CoA ligase in *Rhodopseudomonas palustris*. *FEMS Microbiol Lett* 83:199–204.

33. Dispensa M, Thomas CT, Kim M-K, Perrotta JA, Gibson J, Harwood CS. 1992. Anaerobic growth of *Rhodopseudomonas palustris* on 4-hydroxybenzoate is dependent on AadR, a member of the cyclic AMP receptor protein family of transcriptional regulators. *J Bacteriol* 174:5803–5813.

34. Perrotta JA, Harwood CS. 1994. Anaerobic metabolism of cyclohex-1-ene-1-carboxylate, a proposed intermediate of benzoate degradation by *Rhodopseudomonas palustris*. *Appl Environ Microbiol* 60:1775–1782.

35. Egland PG, Pelletier DA, Dispensa M, Gibson J, Harwood CS. 1997. A cluster of bacterial genes for anaerobic benzene ring biodegradation. *Proc Natl Acad Sci U S A* 94:6484–6489.

36. Gibson J, Dispensa M, Harwood CS. 1997. 4-Hydroxybenzoyl-CoA reductase (dehydroxylating) is required for anaerobic degradation of 4-hydroxybenzoate by

Rhodopseudomonas palustris and shares features with molybdenum-containing hydroxylases. *J Bacteriol* 179:301–309.

37. Egland PG, Gibson J, Harwood CS. 1995. Benzoate-coenzyme A ligase, encoded by *badA*, is one of three ligases able to catalyze benzoyl-coenzyme A formation during anaerobic growth of *Rhodopseudomonas palustris* on benzoate. *J Bacteriol* 177:6545–6551.

38. Egland PG, Harwood CS. 1999. BadR, a new MarR family member, regulates anaerobic benzoate degradation by *Rhodopseudomonas palustris* in concert with AadR, an Fnr family member. *J Bacteriol* 181:2102–2109.

39. Pelletier DA, Harwood CS. 1998. 2-Ketocyclohexanecarboxyl coenzyme A hydrolase, the ring cleavage enzyme required for anaerobic benzoate degradation by *Rhodopseudomonas palustris*. *J Bacteriol* 180:2330–2336.

40. Egland PG, Harwood CS. 2000. HbaR, a 4-hydroxybenzoate sensor and FNR-CRP superfamily member, regulates anaerobic 4-hydroxybenzoate degradation by *Rhodopseudomonas palustris*. *J Bacteriol* 182:100–106.

41. Pelletier DA, Harwood CS. 2000. 2-Hydroxycyclohexanecarboxyl coenzyme A dehydrogenase, an enzyme characteristic of the anaerobic benzoate degradation pathway used by *Rhodopseudomonas palustris*. *J Bacteriol* 182:2753–2760.

42. Harrison FH, Harwood CS. 2005. The *pimFABCDE* operon from *Rhodopseudomonas palustris* mediates dicarboxylic acid degradation and participates in anaerobic benzoate degradation. *Microbiology* 151:727–736.

43. Peres CM, Harwood CS. 2006. BadM is a transcriptional repressor and one of three regulators that control benzoyl coenzyme A reductase gene expression in *Rhodopseudomonas palustris*. *J Bacteriol* 188:8662–8665.

44. Crosby HA, Heiniger EK, Harwood CS, Escalante-Semerana JC. 2010. Reversible N-lysine acetylation regulates the activity of acyl-CoA synthetases involved in anaerobic benzoate catabolism in *Rhodopseudomonas palustris*. *Mol Microbiol* 75:1007–1020.

45. Larimer FW, Chain P, Hauser L, Lamerdin J, Malfatti S, Do L, Land ML, Pelletier DA, Beatty JT, Lang AS, Tabita FR, Gibson JL, Hanson TE, Bobst C, Torres JL, Peres C, Harrison FH, Gibson J, Harwood CS. 2004. Complete genome sequence of the metabolically versatile photosynthetic bacterium *Rhodopseudomonas palustris*. *Nat Biotechnol* 22:55–61.

46. Oda Y, Larimer FW, Chain PS, Malfatti S, Shin MV, Vergez LM, Hauser L, Land ML, Braatsch S, Beatty JT, Pelletier DA, Schaefer AL, Harwood CS. 2008. Multiple genome sequences reveal adaptations of a phototrophic bacterium to sediment micro-environments. *Proc Natl Acad Sci U S A* 105:18543–18548.

47. Oda Y, Samanta SK, Rey F, Wu L, Liu X-D, Yan T-F, Zhou J, Harwood CS. 2005. Functional genomic analysis of three nitrogenase isozymes in *Rhodopseudomonas palustris*. *J Bacteriol* 187:7784–7794.

48. Rey FE, Oda Y, Harwood CS. 2006. Regulation of uptake hydrogenase and effects of hydrogen utilization on gene expression in *Rhodopseudomonas palustris*. *J Bacteriol* 188:6143–6152.

49. Rey FE, Heiniger EK, Harwood CS. 2007. Redirection of metabolism for biological hydrogen production. *Appl Environ Microbiol* 73:1665–1671.

50. Gosse JL, Engel BJ, Rey FE, Harwood CS, Scriven LE, Flickinger MC. 2007. Hydrogen production by photoreactive nanoporous latex coatings of nongrowing *Rhodopseudomonas palustris* CGA009. *Biotechnol Prog* 23:124–130.

51. Huang JJ, Heiniger EK, McKinlay JB, Harwood CS. 2010. Production of hydrogen gas from light and the inorganic electron donor thiosulfate by *Rhodopseudomonas palustris*. *Appl Environ Microbiol* **76**:7717–7722.

52. Gosse JL, Engel BJ, Hui JC, Harwood CS, Flickinger MC. 2010. Progress toward a biomimetic leaf: 4,000 h of hydrogen production by coating-stabilized nongrowing photosynthetic *Rhodopseudomonas palustris*. *Biotechnol Prog* **26**:907–918.

53. Heiniger EK, Oda Y, Samanta SK, Harwood CS. 2012. How posttranslational modification of nitrogenase is circumvented in *Rhodopseudomonas palustris* strains that produce hydrogen gas constitutively. *Appl Environ Microbiol* **78**:1023–1032.

54. Adessi A, McKinlay JB, Harwood CS, DePhilippis R. 2012. A *Rhodopseudomonas palustris nifA** mutant produces H_2 from NH_4^+-containing vegetable wastes. *Int J Hydrogen Energy* **37**:15893–15900.

55. McKinlay JB, Oda Y, Rühl M, Posto AL, Sauer U, Harwood CS. 2014. Non-growing *Rhodopseudomonas palustris* increases the hydrogen gas yield from acetate by shifting from the glyoxylate shunt to the tricarboxylic acid cycle. *J Biol Chem* **289**:1960–1970.

56. Schaefer AL, Greenberg EP, Oliver CM, Oda Y, Huang JJ, Bittan-Banin G, Peres CM, Schmidt S, Juhaszova K, Sufrin JR, Harwood CS. 2008. A new class of homoserine lactone quorum-sensing signals. *Nature* **454**:595–599.

57. Hirakawa H, Oda Y, Phattarasukol S, Armour CD, Castle JC, Raymond CK, Lappala CR, Schaefer AL, Harwood CS, Greenberg EP. 2011. Activity of the *Rhodopseudomonas palustris p*-coumaroyl-homoserine lactone-responsive transcription factor RpaR. *J Bacteriol* **193**:2598–2607.

58. Lindemann A, Pessi G, Schaefer AL, Mattmann ME, Christensen QH, Kessler A, Hennecke H, Blackwell HE, Greenberg EP, Harwood CS. 2011. Quorum sensing in the soybean root-nodulating bacterium *Bradyrhizobium japonicum*: identification of isovaleryl-homoserine lactone, an unusual branched-chain signal. *Proc Natl Acad Sci U S A* **108**:16750–16770.

59. Ahlgren NA, Harwood CS, Schaefer AL, Giraud E, Greenberg EP. 2011. Aryl-homoserine lactone quorum sensing in stem-nodulating photosynthetic bradyrhizobia. *Proc Natl Acad Sci U S A* **108**:7183–7188.

60. Hirakawa H, Harwood CS, Pechter KB, Schaefer AL, Greenberg EP. 2012. Antisense RNA that affects *Rhodopseudomonas palustris* quorum-sensing signal receptor expression. *Proc Natl Acad Sci U S A* **109**:12141–12146.

61. Schaefer AL, Lappala CR, Morlen RP, Pelletier DA, Lu TY, Lankford PK, Harwood CS, Greenberg EP. 2013. LuxR- and luxI-type quorum-sensing circuits are prevalent in members of the *Populus deltoides* microbiome. *Appl Environ Microbiol* **79**:5745–5752.

62. Schaefer AL, Oda Y, Coutinho BG, Pelletier DA, Weiburg J, Venturi V, Greenberg EP, Harwood CS. 2016. A LuxR homolog in a cottonwood tree endophyte that activates gene expression in response to a plant signal or specific peptides. *mBio* **7**:e01101-16.

63. Alvarez-Ortega C, Harwood CS. 2007. Responses of *Pseudomonas aeruginosa* to low oxygen indicate that growth in the cystic fibrosis lung is by aerobic respiration. *Mol Microbiol* **65**:153–165.

64. Starkey M, Hickman JH, Ma L, Zhang N, De Long S, Hinz A, Palacios S, Manoil C, Kirisits MJ, Starner TD, Wozniak DJ, Harwood CS, Parsek MR. 2009. *Pseudomonas aeruginosa* rugose small-colony variants have adaptations that likely promote persistence in the cystic fibrosis lung. *J Bacteriol* **191**:3492–3503.

Women in Microbiology
Edited by Rachel J. Whitaker and Hazel A. Barton
© 2018 American Society for Microbiology. All rights reserved.
doi:10.1128/9781555819545.ch20

Marian Johnson-Thompson: Lifelong Mentor

20

Crystal N. Johnson[1]

Marian Johnson-Thompson earned her bachelor's and master's degrees from Howard University and her Ph.D. from Georgetown University Medical School. After spending the majority of her career at the University of the District of Columbia (UDC), she retired and was conferred the title Professor *Emerita* of Biological and Environmental Sciences. Following her time at UDC, she was Director of Education and Biomedical Research Development at the National Institute of Environmental Health Sciences (NIEHS) of the National Institutes of Health (NIH). Since 2001, she has been an adjunct professor in the Department of Maternal and Child Health, School of Public Health, University of North Carolina at Chapel Hill.

Dr. Johnson-Thompson's publications have characterized the mechanisms of action of the antiviral drug azacytidine against simian virus 40 (SV40) (1, 2), the impact of vacuum UV laser treatment on the stability of the viral DNA (3), and the use of azapyrimidine to treat SV40-infected eukaryotic cells *in vitro* (4). Her 2000 paper titled "Ongoing Research To Identify Environmental Risk Factors in Breast Carcinoma" investigated health disparities in the diagnosis and treatment of breast cancer and was one of the first studies to find that women of color may exhibit higher susceptibility to environment-associated breast cancer (5). In another study 3 years later (6), Dr. Johnson-Thompson reported the need for minority scientists in the clinical realm, as this population is grossly underrepresented and the lack of minority clinicians is associated with a paucity of those qualified to treat underserved patients; thus, an increase in doctors sensitive to these

[1]Department of Environmental Sciences, Louisiana State University, Baton Rouge, LA 70803

important issues would create an increase in treatment of minority patients. In 2013, she was selected for the Science History Makers (7). The History Makers is the largest collection of African-American oral history interviews in the United States, and its interviewees include President Barack Obama. Science History Makers lists her favorite quote as, "To whom much is given, much is required," a sentiment that Dr. Johnson-Thompson has exemplified throughout her career.

Moreover, in the midst of an illustrious career, Dr. Johnson-Thompson also maintained a long-term marriage and raised two healthy and successful sons during a time when working outside the home was not always the norm for women. She is the quintessential picture of "having it all" (Fig. 1). She was elected in 1998 as a Fellow of the American Academy of Microbiology (AAM) and in 2004 as a Fellow of the American Association for the Advancement of Science (AAAS). She personifies the importance of mentoring young scientists at early years and throughout their careers.

Dr. Johnson-Thompson has mentored numerous underrepresented minority students who went on to earn graduate and/or medical degrees. While at UDC, she also taught courses at Howard University. She has directly mentored 14 undergraduate and master's-level students who have gone on to receive a Ph.D. and an additional 17 students at the Ph.D. level. In 2001 she was named the Meyerhoff Scholars Mentor of the Year, and she continues to mentor Meyerhoff Scholars at Duke University and the University of North Carolina at Chapel Hill.

In 2015, I earned tenure in the Louisiana State University Department of Environmental Sciences and was promoted to associate professor. This would never have been possible without long-time mentoring by

FIGURE 1 Dr. Marian Johnson-Thompson at the microscope. Reprinted from reference 8, with permission.

Dr. Johnson-Thompson. I had a rough start in college, even ending up on academic probation with a 0.5 GPA. In 1994, I was doing undergraduate research in the Environmental Endocrinology Lab of the Tulane/Xavier Center for Bioenvironmental Research, which was headed by Dr. John McLachlan. I was shy, insecure, and directionless. I was reasonably intelligent, but I had no idea how to study for college courses or deal with adversity.

I knew that I was excited about science, but I had no idea what came next or how to get there, especially after such a rough start. Dr. McLachlan introduced me to Dr. Marian Johnson-Thompson, who mentored me by phone and by email in the very early formative years of distance mentoring. Throughout my undergraduate years, Dr. Johnson-Thompson was only a phone call or an email away. As a poor black girl who grew up on welfare and food stamps in the small town of Moss Point, MS (population, 13,000), I never knew any scientists, let alone a scientist who was both black and female. In my family, if you were considered smart, you became either a doctor or a lawyer; becoming a scientist was never an option, likely because we did not know any. Not only did Dr. Johnson-Thompson take an interest in my personal success as an undergraduate, but also she represented the very essence of what I discovered I wanted to become, i.e., a scientist. Her mere existence, and the implication that I could possibly do what she did too, kept me focused when I earned low grades in my classes and even had to repeat some classes. She helped me understand that it was okay to fall flat on my face as long as I kept my eyes on the prize. I eventually earned my bachelor's degree in cell and molecular biology from Tulane University in 1997. Two decades later, I would discover that falling and getting back up actually make for a better researcher than getting it perfect the first time.

When it was time to apply to graduate schools, Dr. Johnson-Thompson was right there helping me with specific aspects, such as emphasizing my accomplishments to strengthen my application package, helping me interpret my GRE scores, and answering my numerous questions (Is it good enough? What is the best approach? What schools are best? What do I say during the interview?). She was particularly valuable when a Tulane professor told me that I was not good enough to earn a Ph.D. from the University of Alabama at Birmingham (UAB), because it was ranked third in U.S. microbiology programs. She put me in touch with another mentor at UAB, Dr. Gail Cassell, who was a member and chair of the UAB microbiology department. Dr. Cassell further championed and mentored me as well. Later, Dr. Johnson-Thompson helped me earn an interview for a postdoctoral opportunity at the NIH NIEHS with Dr. Jan Drake, and during this visit she hosted me in her home.

Her eternal patience and willingness to share her knowledge have made me a more conscientious and patient mentor. I ask my students the same questions she asked me 19 years ago, and I encourage them to stay focused, work hard, and keep paying it forward. When I had the opportunity to attend the American Society for Microbiology (ASM) General Meeting in Chicago in 1999, 5 years after Dr. Johnson-Thompson began mentoring me, I finally met her. It was like meeting a Nobel laureate. I was humbled, honored, and a bit star-struck to meet the scientist who had been mentoring me remotely and helping me through many rough periods. Throughout the years she has continued this and remains a mentor to me today, even as a tenured college professor. She has helped me apply for funding, choose which undergraduate and graduate courses to teach, identify research directions, deal with difficult personalities, and deal with perpetually being the only African-American and/or female scientist in the room. She helped me choose a destination for my postdoc position, earn my first appointment to an ASM committee, interview for and earn a tenure-track position, deal with balancing personal and professional aspirations, and forever keep my eyes on the prize.

Dr. Johnson-Thompson's commitment to scholarship is exhibited by her assistance with undergraduate and graduate course selection, encouragement in the face of academic struggles, and assistance with identification of corrective courses of action when necessary. She grew up in segregated schools that had limited resources for black students and depended on inferior and outdated leftover resources from other schools (8). She earned her Ph.D. in molecular virology at the Georgetown University School of Medicine when only a few years before, African-Americans were attending segregated schools throughout the country. She was the first U.S.-born black person to graduate from that program, even mentoring other graduate students during her own graduate studies. Earning her Ph.D. during such a volatile era in our history has made her particularly effective as a mentor, because her fight to the top was disproportionately difficult.

She has demonstrated to me, verbally and by example, that it was never enough to simply be a minority microbiologist; I still had to go out and publish, raise funds for research with a passion, become an active contributor in any scientific group of which I was a member, and make all the right moves to become a well-published and well-funded scientist, all while giving back, paying it forward, and never forgetting my roots. Setting the example, she presented her research on antiviral drugs at numerous conferences and via numerous peer-reviewed publications, including *Virology* and *Antimicrobial Agents and Chemotherapy*, while at UDC. At NIEHS, she continued to publish, specifically on breast cancer policy, mentoring, and academic pipeline issues, while serving on countless panels, committees, and review boards, establishing novel programs, receiving numerous honors, arranging and sustaining mentorships, and introducing and supporting students all along the STEM (science, technology, engineering, and mathematics) research pipeline. In 2010, Dr. Johnson-Thompson was interviewed by *Black Enterprise* magazine for STEM Spotlight Week, and she was asked for actions that would encourage girls to pursue an interest in science. Her responses included the identification of role models, an encouragement of girls to master math early, the use of hands-on activities, and establishment of a supportive environment. More recently, she has lobbied on Capitol Hill and served on multiple Susan G. Komen review panels, which make important decisions on funding applications to the Susan G. Komen Breast Cancer Foundation, the largest and most heavily funded breast cancer organization in North America.

Dr. Johnson-Thompson's commitment to activism is exhibited by her numerous decades of participation in STEM-related activities, particularly those that contribute to the training of underrepresented minorities, including serving on and chairing numerous NIH, ASM, National Science Foundation (NSF), and AAAS advisory boards and panels. She has chaired the NIH Task Force on the Advancement of Minorities in the Sciences (9), Minority Programs Evaluations, and Women Scientists committees. She was an inaugural member of the Roadmap 1.5: Human Microbiome Working Group, which developed the NIH's Human Microbiome Initiative. She was the originator of the William A. Hinton Research Training Award, which honors those who contribute to the training and retention of minority microbiologists, and she has served on the Nomination Committee since 2008, further evidence of her "pay it forward" style of multitiered mentoring. As a result of her work in bringing recognition to Dr. Hinton's contributions, several other organizations have honored him. In 2012, she represented ASM at the National Academy of Sciences (NAS) meeting "Seeking Solutions: Maximizing American Talent by Advancing Women of Color in Academia" at NAS headquarters. For several years, she served as principal investigator of the ASM Minority Travel Program, funded by National Institute of Allergy and Infectious Diseases (NIAID)/ NIH.

In 1997, Dr. Johnson-Thompson established the Bridging Education Science and Technology Program at Hillside High School in Durham, NC, for which the NIEHS equipped the molecular biology lab with supplies and NIEHS scientists trained teachers on how to instruct students. This marked the introduction of students in this high school to hands-on molecular biology experiences, a practice that to this day remains a rarity nationwide.

She has served on the North Carolina School of Science and Mathematics Foundation Board and has written recommendation letters for many students over the years. In 1994, she established the Johnson-Thompson Taylor Endowed Scholarship Fund at Howard University for underrepresented female science students. The scholarship was established at her 25th anniversary at Howard University and was named in honor of her mentor, Dr. Marie Taylor, who was the first female of any race to earn a Ph.D. at Fordham University.

In addition, she has nominated and promoted countless unsung and otherwise unrecognized minorities for very prestigious accolades, including nomination to AAM fellowship. AAM fellows nominated by Dr. Johnson-Thompson include Henry Williams, an African-American male who earned his Ph.D. in microbiology from the University of Maryland at Baltimore in 1979 and is now the director of the Florida A&M University Environmental Sciences Institute; Lizzie Harrell (retired), the first African-American woman to earn her Ph.D. in microbiology from North Carolina State University (in 1978); Karen Nelson, who earned her Ph.D. at Cornell University and is president of the J. Craig Venter Institute; and David Satcher, who earned his Ph.D. in cytogenetics and was the 16th Surgeon General of the United States.

She has also mentored, guided, and advised countless doctoral degree recipients who have benefitted from her actions, including participation in her summer programs, review of graduate fellowships, NIH and NIEHS recruitment efforts, and nominations for prestigious awards. This list includes several individuals. Ashalla Magee Freeman is an African-American woman who earned her Ph.D. in microbiology from the University of Alabama at Birmingham and is now the Director of Diversity Affairs at the University of North Carolina at Chapel Hill, Office of Graduate Education. Kenneth Gibbs earned his Ph.D. at Stanford University and a master of public health degree from Johns Hopkins University and is a program analyst at the National Institute of General Medical Sciences, NIH. Elena Braithwaite is an African-American woman who earned her Ph.D. in toxicology at the University of Kentucky and is now a toxicologist at the FDA in

Rockville, MD. Sherilynn Black is an African-American woman who earned her Ph.D. in neurobiology at Duke University and is currently an assistant professor at Duke University School of Medicine. Pocahontas Jones is an African-American woman who earned her Ph.D. in microbiology/immunology at Howard University, and Dr. Johnson-Thompson served on her dissertation committee. Dr. Jones went on to have an outstanding career at Halifax Community College in North Carolina, with honors including service as a dean and receipt of numerous teaching awards. During this time, Dr. Jones spent a summer engaged in laboratory research at the NIEHS.

Dr. Johnson-Thompson has been invited to give inspirational presentations to a variety of programs beginning at the level of kindergarten to grade 12 (K–12). At the NIEHS, she developed the first K–12 environmental health science education program that was a request for applications to fund K–12 curriculum development and teacher professional development. Other activities included outreach to North Carolina historically black colleges and universities by ensuring student hires in summer programs, group visits to the NIEHS (where students were exposed to current research technologies), and inspirational presentations by underrepresented NIEHS scientists. She developed the Advanced Research Cooperation in Environmental Health extramural program, a collaboration between a minority-serving institution (MSI) and a research-intensive university (RIU). The focus was on providing infrastructure support (both teaching and research) to ensure the flow of Ph.D. candidates from MSIs to RIUs. The funded MSIs included Meharry Medical College, Xavier University, Southern University, and Florida International University.

During her career at the NIEHS, Dr. Johnson-Thompson also ensured recruitment of underrepresented postdoctoral fellows and permanent scientific staff in both intramural and extramural programs. Similarly, during her service as reviewer for many predoctoral funding opportunities, including those offered by the NSF, Department of Defense, NIH, Susan G. Komen for the Cure, and American Cancer Society, she took actions to ensure that underrepresented applicants received fair review. She served as chair for the Burroughs Wellcome Fund Student Science Enrichment Program Advisory Committee, which provided funding that supported programs that enabled primary and secondary school students in North Carolina to participate in hands-on scientific activities and pursue inquiry-based explorations.

Dr. Johnson-Thompson's commitment to community building is also clear. As a founder of Minority Women in Science (MWIS) and later president of the District of Columbia chapter and national chair, she was a key person in the establishment of the Summers of Discovery Program (1979), a collaboration between UDC and Howard University to interest middle school students in science and science careers. This program is still in place to this day, has been one of the first of its kind, and has served as a model for the many similar programs developed since its establishment. She also established the MWIS Christmas Store, in which elementary school students earned currency to allow them to purchase scientific books donated by the AAAS.

She established and chaired the ASM Committee for Microbiological Issues Impacting Minorities, and a major product of this group was a newsletter spotlighting actively publishing underrepresented minority scientists, which facilitated the identification and recruitment of underrepresented groups to serve on review panels and editorial boards. This newsletter, *The Minority Microbiology Mentor*, as of 2 November 2017, has featured >120 minority microbiologists since 2006 (10). She also established the ASM Heroes and Heroines posters, which have featured underrepresented minorities in the sciences, including African-Americans, Hispanics, and American Indians who have represented firsts in their respective fields. She is exemplary in her efforts to highlight the contributions of scientists, especially those otherwise underrepresented or too shy to promote themselves. She has written and lectured extensively on the contributions of U.S.-born black persons to microbiology, and these narratives have appeared in *ASM News* and *Microbe* (11, 12). To see Dr. Johnson-Thompson discussing early African-American microbiologists at the 2017 ASM Microbe meeting, go to http://bit.ly/Johnson-Thompson.

Dr. Johnson-Thompson and I have served jointly on two review panels for the NSF Graduate Research Fellowship Program, and we worked to ensure the selection of qualified minority candidates among applicants. Dr. Johnson-Thompson served on the congressionally mandated NSF Committee on Equal Opportunities in Science and Engineering, which initiated the use of the Broader Impacts component now required by all NSF grant proposals. She has chaired the Trans-NIH Conference on At-Risk Children Planning Committee, and she was a member of the North Carolina Association of Biomedical Research Advisory Board.

Dr. Johnson-Thompson's commitment to identifying, encouraging, training, and mentoring underrepresented STEM scientists transcends age and discipline. I believe it is important to recognize those who are active in preparation for Ph.D. careers, mentorship throughout dissertation studies, and retention and promotion following the dissertation defense. Without the aid of a Ph.D. program, she is one of very few federal employees whose career has held such strong and consistent academic ties and whose activities fed so directly in the STEM Ph.D. pipeline. Specifically, she has supported the research of countless scientists through the years and is responsible for the attainment of many graduate degrees and the publication of numerous publications in peer-reviewed journals. She essentially established her own unique parallel career in academia, on her own time, and the students who she mentored benefitted from this dichotomy because it makes her uniquely equipped to advise students in a way that college professors may not. Dr. Johnson-Thompson is a lifelong full-service mentor and role model, a metric by which others are measured. She is great, and she makes those around her great. I would not be a tenured associate professor today without her as my mentor.

CITATION
Johnson CN. 2018. Marian Johnson-Thompson: lifelong mentor, p 183–192. *In* Whitaker RJ, Barton HA (ed), *Women in Microbiology*. American Society for Microbiology, Washington, DC.

References

1. Johnson-Thompson M, Rosenthal LJ. 1979. Effect of azacytidine on simian virus 40 nucleoprotein complexes. *Antimicrob Agents Chemother* 16:667–673.
2. Johnson-Thompson M, Rosenthal LJ. 1979. Inhibition of SV40 replication by 5-azacytidine: effect on DNA synthesis and conformation. *Virology* 93:605–608.
3. Johnson-Thompson M, Halpern JB, Jackson WM, George J. 1984. Vacuum UV laser induced scission of simian virus 40 DNA. *Photochem Photobiol* 39:17–24.
4. Johnson-Thompson M, Albury D. 1988. Azapyrimidine analogues: inhibition of viral DNA synthesis and protein synthesis in SV40 infected BSC-1 cells. *In Vitro Cell Dev Biol* 24:1114–1120.
5. Johnson-Thompson MC, Guthrie J. 2000. Ongoing research to identify environmental risk factors in breast carcinoma. *Cancer* 88:1224–1229.
6. Newman LA, Pollock RE, Johnson-Thompson MC. 2003. Increasing the pool of academically oriented African-American medical and surgical oncologists. *Cancer* 97: 329–334.
7. The History Makers. *ScienceMakers. Marian Johnson-Thompson.* http://www.thehistory makers.com/biography/marian-johnson-thompson. Accessed 22 March 2017.
8. Matyas ML. 1997. Marian Johnson-Thompson, molecular virologist, 1946–present, p 241–256. *In* Matyas ML, Haley-Oliphant AE (ed), *Women Life Scientists: Past, Present, and Future—Connecting Role Models to the Classroom Curriculum*. American Physiological Society, Bethesda, MD.
9. Johnson-Thompson M, Sullivan C, Olden K. 1996. NIEHS/AACR Task Force on the Advancement of Minorities in Science: vision for a model program. *Cancer Res* 56: 3380–3386.

10. ASM Committee on Microbiological Issues Impacting Minorities (CMIIM). *The Minority Microbiology Mentor*. https://www.asm.org/index.php/publicpolicy-2/135-policy/documents/newsletters/minority-microbiology-mentor-newsletter. Accessed 16 January 2018.
11. Johnson-Thompson MC, Jay J. 1997. Ethnic diversity in ASM: the early history of African-American microbiologists. *ASM News* **63:**77–82.
12. Johnson-Thompson M. 2007. Revisiting the contributions of African American scientists to ASM. *Microbe* **2:**82–87.

Women in Microbiology
Edited by Rachel J. Whitaker and Hazel A. Barton
© 2018 American Society for Microbiology. All rights reserved.
doi:10.1128/9781555819545.ch21

Carol D. Litchfield: Salt of the Earth

21

Bonnie K. Baxter[1] and Kendall Tate-Wright[1]

Microbiologist and biochemist Carol Ann Darlene Litchfield, née Ross, was born in Cincinnati, OH, in 1936. She passed away at her home in Virginia in 2012. Her contributions to the microbiology of oceans, hypersaline systems, salt production ponds, and other extreme environments were noteworthy and meaningful (1). She was also an avid scuba diver and a salt industry historian. As a woman who worked on boats with only male scientists early in her career, she became an active mentor of young women and a strong role model for all of her students.

BECOMING A SCIENTIST

In her youth, Carol had a passion for science, but also for softball (Fig. 1, left panel). Her cousin Randy Ross remembers, "One thing I know was that she loved playing softball and was very good at it. I even have her cleats and glove, which she saved. Whenever she went to a baseball game she got a scorecard and filled it in. I have a bunch of those too. She was a Cincinnati Reds fan for life." Indeed, she followed baseball and often wore her Reds cap in the field.

Carol went to college at the University of Cincinnati, earning a bachelor's degree in medical technology and then a master's in microbiology in 1960 (Fig. 1, right panel). That same year Carol married her husband, Carter Litchfield, a fellow scientist and fellow lover of books and travel. Carter was offered a job as a research scientist at Texas A&M, and the couple moved south.

---

[1]Great Salt Lake Institute, Westminster College, Salt Lake City, UT 84108

Figure 1 **(Left)** Carol Darlene Ross, avid science student and softball player, in grade school. **(Right)** Carol after college, circa 1960, beginning her life in science and about to be married. Photos provided by R. Ross.

A WOMAN IN THE FIELD AND IN THE LAB

After a few years of working in Texas, Carter enrolled in a Ph.D. program there in the organic chemistry department. The A&M Board of Directors did not allow women to be admitted, but there was a provision if they were wives of students and employees. This is how Carol became the first woman to enroll in a Ph.D. program in organic biochemistry at Texas A&M. During this time she began working in microbiology (2). One of her coauthors from her time at Texas A&M was Rita Colwell, then at George Washington University, who would go on to become the first female director of the National Science Foundation. This is how Rita remembers that time: "I worked with Carol when we were both starting our careers in marine microbiology. Carol was at Texas and asked if I would analyze some of her isolates by incorporating them into my numerical taxonomy studies. That launched a wonderful friendship, and our careers continued to intersect thereafter."

With freshly minted doctorates, Carol and Carter left Texas. She accepted a postdoctoral position at the Bangor Marine Science Laboratories at the University College of North Wales. She continued to publish in marine microbiology (3) while she and Carter explored Europe and its bookstores, museums, and antiques. She collected Dutch tiles from the 18th century, British tiles excised from crypts, and Danish ceramics. They enjoyed their time abroad and began applying for jobs back home.

Carol and Carter accepted faculty positions at Rutgers State University of New Jersey. Carter studied soluble oils in marine mammals, and in 1979 left Rutgers to pursue a new career researching and publishing works on the history of fatty oils. Carol held a joint appointment in the Microbiology Department and the Center for Coastal and Environmental Studies. She continued in the field of oceanographic microbiology, in which there were still few women. She served on significant advisory panels, including the New Jersey Governor's Panel on Coastal Waters, and was a member of the American Society for Microbiology. One very important contribution during this time was an edited volume, *Marine Microbiology*, in a series called *Benchmark Papers in Microbiology* (4). Carol selected important papers in this field and bound them alongside her annotations about the significance of each manuscript. With several graduate students, she worked on biomass determination in ocean sediments (5).

While at Rutgers, Carol was interested in halophilic microorganisms, and she began traveling to sample from natural and man-made hypersaline field sites. Rita Colwell recalls, "Carol went way further down the road to become a halophile expert, while I continued but mostly as what I would call a dabbler in the field. We were both intrigued by the nature of the requirement for salt for growth, physiological functioning, and main-tenance of DNA structure." Carol became a mentor to undergraduate and graduate students who took courses from her in microbiology, and the lucky ones were able to sign up for sampling cruises and fieldwork (Fig. 2, left panel). Never too far from the ocean, she continued honing her scuba diving skills.

With her student Russell Vreeland, she discovered a novel genus of moderately halophilic bacteria, *Halomonas* (6), from a sampling trip in the 1970s to a saltern pond on Bonaire, a Dutch island off the coast of Venezuela (Fig. 2, right panel). In their paper, they described one of their isolates, *Halomonas elongata*, which later became the model organism for the study of moderate halophiles. *H. elongata* produces the compatible solute ectoine, which has biotechnological applications. She continued her fascination with the biology of solar salterns and developed an interest in industrial micro-biology (see, e.g., reference 7).

Carol was at Rutgers for 10 years before she decided to leave academia for industry. She held leadership positions at Haskell Laboratory (DuPont), in the environmental toxicology laboratory and its bioremediation sub-sidiary. Carol then founded her own consulting company specializing in environmental and industrial microbiology, Microbial Solutions.

Figure 2 (Left) Carol on the "RV Rutgers" with the sediment grab sampler, 1977. She was investigating the New York City sewage sludge dump site with her students. Photo credit H. Edenborn. **(Right)** Carol and graduate student Russell Vreeland on a sampling trip to Bonaire, circa 1974. Photo provided by R. Vreeland.

Bioremediation, the use of microorganisms in environmental cleanup of industrial waste, was her focus during this time (8–12). Carol served on the U.S. Department of Energy's Environmental Biotechnology Working Group in 1989.

Carol always relished the applied work (13–15), but she missed students and academia. In 1993, she accepted a faculty position at George Mason University (GMU), beginning in the Biology Department and later joining the Environmental Science and Policy Department. She began to shift her work entirely towards salt industry and to hypersaline systems where life exists at salt saturation. She became a fixture in the halophile community. In the early 1990s, she attended her first International Halophiles Conference, an association that continued the rest of her life (Fig. 3).

She was an active member of the International Committee on Systematics of Prokaryotes Subcommittee on the Taxonomy of Halobacteriaceae (e.g., reference 16). Carol remained active in the Society of Industrial Microbiology and Biotechnology (SIMB), serving as president from 2007 to 2008. She won their Charles Porter prize in 2012 for her achievements in applied microbiology. The SIMB organization gives a prize each year, the Carol D. Litchfield Best Student Oral Presentation, at their national conference.

Carol's research interests in halophilic microbiology maintained a chemistry slant, harkening back to her early training. For example, she utilized polar lipid analysis and (carotenoid) pigment analysis as tools when

Figure 3 Carol with Aharon Oren and Bonnie Baxter at the International Halophiles Conference in Essex, UK, 2007. Photo provided by B. K. Baxter.

characterizing the biota of saltern ponds and salt lakes (17). Carol remained interested in the application of science in hypersaline environments to space science, in particular postulating that halophilic archaea are good models for pursuing questions about life on Mars (18). Her lab excelled at traditional microbiology approaches, but she welcomed new technologies that augmented cultivation with molecular, cultivation-independent, techniques (19, 20). She believed in a combined approach: while cultivation allows exploration of the microbial physiology aspects, molecular surveys can tell more about the whole population (21–24).

One of the authors of this chapter, Bonnie K. Baxter, first met Carol in 2002. Her recollections follow. "I was a biochemist who moved to Salt Lake City for a job at a liberal arts institution, Westminster College. I studied DNA repair, and given my proximity to Great Salt Lake, I wanted to use halophiles from the lake as lab models for my DNA damage and repair studies. The problem was that there was no one working on the microbiology of this iconic lake. This meant I had to learn to field sample and to cultivate from the environment, skills I did not have. I contacted Carol and asked if she could help. Her passion was evident as she told me, 'Great Salt Lake has been ignored, and I will help you do this work.' Within a month, she flew out to sample (Fig. 4). She taught me everything I know about microbiology in extreme environments. I personally experienced not only her dedication to knowledge creation but also her active mentorship style. Our work resulted in many field days and presentations and a couple of joint publications (24, 25)."

Figure 4 Carol, center, field sampling at the north shore of Great Salt Lake with students from Westminster College in Utah, June 2002. Photo credit B. K. Baxter.

A SPECIAL MENTORING STYLE

Carol was a good scientist, thoughtful and thorough, and she wanted each of her students to know all aspects of the job. While working on her dissertation, Swati Almeida-Dalmet noted about Carol, "As a mentor, she gave freedom to her students to explore all the options in research. Sometimes my decisions were wrong, but from those I learned my lessons. In the lab she allowed me to do all the things from scratch, from buying my supplies to analyzing the data using bioinformatics."

Carol had high standards and asked a lot of her students as well as herself. A colleague of Carol's, Dr. Aharon Oren, who is renowned for his work on the halophiles of the Dead Sea, worked with Carol and her students on the diversity and biochemistry of solar salterns (26, 27). Aharon remembers Carol's laboratory rules for her students. "In the lab in Fairfax, she demanded that her students clean up their desks at the end of the working day so that after they left, no one could see that anybody had worked there during the day. I sometimes referred to this as the 'Litchfield Principle' but I have never been able to convince my students and technician to implement the 'Litchfield Principle' in my lab in Jerusalem."

Carol's students knew that professionalism was expected. Swati Almeida-Dalmet, who was Carol's last graduate student at GMU, remembers that her mentor maintained a long list of duties and responsibilities of each student, especially time together for presenting and discussing: "She was very particular about lab meetings and made sure that everyone attended the meeting." Decades before, Hank Edenborn, now at the U.S. Department of Energy, was her last graduate student at Rutgers. He worked alongside Carol on boats and in the lab and recently noted, "Carol both commanded and demanded our respect. Always 'Dr. Litchfield' to her face."

Carol was not just demanding; she was also an inspiration to her students throughout her career. Former Rutgers student and current Biotech executive Joe Zindulis says, "Carol helped me to fall in love with microbiology, and I quickly learned that she would go out of her way to answer any micro question even though we had a grad student who was teaching our section. She did not even have to be there, but she continued to appear in the lab." Another student of Carol's at Rutgers who works at the university today, Monica Devanas, shared the significance of this inspiration: "Carol mentioned that when you are really enthralled by your work, you think of it all the time: when you are doing dishes, when you are walking across campus. At that time in my young life, research was compartmentalized as 'in lab' in my mind. But now when I am washing the dishes but thinking about my latest project, I remember Carol and think, yes, you were so right. When you love what you are doing it is always on your mind. And at those special times, so is Carol."

Despite the rigid structure she created in the work environment, Carol had an informal and warm style of mentoring that went beyond the formality. She created a sense of family, which she cultivated even as each student navigated their own path and careers. Of his time at Rutgers, Russell Vreeland said, "Carol had an interesting view of students: we were her kids." Russell became her scientific progeny as well, as he spent his entire career in the field of halophilic microorganisms, recently retiring from West Chester University. Joe Zindulis confirmed this family atmosphere: "Although we were Carol's students, we were really Carol's family. We began our relationship with her as her upstart teenagers. She encouraged each of us to grow in our own ways in science and beyond. She opened doors for each of us throughout the entirety of our careers. Some of us stayed close, and some of us moved a little farther away. However, all of us know what she did for us,

Figure 5 Carol and her GMU graduate student, Swati Almeida-Dalmet. Photo provided by S. Dalmet.

and all of us will remember her for the rest of our lives." Carol's philosophy was that if you were not going home for a major holiday, you were coming to her house because you were her family. Swati Almeida-Dalmet said, "She was a very outgoing and social person (Fig. 5). Every year she kept a dinner party for her students and friends at her residence. She was an incredibly good host and always gave a warm welcome to everyone."

Monica Devanas remembers that Carol's approach was indeed special, going beyond the lab. "Just as I was completing my doctoral research with Carol, she took an exciting offer from DuPont to head up their ground water remediation project. That meant that those of us still in the lab needed to send our dissertation drafts to her and visit her at her home to work on each round of edits and new sections. The drive to her new home was planned to always arrive in time for Friday night dinner and an overnight visit. My colleague Linda Kiefer and I would drive down; one of us would take the after-dinner time slot and the other the Saturday morning session. There would also be some outing, maybe to Kennett Square for mushrooms for lunch or a farmers' market. Then off we would go, back to Rutgers for the next round of confirming experiments or draft writing."

Monica also remembers the day she became Carol's friend and not just her student. There was a softball game between microbiologists during a break at a Gordon Conference and Carol, proficient at the sport, was playing. "When she came up to bat, she hit quite a long shot beyond the infield and we all screamed for her to go on to second base, 'Go, Carol! Run, Carol! Way to go, Carol!' It certainly seemed sacrilegious, to use her first name, but one simply cannot cheer, 'Go, Dr. Litchfield!'" Carol later told Monica that they were now friends because Monica was able to call her by her first name!

Hank Edenborn said he never did call her by her first name in graduate school, but they did call Carol "CDL" behind her back as some form of rebellion. "A favorite recreational sport of mine was dubbed Marine Science Hockey, played using a plastic coffee can lid like a Frisbee, with an open doorway for a goal." Apparently, Carol never caught him since the student on watch was given the task of issuing the "CDL alert!!" as she approached the lab. Likely Carol would have both chastised them and rejoiced in the laughs her students shared.

Russell Vreeland paints a lovely picture of his graduate mentor that echoes her ability to be demanding, critical, and serious but simultaneously warm and humorous. After sampling at the Bonaire site, Russ left to bring the samples back to the lab. "Carol packed the samples in a suitcase, and she marked the sleeves of petri dishes 'photosynthetic media.' I was a long-hair hippie, and customs was suspicious of me. The customs official knew enough science to know that photosynthetic meant 'plants.' He accused me of smuggling plants into the U.S. I replied, 'That media contains 25% sodium chloride. Do you think anything can live in there?' Later Carol scolded me, saying I misled the agent since we actually studied things that lived in high salt!" But later he got her back. "We were sampling in Bonaire another time, and I convinced the sewage truck to dump in the solar saltern ponds. Carol had to sample the poo as a control, and she fell in! I was in stitches, and I told her that it was 25% salt and it would kill anything!" Later, they were both colleagues in the field of halophiles, and after reading a manuscript he wrote, Carol told Russ she thought he was an excellent writer. She added, "I never would have believed it, based on your thesis."

This is the mentor all of her students describe. Carol actively worked to make them all well-rounded scientists who are dedicated to the team and the work, but she clearly had another mission of building a community of human beings that she thought of as her family.

HISTORIAN AND ARCHIVIST

Carol had a love of antiques and especially old books. She amassed an incredible collection of microbiology texts, which she donated to GMU upon her official retirement in 2008. The Carol D. Litchfield Microbiology Collection also "contains journal publications, government studies, microfiche, and research data, both published and personal" (12).

Perhaps unusual among scientists, Carol also was an amateur (salt industry) historian, and she often presented at the Society of Industrial Archaeology conferences. She traveled the world visiting historic salt production sites. She coedited a book of essays on the history of salt and salt making (28). In New Jersey, Carol and Carter each had a dedicated space for their collections. Rita Colwell remembers that Carol "was the only person that I knew who had a professional library in her home!" When Carol and Carter settled in Virginia, they bought a condominium but rented a separate two-bedroom apartment for their "library." Both bedrooms and the living room were outfitted with rows of library shelving. Carter maintained a collection of books and artifacts on oils and an office space for his publication business. Carol's half of the shelving was filled with books, maps, and samples, including unusual items such as a camel strap for carrying salt across the desert. Carol had a desk and computer there where she accessioned each item and collected information on its authenticity and value. As requested in her will, her extensive collection is now at the Hagley American Industry Museum in Delaware (29). Hagley says this about the Carol Litchfield Collection on the History of Salt: "She combed the earth for anything and everything related to salt: salt sacks, salt specimens, letters, account books, broadsides, maps, stamps, photographs, postcards, trade cards—even blank sheets of corporate letterhead, so long as it came from a salt company…The result is a world-class collection on the subject of salt, spanning over four centuries and written in eighteen languages." Hagley is also the custodian of the Carter Litchfield Collection on the History of Fatty Materials, which was established after Carter's death in 2007. In their estate planning, Carol and Carter established an endowment at the museum to support the maintenance of their collections.

FINAL CHAPTER

In October of 2011, for her 75th birthday, Carol Litchfield wanted a party with her closest friends. She hosted a weekend of events in Washington, DC (Fig. 6, left panel). Rita Colwell remembers, "Carol invited her former students (whom she dearly loved) and close friends for a memorable Potomac River boat cruise. It was a gala evening and everybody had a superb time. That was classic Carol Litchfield: friend, scientist, entrepreneur, music and art lover, and the kindest of human beings and the best of friends."

At that time she did not know that in about 4 months, she would be diagnosed with pancreatic cancer. That party was full of joyful shared memories, and her impact on those around her was evident. It was a fitting tribute. She passed away 3 April 2012. Russell Vreeland reflects today, "I still want to call Carol all the time to see what she thinks about something." Rita Colwell says, "I still miss her and our intermittent dinners at our respective homes with our husbands, enjoying good food and wine…and

Figure 6 (Left) Carol celebrating her 75th birthday with friends, October 2011. Photo credit B. K. Baxter. **(Right)** One of Carol's photographs from a scuba trip to Turks and Caicos, 2005. Photo credit C. D. Litchfield.

excellent conversations, especially to devise new experiments and share new ideas."

During the time she was in hospice care, Carol planned to be become part of the ocean she had studied and explored. She asked for her cremation ashes to be made into a "pod" that would be deployed into the ocean. The mission of this project not only is focused on providing a location one's ashes can be delivered but also creates a man-made system that would support reef recovery and growth of the natural marine environment (30). The location of Carol's pod (25° 57.780 north, 80° 05.886 west) can be visited by divers. A couple of important items were included with Carol's ashes: her Cincinnati Reds visor and her license plate, which read: "SALT BUG." As a tribute to those salty bugs, her family also had a smaller pod made of the same material, and it now lies in the north arm of Great Salt Lake, no doubt inhabited by halophiles.

Carol Litchfield explored salty environments and explained them to us. Her legacy is not only in science, where she created volumes of literature on the subject, but also in history. Carol's archives in microbiology (12) and in salt industry history (29) will provide many resources for future researchers. Those who worked with her understand that her greatest gift was her collaborative spirit. A longtime colleague from Australia, Mike Dyall-Smith, recently named a microorganism in Carol's honor: *Halohasta litchfieldiae*, which was found in the salty waters of Deep Lake, Antarctica (31). This is perhaps the finest tribute for a person dedicated to cultivation, molecular biology, physiology, and taxonomy of halophiles.

ACKNOWLEDGMENTS

We thank Carol's friends, family, students, and colleagues (most would say that Carol made no distinction between those categories). Randy Ross, her cousin whom she called her "little brother," graciously provided photos and family information. Rita Colwell and Aharon Oren reflected on Carol as a colleague. The voices of Carol's students were represented by Swati Almeida-Dalmet, Hank Edenborn, Monica Devanas, Russell Vreeland, and Joe Zindulis. We are grateful for the thoughts they shared to make this biography a real tribute.

CITATION

Baxter BK, Tate-Wright K. 2018. Carol D. Litchfield: salt of the earth, p 193–206. *In* Whitaker RJ, Barton HA (ed), *Women in Microbiology*. American Society for Microbiology, Washington, DC.

References

1. **Baxter BK, Gunde-Cimerman N, Oren A.** 2014. Salty sisters: the women of halophiles. *Front Microbiol* 5:192.
2. **Litchfield CD, Colwell RR, Prescott JM.** 1969. Numerical taxonomy of heterotrophic bacteria growing in association with continuous-culture *Chlorella sorokiniana*. *Appl Microbiol* 18:1044–1049.
3. **Litchfield C.** 1974. Non-equivalence of proteins from marine and conventional sources for the cultivation of marine bacteria, p 354–362. *In* Colwell RR, Morita RY (ed), *Effect of the Ocean Environment on Microbial Activities; Proceedings*. University Park Press, Baltimore, MD.
4. **Litchfield CD.** 1976. *Marine Microbiology.* Dowden, Hutchinson & Ross, New York, NY; distributed by Halsted Press, Stroudsburg, PA.
5. **Litchfield CD, Devanas MA, Zindulis J, Carty CE, Nakas JP, Martin EL.** 1979. Application of the 14C organic mineralization technique to marine sediments, p 128–147. *In* Litchfield CD, Seyfried PL (ed), *Methodology for Biomass Determinations and Microbial Activities in Sediments*. ASTM STP 673. American Society for Testing Materials, West Conshohocken, PA.
6. **Vreeland RH, Litchfield CD, Martin EL, Elliot E.** 1980. *Halomonas elongata*, a new genus and species of extremely salt-tolerant bacteria. *Int J Syst Bacteriol* 30:485–495.
7. **Litchfield CD.** 1991. Red—the magic color for solar salt production, p 403–412. *In* Hocquet J-C, Palme R (ed), *Das Salz in der Rechts-und Handelsgeschichte*. Berenkamp, Hall in Tirol, Austria.

8. Litchfield CD. 1979. *Microbial Contributions to Organic Mineralization in the Waters and Sediments of the New York Bight: Final Report for 1977–78, NOAA-MESA-New York Bight Studies.* Center for Coastal and Environmental Studies—Marine Sciences Division, Rutgers, the State University, New Brunswick, NJ.

9. Litchfield CD. 1982. Benthic and planktonic bacteria. *In SINC Compendium Report.* http://sca.gmu.edu/finding_aids/litchfield.html. Accessed 10 October 2017.

10. Malone TC, Chervin MB, Garside C, Litchfield CD, Stepien JC, Thomas JP. 1982. *Synoptic Investigation of Nutrient Cycling in the Coastal Plume of the Hudson and Raritan Rivers: Plankton Dynamics.* Hudson River Environmental Society, Inc, New York, NY.

11. Litchfield CD. 1990. *In Situ Bioremediation at the Victoria West Landfill (Final Report).* http://sca.gmu.edu/finding_aids/litchfield.html.

12. George Mason University Libraries. 2016. *Carol D. Litchfield Microbiology Collection. Collection C0047, Special Collections and Archives.* http://sca.gmu.edu/finding_aids/litchfield.html.

13. Litchfield CD. 2004. Evolution of extremophiles from laboratory oddities to their practical applications: a selective review of the patent literature. *SIM News* 54:245–253.

14. Litchfield CD. 2005. Thirty years and counting: bioremediation in its prime? *Bioscience* 55:273–279.

15. Litchfield CD. 2011. Potential for industrial products from the halophilic Archaea. *J Ind Microbiol Biotechnol* 38:1635–1647.

16. Arahal DR, Vreeland RH, Litchfield CD, Mormile MR, Tindall BJ, Oren A, Bejar V, Quesada E, Ventosa A. 2007. Recommended minimal standards for describing new taxa of the family *Halomonadaceae. Int J Syst Evol Microbiol* 57:2436–2446.

17. Litchfield CD, Irby A, Kis-Papo T, Oren A. 2000. Comparisons of the polar lipid and pigment profiles of two solar salterns located in Newark, California, USA, and Eilat, Israel. *Extremophiles* 4:259–265.

18. Litchfield CD. 1998. Survival strategies for microorganisms in hypersaline environments and their relevance to life on early Mars. *Meteorit Planet Sci* 33:813–819.

19. Litchfield CD, Gillevet PM. 2002. Microbial diversity and complexity in hypersaline environments: a preliminary assessment. *J Ind Microbiol Biotechnol* 28:48–55.

20. Litchfield CD, Sikaroodi M, Gillevet PM. 2006. Characterization of natural communities of halophilic microorganisms. *Methods Microbiol* 35:513–533.

21. Litchfield CD. 2004. Microbial molecular and physiological diversity in hypersaline environments, p 49–61. *In* Ventosa A (ed), *Halophilic Microorganisms.* Springer, Berlin, Germany.

22. Litchfield CD, Sikaroodi M, Gillevet PM. 2005. The microbial diversity of a solar saltern on San Francisco Bay, p 59–69. *In* Gunde-Cimerman N, Oren A, Plemenitaš A (ed), *Adaptation to Life at High Salt Concentrations in Archaea, Bacteria, and Eukarya.* Springer, Dordrecht, the Netherlands.

23. Pesenti PT, Sikaroodi M, Gillevet PM, Sánchez-Porro C, Ventosa A, Litchfield CD. 2008. *Halorubrum californiense* sp. nov., an extreme archaeal halophile isolated from a crystallizer pond at a solar salt plant in California, USA. *Int J Syst Evol Microbiol* 58:2710–2715.

24. Almeida-Dalmet S, Sikaroodi M, Gillevet PM, Litchfield CD, Baxter BK. 2015. Temporal study of the microbial diversity of the North Arm of Great Salt Lake, Utah, U.S. *Microorganisms* 3:310–326.

25. **Baxter BK, Litchfield CD, Sowers K, Griffith JD, DasSarma PA, DasSarma S.** 2005. Microbial diversity of great salt lake, p 9–25. *In* Gunde-Cimerman N, Oren A, Plemenitaš A (ed), *Adaptation to Life at High Salt Concentrations in Archaea, Bacteria, and Eukarya.* Springer, Dordrecht, the Netherlands.
26. **Litchfield CD, Irby A, Kis-Papo T, Oren A.** 2001. Comparative metabolic diversity in two solar salterns. *Hydrobiologia* **466**:73–80.
27. **Litchfield CD, Oren A.** 2001. Polar lipids as biomarkers for the study of the microbial community structure of solar salterns in different geographic locations. *Hydrobiologia* **466**:81–89.
28. **Litchfield CD, Palme R, Piasecki P (ed).** 2001. *Le monde du sel. Melanges offerts a Jean-Claude Hocquet.* Berenkamp, Hall in Tirol, Austria.
29. **Hagley Museum and Library.** 19 August 2016. *Introducing the Carol Litchfield Collection on the History of Salt.* http://www.hagley.org/librarynews/introducing-carol-litchfield-collection-history-salt.
30. **Eternal Reefs.** 2016. *The Eternal Reefs Story.* http://eternalreefs.com/the-eternal-reefs-story/.
31. **Mou YZ, Qiu XX, Zhao ML, Cui HL, Oh D, Dyall-Smith ML.** 2012. *Halohasta litorea* gen. nov. sp. nov., and *Halohasta litchfieldiae* sp. nov., isolated from the Daliang aquaculture farm, China and from Deep Lake, Antarctica, respectively. *Extremophiles* **16**:895–901.

Women in Microbiology
Edited by Rachel J. Whitaker and Hazel A. Barton
© 2018 American Society for Microbiology. All rights reserved.
doi:10.1128/9781555819545.ch22

Ruth E. Moore: The First African-American to Earn a Ph.D. in the Natural Sciences

22

Candace N. Rouchon[1]

During a time of gender and racial discrimination, Dr. Ruth E. Moore became the first African-American to earn a Ph.D. in the natural sciences. Ruth Ella Moore was born in Columbus, Ohio, on 19 May 1903 (1, 2). She completed her undergraduate and graduate studies at Ohio State University, where she earned a B.S. in 1926 and an M.A. in 1927 (2). After receiving her master's degree, Dr. Moore continued her graduate education at Ohio State University and earned a Ph.D. in bacteriology in 1933 (2). Her Ph.D. dissertation was titled "Studies on Dissociation of *Mycobacterium tuberculosis*: A New Method of Concentration on the Tubercle Bacilli as Applied to Sputum and Urine Examination" (2). During graduate school, she also gained teaching experience by working as an instructor at Tennessee State University (1).

In 1940, Dr. Moore began a long and illustrious career at Howard University Medical College (2), a historically black university located in Washington, DC (Fig. 1). She was initially hired as an assistant professor of bacteriology and held that position for several years (2). While at Howard University, Dr. Moore participated in research on a diverse array of scientific topics, including *Enterobacteriaceae* and blood grouping of individuals from different backgrounds (2). Beginning in 1952, Dr. Moore served as the chair of the bacteriology department, and she was subsequently promoted to the position of associate professor (2). During this time, she published her findings on blood groupings in African-American populations in journals such as *American Journal of Physical Anthropology* and *Journal of the American*

[1]Department of Microbiology and Immunology, Uniformed Services University of the Health Sciences, Bethesda, MD 20814

Figure 1 Dr. Ruth E. Moore in the bacteriology laboratory at Howard University Medical College in 1949. Image from the Howard University yearbook, *The Bison: 1949* (6).

Medical Association (3, 4). In addition to teaching bacteriology, Dr. Moore served as chair of both the scholarship and loans committee and the student guidance committee (5). In 1957, she resigned from her position as department head, but she continued to teach bacteriology at Howard University part-time (2). Dr. Moore undoubtedly made a lasting impression on her students, as demonstrated in the 1949 publication of the Howard University College of Medicine yearbook, called *The Bison* (6). In the yearbook, the long-time educator was minted the "most punctual" faculty member, as voted by her students. This superlative was likely well-deserved, as the students also described her tendency to dismiss students who were late for class. When asked to describe Dr. Moore's personality, the students commented on her ability to design unique practical exams for class and her zest for testing the knowledge of her students through surprise quizzes. Dr. Moore seemed to be a scientist who was not only focused on teaching microbiology but also interested in molding the character of the future scientists and physicians whom she was educating.

Dr. Moore retired from Howard University in 1973 and settled in Rockville, MD, for the remainder of her life (5). She died on 19 July 1994 at the age of 91. Her obituary in the *Washington Post* highlighted many of her additional achievements (5). Included among these accomplishments are the honorary doctorate that she received from Gettysburg College in 1989, the Centennial Award for Distinguished Alumni from Ohio State University in 1970, and the Magnificent Professor Award from Howard University. She was also a member of the American Society for Microbiology and other professional organizations (1, 5). In addition, Dr. Moore was active in her personal life within her church community and served as a member of the National Council of Churches and the Augustana Lutheran Church in Washington, DC (5).

As a pioneer, educator, and scientist with a career that stretched over decades, Dr. Ruth Moore's contributions to the United States have been revered both within and outside the scientific community. She is honored every year by Howard University through the Ruth E. Moore and Lloyd H. Newman Service Award. This donor-based scholarship is awarded to students exhibiting selflessness towards their classmates and through community service. Indeed, this award exemplifies Dr. Moore's commitment to serving her scientific and educational community for the majority of her life. Her influences have been felt even as far as the U.S. government: members of Congress have lobbied for acknowledgment of her scientific contributions. On 15 March 2005, Congresswoman Eddie Bernice Johnson of Texas introduced a bill entitled "The Significance of African-American Women in the United States Scientific Community" into the House of Representatives (7). Ms. Johnson recommended that Congress acknowledge the contributions of female scientists, as many of these women were able to establish and maintain successful careers despite facing such major obstacles as discrimination and prejudice. Congresswoman Johnson also emphatically argued that by celebrating these innovators, new generations of female and minority scientists would be inspired to pursue similar aspirations, thus continuing to diversify and enrich the scientific community. Dr. Moore was the first among the scientists listed in the bill to be honored for their contributions to science and to the United States. The bill passed the House of Representatives on 26 April 2005 (7).

There is very little knowledge of the specific steps that Dr. Moore took to earn her doctorate and to obtain her faculty positions. Most of the literature about her primarily describes her accomplishments after starting her position at Howard University (1, 2, 6, 8). However, she likely faced many obstacles as both a female and a minority scientist. Dr. Moore earned her doctorate and began to teach as a professor in the 1930s and 1940s, a time during which it was difficult for women to obtain advanced degrees and to maintain careers. In "Always the Exception: Women and Women of Color Scientists in Historical Perspective," Douglas M. Haynes describes these years as a time when women were pressured by society and/or their families to make the difficult choice between raising children and pursuing the profession that they desired (9). For many women, the responsibility of caring for their families often deterred or delayed the completion of their graduate programs and the progression of their careers. In some cases, women were not hired due to the fear that they would get married and not be able to fulfill their duties as faculty members.

In the 1940s, prestigious universities were reluctant to hire women, and the faculty at these schools was predominantly male (8). Due to the example set by these universities, less renowned colleges perceived that retaining women within their institutions was a demonstration of inferior quality and standards. Therefore, many women were fired from positions at smaller universities and women's colleges or simply not even considered for faculty positions at these schools. Women who were successful in obtaining faculty positions within academia faced additional challenges: they were often not promoted from their initial position or not considered for higher-ranking titles. Unfortunately, female scientists were frequently not supported in their professional ventures and not encouraged to utilize their degrees for career advancement. Some women within academia did recognize that their female colleagues were being treated unfairly, but they were afraid to speak up due to the fragility of their own positions. Even organizations that advocated for women in the scientific community lacked the proper backing to aid female scientists. For example, the American Association of University Women began to take issue with the way that women were being treated in academia in the 1940s; however, the organization was attacked for being discriminatory itself. It was not until the late 1960s and early 1970s that institutions and scientific organizations began to acknowledge and take action towards ameliorating the employment disparities between women and men within the sciences (8).

During such a difficult time for female scientists, Dr. Moore was a glimmering example of perseverance. Although she was successful in acquiring a faculty position at Howard University, it is clear that she faced barriers as she progressed through her career. While serving as chair of the bacteriology department, Dr. Moore only held the title of assistant professor and was not promoted to associate professor until several years later. Furthermore, she continued to teach at the university for several decades, but there is no evidence that she ever became a tenured professor (2). Additionally, Dr. Moore may have had to overcome racial prejudice while searching for positions before landing at Howard University. At that time there were limited positions for minority scientists and educators due to rampant discrimination. Minority women were commonly limited to applying for positions at universities where faculty members were willing to accept them or to institutions where the faculty and students consisted of primarily underrepresented minorities (9). Despite these obstacles, Dr. Moore established a successful career and made her mark on both Howard University and the rest of the scientific community. She not only paved the way for herself as a scientist but also dedicated her life to educating and supporting others.

Female scientists today still face many of the same challenges; however, Dr. Moore is an inspiration for those who struggle to reach their goals in the face of doubt or discrimination. Her story serves as a source of encouragement not only for the next generation of female researchers but also for male scientists. She also shows us that service can and should be an important piece of the educational and research cultures, as she continued to serve both her personal community and academic institution even after retiring. Dr. Moore demonstrated that being a scientist is not only about what we achieve as individuals but also about how we impact the lives of others.

CITATION

Rouchon CN. 2018. Ruth E. Moore: The first African-American to earn a Ph.D. in the natural sciences, p 207–212. *In* Whitaker RJ, Barton HA (ed), *Women in Microbiology*. American Society for Microbiology, Washington, DC.

References

1. Harvey JOM. 2000. *The Biographical Dictionary of Women in Science: Pioneering Lives from Ancient Times to the Mid-Twentieth Century*. Taylor & Francis, New York, NY.
2. Warren W. 1999. *Black Women Scientists in the United States*. Indiana University Press, Bloomington, IN.
3. Moore RE. 1955. Distribution of blood factors, ABO, MN and Rh in a group of American Negroes. *Am J Phys Anthropol* **13**:121–128.
4. Moore RE. 1957. Occurrence of Rh antigen V in a group of American Negroes. *J Am Med Assoc* **163**:544–545.
5. Anonymous. 5 August 1994. Microbiologist Ruth Moore dies at 91. Washington Post, Washington, DC. https://www.washingtonpost.com/archive/local/1994/08/05/microbiologist-ruth-moore-dies-at-91/385c5dee-20a6-46b0-ad2b-bb95cd02075b/.
6. Howard University. 1949. *The Bison: 1949*, vol 118. Howard University, Washington, DC.

7. Congress. 2005. *Congressional Record: Proceedings and debates of the 109th Congress, First Session*, vol 151, p 7786–7787. US Government, Washington, DC.
8. Rossiter MW. 1995. *Women Scientists in America: Before Affirmative Action, 1940–1972*, vol 2. The Johns Hopkins University Press, Baltimore, MD.
9. Haynes DM. 2014. Always the exception: women and women of color scientists in historical perspective. *Peer Rev* 16:25.

Women in Microbiology
Edited by Rachel J. Whitaker and Hazel A. Barton
© 2018 American Society for Microbiology. All rights reserved.
doi:10.1128/9781555819545.ch23

Nancy A. Moran: The Winding Path of a Brilliant Scientific Life

23

John P. McCutcheon[1]

Nancy A. Moran is the Leslie Surginer Endowed Professor of Integrative Biology at the University of Texas at Austin (UT). She was a professor at the University of Arizona from 1986 to 2010 and at Yale University from 2010 to 2013, and she has been at UT since 2013. Nancy is a member of the American Academy of Microbiology, the American Academy of Arts and Sciences, and the National Academy of Sciences, and she has won numerous awards over her career. She is interested in how biological complexity arises in associations of organisms, with a particular focus on symbioses between insects and microbes. In writing this chapter, I wanted to explore the questions "How did she get to where she is now?" and "When was her career path clear to her?" The answers to these questions not only reveal the (perhaps surprising) way that Nancy became a scientist but also show how the structure of modern education might hurt the chances of a young person who finds herself in a similar situation today.

Nancy was raised in Dallas, TX, as one of eight siblings in a large Catholic family. Her father owned a drive-in movie theater, and her mother was a homemaker who had grown up on an Oklahoma farm during the Great Depression. As a child, Nancy had an innate affinity for the natural world. As she puts it, "I was definitely the little girl with bugs in jars." However, the idea that one could catch and study insects as a profession was completely unknown to her, as very few of her relatives had attended college.

Nancy went to a Catholic high school where, despite her early interests in insects and nature, she hated biology class. It was taught by an uninspiring

---

[1]Division of Biological Sciences, University of Montana, Missoula, MT 59812

Figure 1 Photo of Nancy A. Moran.

and slovenly monk—nuns didn't teach biology—and the course left her with the impression that biology was gross. She realized in February of her junior year that she had enough credits to graduate in May, so she did. However, she left with what she describes as "no plan at all."

Through some of her mother's connections in the Catholic Church, she was admitted to the University of Dallas as an art major. While there, she took a challenging art course that made her realize how difficult life would be as a professional artist. She was also tiring of her religion-based education, so she transferred to UT-Austin after her freshman year at Dallas. At UT, Nancy started with an emphasis in philosophy, but again with what she describes as "essentially no direction at all."

While continuing to take art and philosophy courses, she found her way into the Plan II Honors Program at UT. This program provided her considerable flexibility in the courses she chose, essentially allowing her to design her own personalized major. While still more or less a philosophy major, she took introductory biology in the spring of her junior year to help fulfill the breadth requirements of Plan II. It was a revelation. Nancy realized that biology was not, in fact, gross but rather fascinating, complex, and intellectually vast. She still fondly remembers the book used in the course, William Keeton's *Biological Science*.

Graduating in Plan II required her to do a senior thesis. Nancy selected the topic of mate choice in pigeons, which led to a broader interest in the evolution of sex. It was during the writing of this thesis that she realized it was possible to "do science" as a career. With no real sense of how to pick a graduate school, she asked students at UT if they had ideas, and several suggested the University of Michigan.

At Michigan, Nancy worked first with Richard Alexander because of her interests in insects, but she was also able to interact with two giants of evolutionary biology, John Maynard Smith and William D. Hamilton, who became her doctoral co-advisor along with Alexander. Even these lofty influences didn't lay the groundwork for what one might call a traditional path after graduate school. Before she started her first faculty position at the University of Arizona, she chose to do a postdoc at Northern Arizona University followed by a National Academy of Sciences exchange fellowship in the former Czechoslovakia. These two choices were made partly because of the science happening in Flagstaff and České Budějovice but mostly, as she says, because she "just wanted to go to those places. I wasn't thinking about making the right career moves."

Even her (now well-known) choice of working on insect endosymbionts was due somewhat to chance. She first studied aphids because of her interests in sexual evolution and phenotypic plasticity, and this work led to her reading the work of Paul Buchner, one of the pioneers of insect symbiosis in the early 20th century. But it was an out-of-the-blue phone call from Paul Baumann, a microbiologist at the University of California, Davis, with the suggestion that they collaborate on aphids and their intracellular bacteria that pushed her in the direction of symbiosis. This aphid endosymbiont work eventually expanded to include other sap-feeding insects and, more recently, to her research into the gut microbiomes of bees.

Reflecting on the nonlinearity of her career path, and her relatively late entry into the formal biology curriculum in college, Nancy worries that the structure of modern high schools and universities has evolved to place too much emphasis on choices made early in life. It isn't clear to her that many girls are necessarily interested in science during their early teenage years, or may have been taught to believe that science "isn't for them," and so the shunting of young people into science, technology, engineering, and math (STEM) paths during middle and high school tends to disfavor girls. By the time these young women get to college, even if they later develop an interest in STEM fields, they have often taken watered-down nonmajor science and math classes as freshman and sophomores. At that point, it becomes logistically and (often) financially difficult to change to a STEM major because many of their completed math and science courses don't count in their would-be new major. It's not difficult to hear Nancy's story and wonder how much talent we are missing today by inadvertently diverting young women away from viable STEM options.

In thinking about my own experience in Nancy's lab, which I will share now, I wonder how her life experiences shaped her approach to science and mentoring students. While my own experience is just one example, I hope it gives a sense of the impact she had on me personally and scientifically. But, of course, I am not the only one. The list of Nancy's successful trainees is long—very long.

When I arrived with my dog to start my postdoc in Tucson, AZ, in the fall of 2006, I was lost. Not physically lost (it's easy to drive from St. Louis to Tucson), not emotionally lost (I was excited to finally live with my wife again after 6 months of living apart), but scientifically lost. I was incredibly lucky to have worked with brilliant and kind people during my undergraduate and graduate training, but these amazing experiences had left me with close to no idea of what I was going to do with my life. The 17-hour drive with my dog brought little further clarity.

I was a collector of scientific techniques. I thought x-ray crystallography and electron microscopy sounded cool, so I did those. I thought computational biology sounded fun, so I tried that. Genomics seemed neat, so, sure, why not dabble? But towards the end of graduate school it occurred to me that I had never actually tried biology. Of course I had always worked on biological problems, but I had never been out in the field, had never seen one of my study organisms in the wild, and had never thought about how interactions between organisms in nature might shape the way that things work in the lab.

I decided to try and do a postdoc with someone who did this kind of work. But what field? I had always loved microbiology, so that was a given. I thought that host-microbe interactions seemed pretty cool, so that was a possibility. One of my committee members asked me if I knew the work of Nancy Moran, and I said (probably something like), "Yeah, sorta." But experience had taught me that when this particular committee member suggested something, I should probably listen. I read Nancy's papers and was blown away. The work was incredible—interesting, creative, diverse, rigorous, and, importantly, the kind of biology I wanted to try.

I wrote what couldn't have been an overwhelmingly impressive email to Nancy asking about the possibility of doing a postdoc ("Dear Nancy, I am an amateur microbiologist, with no training in evolutionary biology or entomology. I am decent with computers, so I think I could be OK at genomics but I've never really done it. Sincerely, John.") But somehow, Nancy said I could come. Looking back now, her positive response to my email was the key turning point in my scientific life. Nancy helped me cobble all of the various techniques I had collected over the years into a semicoherent theme. She patiently helped me with my (vast) misunderstanding of evolutionary biology. She showed me how to write a paper, how to give a talk, how to write a grant, and how to think about scientific problems. She gave me the direction I had been looking for, and by doing so, she gave me my career.

If I had been her only postdoc, or her only student, one might be able to put my turning out okay down to luck. However, when I look at the list of my fellow lab alumni, I think about the amount of time she has put into these people. Could I have been the only one who needed her insight, her patience, her high expectations, her experience, her clear thinking, and her cool head? Could I have been the only one who benefitted from the numerous professional opportunities she so generously gave? Of course not. Add this output of successful scientists to her impressive output of publications and it becomes rather easy to see why Nancy is one of the most highly regarded microbiologists of the modern era.

CITATION
McCutcheon JP. 2018. Nancy A. Moran: the winding path of a brilliant scientific life, p 213–217. *In* Whitaker RJ, Barton HA (ed), *Women in Microbiology*. American Society for Microbiology, Washington, DC.

Women in Microbiology
Edited by Rachel J. Whitaker and Hazel A. Barton
© 2018 American Society for Microbiology. All rights reserved.
doi:10.1128/9781555819545.ch24

Flora Patterson: Ensuring That No Knowledge Is Ever Lost

24

Hannah T. Reynolds[1]

Flora Patterson, the first woman mycologist at the U.S. Department of Agriculture (USDA), exemplifies the tenacity, audacity, and perspicacity of a true scientific visionary (Fig. 1). She entered her tenure at the USDA in 1896, in many ways an anomaly, following what even today would be considered a nontraditional career path. She attended several colleges and universities, earning "Mistress of Arts" degrees which were awarded for her scientific contributions rather than the 4-year curriculum typical of today's universities (1). She began her pursuit of science as a hobby and found herself drawn into the study of botany and fungi in college. She had sought education early in life but ended her formal studies when she married. After her husband died, she took a plant pathology course at Iowa State University, returning to scientific study "on the theory that no knowledge is ever lost" (2). She continued her education at Radcliffe College and worked in the Harvard Gray Herbarium (1). After she applied to a government position at the suggestion of a friend, with no expectation of success, her life changed when she was hired as an assistant pathologist based on her high score on the civil service entrance exam. This first job led to her heading the U.S. National Fungus Collection with the official title of Mycologist in Charge of Mycological and Pathological Collections. Over her years of service, Patterson expanded the department and the National Fungus Collection, cataloguing tens of thousands of specimens (3). This collection allowed her to identify multiple diseases that could harm agricultural and wild plants in

[1]Department of Biological & Environmental Sciences, Western Connecticut State University, Danbury, CT 06810

Figure 1 Flora Patterson poised at the microscope. Photo credit: Library of Congress, Prints & Photographs Division, reproduction number LC-DIG-npcc-30944.

the United States, including fungal diseases on the first set of cherry trees sent to Washington, DC.

LIVING IN FLORA PATTERSON'S OHIO

Flora Patterson was born Flora Wambaugh in Columbus, OH, in 1847 and spent her early childhood there with her parents (A.B. and Sarah) and older sister Mary (4). She and her family lived in the Franklinton neighborhood in 1850, when she was 2 years old (4). When I moved to Columbus for postdoctoral research, I enjoyed visiting the historic brick homes in that neighborhood, some of which played important roles in the Underground Railroad. I could easily imagine how Patterson's life in Ohio inspired her to study mycology and plant diseases. The frequent rains lead to numerous mushrooms in and outside the city, and the warm, humid summers make the Ohio River Valley a hotbed of plant diseases. The forests I have explored contain a different assemblage of trees from those Patterson would have seen, due in large part to the loss of the American chestnut as the major canopy tree in the eastern United States. The dramatic reconfiguration of American forests from chestnut blight disease occurred early in the 20th century, when Patterson worked at the USDA. This invasive disease was one of the first to be well studied and made a compelling case for the necessity of preventing and managing invasive plant diseases. First noted in the New York Botanical Garden in 1904, the chestnut blight fungus, *Cryphonectria parasitica*, continues to infect and kill chestnut sprouts (5). The chestnut trees that Patterson would have seen as a child in Columbus are now replaced by oaks, maples, and crabapples—themselves subject to their own arrays of fungal and insect pests.

The dramatic decline of the American chestnut, which I have never seen in its maturity, is replicated by numerous invasive pests and diseases that have followed. Ragged, broken trunks of ash trees are sprinkled through the forests here, looking like lightning-struck trees. The invasive insect emerald ash borer is responsible, currently eradicating ash trees from North America, while the ash dieback disease fungus *Hymenoscyphus fraxinea* has killed so many ash trees in Europe that they face continent-wide extinction (6). There is some hope that resistant trees in both continents could eventually regenerate the ash populations (6, 7); however, introduction of ash dieback disease into North America would exacerbate an already devastating infestation (8). Both European and North American scientists are monitoring the health of ash trees closely, working to ameliorate their local problems while trying to prevent the entry of the threats plaguing the other continent. A dying ash extends over my street, a mix of healthy-looking branches and leaves and broken snags; it likely has less than 5 years to live and has already dropped several large branches onto the pavement. I avoid parking underneath it.

Every newly invasive pathogen presents a new set of dangers and scientific challenges, as even the most basic facts of its life cycle and ecology may be unknown at the beginning of an outbreak. As a graduate student, I became fascinated by both historic and current fungal epidemics. My interest led me to study the invasive fungus causing white-nose syndrome, which has killed several million North American bats (9). Like *C. parasitica*, the white-nose syndrome fungus, *Pseudogymnoascus destructans*, threatens to overturn the established dominance of its most widespread host species, the little brown bat (*Myotis lucifugus*) (10). The history of fungal epidemics underscores the importance of knowledge of the life cycles, hosts, and persistence of pathogens. Basic research on the plant pathogen biology forms the necessary foundation for managing these diseases. Patterson established a foundation for plant inspections at the USDA, working to identify plant pathogens and infected plants even before the plant quarantine laws we rely on were written.

GAINING KNOWLEDGE: EARLY LIFE AND EDUCATION

By 1860 Flora's family, which now included her younger brother Eugene, had moved to Yellow Springs, OH, which now forms part of the Dayton metropolitan area (11). There, they shared a home with another set of Wambaughs (parents, two children, and a 19-year-old "domestic"). Primarily educated by tutors (12), she attended the preparatory school at Antioch College for 1 year (1860–1861), when she was 13 years old (13). The next records of her formal education show her entering Cincinnati Wesleyan Female College in 1864 as a senior (14). Her college charged $30 tuition for each 20-week term, and the senior class in 1864–1865 consisted of 15 women. Although Flora was only 18 years old, she was able to graduate in 1865 with her class, having completed the requirements for the Classical Course and earning an MLA (Mistress of Liberal Arts). Students following the Classical Course of studies would have taken geology in their senior year, along with courses in Latin and philosophy. Other science courses required to graduate included astronomy, zoology, chemistry, and botany; Flora's ability to matriculate at this advanced level and graduate within the year indicates her high level of academic achievement and passion for science at an early age.

In 1870, Flora lived with her parents in Springfield, OH, directly before her marriage at age 23 to steamboat captain Edwin Patterson (15). Ten years later, the couple resided in Ripley, OH, a village on the banks of the Ohio River famous for its role in the Underground Railroad (16), along with their two young sons and Flora's mother (17). After her marriage, Flora earned another degree *in cursu* from the Cincinnati Wesleyan Female College: the A. M., or Master of Arts. This degree was awarded to alumnae of the college who had engaged in scientific or literary pursuits for at least 3 consecutive years (18). Flora earned this degree in 1883 (1), but unfortunately, there is no record of the details of the achievements that served as the basis for her degree (C. Holliger, personal communication). A few years afterward (9 July 1889), her husband died after a long convalescence following a steamboat explosion, and Flora moved to Cincinnati to live with her younger brother, Eugene Wambaugh (1). Eugene was hired as a law professor at the State University of Iowa, and Flora recommended her career in science, studying plant pathology and earning another A. M. degree. When her brother moved to work at Harvard Law School, she planned to study botany at Yale University, but she discovered to her dismay that women were not permitted entry; she responded by attending the rival institution. From 1892 to 1895, Patterson studied at the Harvard Annex, later named Radcliffe College (19), and worked at the Gray Herbarium of Harvard University. This experience led to her applying to and joining the USDA as an assistant pathologist.

EXPANDING KNOWLEDGE: PLANT PATHOGEN RESEARCH AT THE USDA

Promoted to mycologist in 1901, Patterson rapidly expanded her department, hiring four other mycologists (Vera Charles, Anna Jenkins, Edith Cash, and William Diehl) to assist her in identifying, investigating, and curating an increasingly massive collection of fungal specimens (20). From early in her tenure, she emphasized the importance of reference specimens in plant pathology; descriptions of fungal species in literature may be ambiguous, whereas comparison with a known specimen permits a more confident identification. She sought multiple such specimens from collections around the world, preserving the knowledge of new species as they were first described (3). She ultimately added over 90,000 specimens to the U.S. National Collection, which remains the largest fungus collection in the world.

Agriculture in the United States depends heavily on the inspection and quarantine of potentially infected plant material. Under Patterson's leadership, the USDA recognized and prevented major threats to agricultural and plant health. This important work, then as now, relied on the routine diagnosis of diseases on both imported plants and local crops. She began a plant inspection program in 1906, 6 years before the passage of the Plant Quarantine Act in 1912. The Plant Quarantine Act, which provides for inspection and quarantine of nursery stock and other plants entering the United States, lists only three specific pathogens: the Mediterranean fruit fly and the pathogens causing potato wart disease and white-pine blister rust (21). The latter two pathogens are fungi, both of which were assessed by Patterson. White-pine blister rust was initially identified in the United States by Patterson (22), and her division placed an embargo on imported potatoes from the United Kingdom to prevent entry of potato wart disease (3). Potato wart disease remains an economically devastating disease that produces masses of unsightly galls on potato tubers and stems and can also prevent the emergence of plants from infected seed plants (23). Because the causal fungus persists in infected soils for decades as hardened masses of fungal tissue known as sclerotia, fields where it is discovered are prohibited from being used for potato production, causing millions of dollars in losses. Through quick identification, Patterson and Charles were able to slow its spread within the country, protecting American agriculture and demonstrating the benefits of effective plant health monitoring. When her department was threatened with a disruptive restructuring that would end the plant inspection program, Patterson presented compelling reasons for preventing the entry of diseased plants into the country (3). She succeeded in defending this program, and the plant pathologists at the USDA continue their important inspection services today.

I learned of the most famous plant quarantine story in U.S. history when I was 6 years old. The story, as I learned it in first grade, goes something like this. The annual Cherry Blossom Festival in Washington, DC, occurs thanks to a gift from the Japanese government in the early 20th century. The original gift was burned, however, when the first set of trees sent to Washington were found to be diseased and infested with pests. Japan graciously replaced the sickly trees with healthy ones, and the Cherry Blossom Festival has delighted DC residents and visitors for over a century. I was 29 when I learned the rest of the story, including the detail that Patterson was one of three USDA inspectors who examined that ill-fated first shipment of trees—numbering over 2,000 (3). Entomologist J. G. Sanders, nematologist A. F. Woods, and mycologist Flora Patterson each wrote a separate letter describing the state of the original trees. The acting chief of the bureau, C. L. Marlott, suggested in a fourth letter that the trees had probably been chosen because of their large size, which would make an immediately pleasing show, but that younger trees would be more likely to be healthy. The trees, even if planted, would likely not survive. Flora's letter details that over 45% were infested with crown gall and that five showed girdling with *Pestalozzia*, now named *Pestalotia*. She explained that because crown gall was already widespread in the United States, it did not pose an economic threat; however, the species could not be identified in such a limited time frame to determine if it was indigenous to the United States or not. *Pestalotia* is a diverse plant pathogen genus that typically forms tiny black bumps on plant stems and tree bark but may also cause dead spots on leaves or rot some varieties of fruits (24). Its segmented spores look like brown peas in transparent pods with hair-like appendages. The original, infected cherry specimens can be found in the USDA National Fungus Collections. I am impressed at the care and attention that must have been required to find this pathogen on 5 out of 2,000 trees in such a short amount of time.

SHARING KNOWLEDGE: OUTREACH TO THE PUBLIC AND PUBLICATIONS

As an additional public service, Patterson identified mushrooms for visitors to the Department of Agriculture building. Mondays were the busiest days for this public service, because people would walk in with mushrooms they had collected over the weekend (2). Her responsibilities in mushroom identification came with the job; in response to a reporter's question about her career path, she "laughingly explain[ed] that she had no idea of specializing in mushrooms, but [...] it was thrust upon her." Patterson clearly excelled at this work, as she published a guide related to mushroom collection and identification of edible and poisonous species. This departure from her typical work on plant pathogens to focus on edible mushrooms was partly motivated by World War I. In the winter of 1915, war was raging in Europe, but the *Lusitania* was still afloat, the Battle of the Somme was over a year away, and the United States would not join the war for another 2 years. One of the concerns in the United States, in fact, was that the supply of mushrooms from France would be diminished, inspiring interest in developing an American mushroom industry. Patterson therefore performed several experiments that winter to select a method she could recommend to the public for mushroom preservation. She included a brief description of mushroom education in both France and Germany, indicating to the American public that in Europe, mushrooms were highly prized: "For some years, certain foreign Governments have been endeavoring to teach their citizens the food value of mushrooms (25)." The authors further describe the popularity of mushroom education and exhibits in France and Germany, even to the point of teaching mushroom identification to schoolchildren. In a second bulletin on mushrooming, again cowritten with Vera Charles, Patterson added an introductory section that listed and debunked numerous, widespread pieces of folk wisdom about how to distinguish edibles from poisonous mushrooms (26). She sternly advised readers not to trust any "simple test," such as cooking mushrooms with a silver spoon ("absolutely no reliance can be placed on this test"), peeling the cap ("equally baseless"), soaking/boiling in salt water ("no foundation in fact"), or taking proof of edibility from insects crawling on or eating the mushroom ("a dangerous supposition"). Her subsequent diagrams, photographs, and descriptions of the common poisonous mushrooms, which could sicken or even kill unwary collectors, were clear and direct. In both bulletins, she included multiple recipes for mushrooms in general and for particular species and preservation methods for those who learned to confidently identify the edible mushrooms —and to collect and eat only those with a clear identity.

Patterson published scholarly works as well. Women scientists working for the USDA during the early 20th century were, unlike many women working in academia, permitted to publish under their own names (27). Patterson's publications cover a wide variety of fungi, which is not surprising given the breadth of diseases her department evaluated. For example, in bulletin 171 of the USDA Bureau of Plant Diseases, she and her assistants, Vera Charles and Frank Veihmeyer, illustrated and described five novel plant diseases (28).

She also published in-depth experiments on the pineapple fungus *Thielaviopsis paradoxa*, which causes a postharvest fruit rot (28). The director of the Hawaii Agricultural Experiment Station had requested that Patterson investigate a possible treatment of the pathogen with formaldehyde gas: how much would be required to kill the fungus, and would the pineapple industry benefit from this application? Patterson presented the results of multiple fumigation tests to determine the level of formalin needed to prevent black rot of pineapples in storage. She designed a specialized fumigation chamber that could allow drawers of fruit or petri dishes to be sealed before adding a mix of formaldehyde and water to crystals of potassium permanganate kept in a large beaker at the bottom of the incubation chamber. The ensuing reaction filled the sealed box with formaldehyde gas. By varying the concentration of formaldehyde and exposure time, Patterson determined that at least 30 minutes of exposure to between 750 and 1,300 cc of formaldehyde per 1,000 cubic feet of air was necessary to prevent fungal growth. After such treatment, fumigated fruit remained firm, while nonfumigated fruit decayed after only 120 hours. The pineapple rot pathogen Patterson studied, *T. paradoxa* (29), currently causes disease on multiple tropical crops (30). As formalin fumigation fell from favor, it was replaced with antifungal dip treatments (31). Although it would be desirable from the viewpoint of consumers to find an alternative to these antifungal treatments, attempts to treat pineapples without the use of antifungals have been unsuccessful in diminishing black rot (32), which remains the most problematic postharvest disease of pineapples (33). Patterson's experiments demonstrate that she was active not only in identifying new pathogens but also in attempting to treat them. She additionally determined the range of temperatures and humidity levels that exacerbated pineapple rot symptoms, that *T. paradoxa* macrospores were more resilient than the microspores, and that symptoms were worse on wound-inoculated fruits (28). Preventing wounding of the fruit during harvest and transport remains critical for minimizing the risk of infection. These discoveries assisted with the marketing of pineapples, permitting a tropical fruit to be stored and sold in temperate regions.

PATTERSON, "WOMAN SAVANT"
Patterson was able to forge a career as a scientist during a time when few women were able to do so. She impressed the world around her, and she

appeared in several news articles about women scientists in the early 1900s with headlines like "Fair Women Seeking Secrets of Plant and Animal Phenomena" and "Women Savants Who Sell Knowledge to the Government" (34–38). As concern about the risk of plant diseases grew during the early 20th century, so did interest in Patterson's work; as one 1914 newspaper stated, "This woman stands as a shield between the American farmer and the plant disease germs of Europe" (39). Some of these early articles, although admiring her authority as a fungal expert, her extensive research, and "great and wonderful collection" of fungi and diseased plants, also hastened to assure readers of her femininity:

"Mrs. Patterson has a gentle dignity, a poise of manner, and a gracious personality which make her perhaps the strongest proof one could have that an absorbing interest in technical scientific subjects does not interfere with a woman's womanliness and charm" (34).

Patterson's own writing style tended to be forthright and eloquent, while she may have been gentle and gracious in person. Furthermore, a later interview regarding following "ridiculous" folklore for mushroom identification, or "unintelligent women" voters negating any advantage of women's suffrage (2), indicates that she prized clarity over charm.

In addition to appearing in the press as an example of a female scientist, Patterson was frequently in the newspapers' social and clubs section due to her community service. Her life in Washington, DC, included leadership in women's organizations: she participated regularly in the College Women's Club (40, 41), served as treasurer and corresponding secretary for her chapter of the Daughters of the American Revolution (42–44), and was president of a section of the Needlework Guild, which sewed for the military during World War I (45). Patterson, as president of the Radcliffe Club, hosted talks about the women's suffrage movement and opportunities for women to study at the Sorbonne (44). Her apartment building (The Decatur), built in 1912 shortly before she moved in, still stands near Dupont Circle in Washington, DC (44, 46). I was surprised to learn that despite hiring and mentoring other women scientists and promoting women's education, Patterson opposed women's suffrage, fearing an increase in the number of "unintelligent" voters (2). Five years following this declaration, women were able to vote for the first time nationwide; whether Patterson had altered her views is uncertain. In any case, as a resident of the District of Columbia, she would have remained unable to vote in national elections.

Patterson contributed to the 1920 work *Careers for Women*, allowing us to see her own views regarding women's opportunities in science—namely, that they were difficult to obtain and insufficiently lucrative (47). At the time, 29 women held pathology or mycology positions in the entire federal government, with salaries ranging from $1,320 to $2,400 (the equivalent to approximately $16,000 to $23,000 in 2017) (48). Patterson commented that "A survey of positions now occupied by women in this line does not form a basis for expectation of financial returns in any way commensurate with the expenditure of time and money involved in this preparation."

Several male professors provided quotes assuring that although no female plant pathologists were currently working in research at their institutions, there existed no apparent obstacle for women in plant pathology. Patterson evidently held a less sunny view of women's opportunities: "The writer sincerely wishes it could be said that for the women pathologists there is always room at the top. There is no such pleasing prospect in view, but the exceptional woman may secure such a coveted position, and time proving woman's efficiency, the old-fashioned prejudice may be overcome."

Patterson was 70 years old when she wrote these words, having worked for the USDA for 25 years. She had overcome much herself, both personally and professionally, to establish and preserve the National Fungus Collection.

CONCLUSION

In a world full of "the old-fashioned prejudice" (47) against women, at a time when many women scientists worked as unpaid lab assistants, Flora Patterson found a route to employment, leadership, and success in science. Patterson also hired and mentored other women scientists, contributing to the emergence of Washington, DC, as a notable center for women in science in the early 20th century. Moreover, her expansion of the U.S. National Fungus Collection, which under her administration became the largest such collection in the world, provided a lasting and invaluable resource for the nation and world.

ACKNOWLEDGMENTS

Several people provided invaluable archival research assistance. I thank Carol Holliger at the Ohio Wesleyan University, who provided the information regarding the curriculum and alumna lists of the Cincinnati Wesleyan Female College; Scott Sanders at Antioch College, who provided information about Patterson's enrollment in the Antioch Preparatory Class;

Susan Reynolds; who provided Census Bureau and newspaper archival access; and the Columbus Historical Society for information about her birthplace and neighborhood.

CITATION

Reynolds HT. 2018. Flora Patterson: ensuring that no knowledge is ever lost, p 219–231. *In* Whitaker RJ, Barton HA (ed), *Women in Microbiology*. American Society for Microbiology, Washington, DC.

References

1. Charles VK. 1929. Flora Wambaugh Patterson. *Mycologia* 21:1–4.
2. The Sunday Star. 30 May 1915. Woman employee in bureau of plant industry is an expert on mushrooms. *The Sunday Star*, 4th ed. Washington, DC.
3. Rossman AY. 2002. Flora W. Patterson: the first woman mycologist at the USDA. *Plant Health Instructor* doi:10.1094/PHI-I-2002-0815-01.
4. US Census. 1850. *Seventh Census of the United States, 1850*. US Census, Washington, DC.
5. Money NP. 2007. *The Triumph of the Fungi: A Rotten History*. Oxford University Press, Oxford, United Kingdom.
6. McKinney LV, Nielsen LR, Collinge DB, Thomsen IM, Hansen JK, Kjaer ED. 2014. The ash dieback crisis: genetic variation in resistance can prove a long-term solution. *Plant Pathol* 63:485–499.
7. Tanis SR, McCullough DG. 2012. Differential persistence of blue ash and white ash following emerald ash borer invasion. *Can J Res* 42:1542–1550.
8. Kowalski T, Bilański P, Holdenrieder O. 2015. Virulence of *Hymenoscyphus albidus* and *H. fraxineus* on *Fraxinus excelsior* and *F. pennsylvanica*. *PLoS One* 10: e0141592.
9. Froschauer A, Coleman J. 17 January 2012. *North American bat death toll exceeds 5.5 million from white-nose syndrome. News release*. US Fish and Wildlife Service, Arlington, VA.
10. Frick WF, Pollock JF, Hicks AC, Langwig KE, Reynolds DS, Turner GG, Butchkoski CM, Kunz TH. 2010. An emerging disease causes regional population collapse of a common North American bat species. *Science* 329:679–682.
11. Bureau of the Census. 1860. *Eighth Census of the United States, 1860*. Bureau of the Census, Washington, DC.
12. Ogilvie M, Harvey J, Rossiter M. 2000. *The Biographical Dictionary of Women in Science: Pioneering Lives from Ancient Times to the Mid-20th Century*. Routledge, London, United Kingdom.
13. R H Burkholder. 1860. *Catalogue of Antioch College of Yellow Springs, Greene Co., Ohio, for the Academical Year 1860-61*. R H Burkholder, Yellow Springs, OH.
14. Alumnae Association. 1882. *Cincinnati Wesleyan College for Young Women*. Cincinnati Wesleyan College for Young Women, Cincinnati, OH.
15. Bureau of the Census. 1870. *Ninth Census of the United States, 1870*. Bureau of the Census, Washington, DC.

16. Hoh JL Jr. 2008. *Places on the Underground Railroad: Ripley, Ohio.* ChedderBrau Publications, Milwaukee, WI.
17. **Bureau of the Census.** 1880. *Tenth Census of the United States, 1880.* Bureau of the Census, Washington, DC.
18. **Cincinnati Wesleyan College for Young Women.** 1882. *Cincinnati Wesleyan College for Young Women, Catalog 1882-1883.* Cincinnati Wesleyan College for Young Women, Cincinnati, OH.
19. **The Harvard Graduates' Magazine Association.** 1897. *The Harvard Graduates' Magazine, Vol 5 1896–1897.* The Harvard Graduates' Magazine Association, Boston, MA.
20. **USDA ARS.** U.S. National Fungus Collections—history. USDA ARS, Beltsville, MD. https://www.ars.usda.gov/northeast-area/beltsville-md/beltsville-agricultural-research-center/mycology-and-nematology-genetic-diversity-and-biology-laboratory/docs/us-national-fungus-collections-bpi/us-national-fungus-collections-history/.
21. **US Government Printing Office.** 1912. *Plant Quarantine Act. 7 U.S.C. 151–164.*
22. **Pierce RG.** 1917. Early discovery of white pine blister rust in the United States. *Phytopathology* 7:224–225.
23. **Franc G.** 2007. Potato wart. *APSnet Features.* doi:10.1094/APSnetFeature-2007-0607.
24. **Maharachchikumbura SS, Hyde KD, Groenewald JZ, Xu J, Crous PW.** 2014. *Pestalotiopsis* revisited. *Stud Mycol* 79:121–186.
25. **Patterson FW, Charles VK.** 1915. *Mushrooms and Other Common Fungi.* USDA bulletin no. 175. Government Printing Office, Washington, DC.
26. **Patterson FW, Charles VK.** 1917. *Some Common Edible and Poisonous Mushrooms.* USDA farmers bulletin no. 796. Government Printing Office, Washington, DC.
27. **Wayne TK.** 2010. *American Women of Science Since 1900.* ABC-CLIO, LLC, Santa Barbara, CA.
28. **Patterson FW, Charles VK.** 1910. *Some Fungous Diseases of Economic Importance. I. Miscellaneous Diseases. II. Pineapple Rot Caused by* Thielaviopsis paradoxa. USDA Bureau of Plant Industry bulletin 171. Government Printing Office, Washington, DC.
29. **de Beer ZW, Duong TA, Barnes I, Wingfield BD, Wingfield MJ.** 2014. Redefining *Ceratocystis* and allied genera. *Stud Mycol* 79:187–219.
30. **Cline E.** Ceratocystis paradoxa *and* Thielaviopsis thielavioides. US National Fungus Collections, ARS, USDA, Bethesda, MD.
31. **Frossard P.** 1978. Control of pineapple rot caused by *Thielaviopsis paradoxa.* Importance of storage temperature and fungicidal treatment. *Fruits* 33:91–99.
32. **Wijeratnam RSW, Hewajulige IGN, Abeyratne N.** 2005. Postharvest hot water treatment for the control of *Thielaviopsis* black rot of pineapple. *Postharvest Biol Technol* 36:323–327.
33. **Yahia E (ed).** 2011. *Postharvest Biology and Technology of Tropical and Subtropical Fruits.* Woodhead Publishing Limited, Cambridge, United Kingdom.
34. **The Washington Post.** 22 October 1905. Fair women seeking secrets of plant and animal phenomena. *The Washington Post*, Washington, DC.
35. **The Topeka State Journal.** 2 December 1905. What becomes of American college women. *The Topeka State Journal*, Topeka, KS.
36. **The Washington Post.** 15 August 1909. Women savants who sell knowledge to the government. *The Washington Post*, Washington, DC.

37. Bache R. 8 May 1910. Hunting for new plants. *Salt Lake Tribune*. Salt Lake City, UT.
38. Indianapolis Star. 7 July 1913. *Indianapolis Star*. Indianapolis, IN.
39. South Bend News Times. 14 July 1914. Sick plants nursed by Uncle Sam's high-salaried 'plant-doctor.' *South Bend News Times*, South Bend, IN.
40. The Washington Post. 25 January 1914. College Women's Club. Clubs and Society. *The Washington Post*, Washington, DC.
41. The Washington Post. 24 January 1915. Social events. *The Washington Post*, Washington, DC.
42. The Washington Post. 23 December 1905. Social and personal. *The Washington Post*, Washington, DC.
43. The Washington Evening Star. 3 May 1907. Daughters engage in partisan strife. *The Washington Evening Star*, Washington, DC.
44. The Washington Post. 25 May 1913. Clubs and societies. *The Washington Post*, Washington, DC.
45. The Washington Post. 5 March 1914. Clubs and society. *The Washington Post*, Washington, DC.
46. The Washington Herald. 1 December 1912. Radcliffe Club. *The Washington Herald*, Washington, DC.
47. Patterson FW. 1920. *Careers for Women*. Houghton Mifflin, The Riverside Press, Cambridge, MA.
48. Bureau of Labor Statistics. CPI inflation calculator. https://data.bls.gov/cgi-bin/cpicalc.pl. Accessed 10 November 2017.

Women in Microbiology
Edited by Rachel J. Whitaker and Hazel A. Barton
© 2018 American Society for Microbiology. All rights reserved.
doi:10.1128/9781555819545.ch25

Felicitas Pfeifer: Creativity through Freedom

25

Christa Schleper[1]

She entered the room in small triple steps, her voluminous white beard hanging somewhat loosely around her cheeks. The jacket in camouflage colors looked somehow massive on her shoulders, but the bicycle helmet had a good fit and was firmly fixed on her head. With her pants pressed tightly to her legs by metal clamps, she was pushing the sports bike into the seminar room, offering an enthusiastic "Good morning!" in the deepest possible voice before breaking into a kind of philippic that could be heard far across the hall on a controversial paper that denied the recognition of *Archaea* as a third domain.

It was with great vigor and fun that Felicitas Pfeifer (Fig. 1) played the role of Max Planck Institute department head Wolfram Zillig in theater plays at our department. The casting was particularly hilarious, because in real life Felicitas's calm and balancing style as a group leader was diametrically opposed to the unpredictable and rumbling temper of Wolfram.

Of course, none of us were professional actors, but we practiced seriously for these events, in which we poked fun at those who were moving on from the Institute or retiring. It was Felicitas who often initiated these performances, and although they were sometimes embarrassing (mostly for those of us performing), they became a really fun way to see our labmates off, independent of whether our audience laughed to tears because of our jokes or because of our bad acting. In retrospect, I believe that these little performances were also a good release, allowing us to make fun of ourselves and our colleagues instead of accumulating unneeded frustration over failed

---

[1]Archaea Biology and Ecogenomics Division, University of Vienna, 1090 Vienna, Austria

Figure 1 Felicitas Pfeifer, Professor of Microbiology at Darmstadt Technical University, Germany.

experiments or even the small grudges that can sometimes build up in intense research environments.

It was in this department at the Max Planck Institute for Biochemistry in Martinsried, near Munich, that I first met Felicitas Pfeifer in 1988. She was a young group leader working with halophilic archaea, developing genetic tools and studying the regulation of gas vesicle synthesis in these organisms. Felicitas had done her Ph.D. work with Werner Goebel in Würzburg on the variable plasmids of halophiles and had spent a 3-year period doing postdoc work in Herb Boyer's lab at the University of California in San Francisco, studying the regulation of bacteriorhodopsin before joining Wolfram's department back in Germany in 1985.

When we met, I was a master's student with Wolfram, and it was a vibrant time in archaeal research. *Archaea*, back then still known as archaebacteria, had been recognized as a third fundamental group of organisms by Carl Woese, and Wolfram's lab had just discovered that the RNA polymerase of archaea is complex and homologous with eukaryotic RNA polymerase. The task of my master's thesis was to confirm that a hyperthermophilic bacterium (*Thermotoga maritima*) has indeed an RNA polymerase like that of bacteria and not a more complex version as in archaea, although the organism is adapted to high temperatures like many archaea. After the biochemical isolation, I had to clone and sequence the genes of the subunits—not such a trivial job back in the 1980s.

I received lots of advice from my mostly male labmates about how to best solve the problem. Every person I asked really took his time to let me in on the secrets of cloning a gene of unknown sequence. There were two stereotypes of postdocs or Ph.D. student labmates: either they were really smart and a bit intimidating because they seemed to harbor infinite wisdom or they were excellent at sports and talked about their weekend adventures in mountain climbing and biking (sometimes equally intimidating). Only a few were incarnations of both types, like Wolfram Zillig himself. In particular, from the smart fraction I got masses of useful advice but was also informed at length about all the possible obstacles and the alternate routes I could pursue. Hearing all these caveats scared me, and I started to wonder how one could ever succeed at cloning a gene. It sounded like really difficult science. Strikingly, the advice I got from different colleagues was never compatible; it was rather contradictory, so I became increasingly confused. It was also a bit suspicious to me that my labmates sometimes spent a lot of effort to explain why the alternative protocol from their colleague would certainly not work.

It was difficult to sort out the advice I received: I had no clue how to get the relevant information on which I would base my decision. In hindsight, I probably followed the protocols of the most attractive rather than of the smartest guy. That is, until I met Felicitas.

Felicitas used a different logic but also a different style and language to communicate and teach. Her explanations were very straightforward, and she demystified science for me. All of a sudden, molecular biology was logical and simple, and I started to really love it. Most importantly, from discussions with her I developed self-confidence and was able to develop my own way of figuring out how to proceed with experiments. Chatting together regularly at social gatherings on Friday afternoons, we also discovered how the different personalities of the experimenters in our lab were reflected in the way they approached science. These discussions were lots of fun, and we collected much good material to use for the next farewell party theater production.

Since our initial time together, Felicitas has gone on to uncover further details of the ecophysiology and the complex gas vesicle synthesis, biofilm formation, and signal transduction of halophilic archaea. She described one of the first "eukaryotic-like" transcriptional regulators, a leucine zipper in an archaeon. In 1994, she became a professor and chair of the microbiology department at Darmstadt University of Technology, where she established a strong teaching program.

When I joined her institute as an independent group leader in 1997 after my postdoc work, I realized how effective and logical Felicitas was in leadership and science politics and how strong she was due to her lack of vanity, staying concentrated on matters of importance to her. Nevertheless, it was probably my earliest encounters with Felicitas that influenced me most. Although it may sound simple now, at the time it was like a revelation: I needed support to establish some self-confidence, but even more, I needed enough freedom to figure out my own strategy for how to approach a problem. This was perhaps key to gaining independence and developing a true passion for science. I hope that every young female scientist finds her own Felicitas to guide her in developing her own approach to science.

CITATION

Schleper C. 2018. Felicitas Pfeifer: creativity through freedom, p 233–236. *In* Whitaker RJ, Barton HA (ed), *Women in Microbiology*. American Society for Microbiology, Washington, DC.

Women in Microbiology
Edited by Rachel J. Whitaker and Hazel A. Barton
© 2018 American Society for Microbiology. All rights reserved.
doi:10.1128/9781555819545.ch26

Beatrix Potter: An Early Mycologist

26

Millicent E. Goldschmidt[1]

The name Beatrix Potter is known worldwide as that of an outstanding writer and illustrator of children's books, including the famous *The Tale of Peter Rabbit* (published in 1902). Potter was also a consummate artist, and museums have exhibited her landscapes; however, she was also a serious scientist with a microscope and produced many detailed drawings of flowers, animals, fossils, insects, and fungi. Beatrix was born Helen Beatrix Potter into a prosperous British family in 1866, during the reign of Queen Victoria. Societal norms meant that Beatrix was trained by a governess at home, where her studies were focused on those skills that would make her an excellent wife and mother, including entertaining and art (Fig. 1) (1). In her journal, Beatrix attributes this lack of a formal education as critical to her willingness to explore the natural world, without the confined thinking of the time (2).

Beatrix's parents were very protective of her and her younger brother, Walter (2). The children spent many lonely years trapped in their home by their parents' fear of communicable diseases from other children. To entertain themselves, Beatrix and Walter were permitted to have a large variety of pets, including mice, rabbits, hedgehogs, snails, lizards, and even bats. Many of these pets would go on to hold prominent positions in Beatrix's stories (1). Art in particular caught Beatrix's attention, as did the artistic style of the time, which was greatly influenced by the art critic John Ruskin. It was Ruskin who argued that the principle of the artist was to find the truth in

---

[1][Professor Emerita] Department of Diagnostic and Biomedical Sciences, University of Texas Health Science Center at Houston, Houston, TX 77030

Figure 1 A young Beatrix Potter
(age 15) holding her dog, Spot.

nature, and the resulting fashion was for art that was beautiful in its detail
of the natural world (3). With the help of her tutors, by the age of 9, Beatrix
had already drawn fully annotated pictures of the caterpillars that she kept as
pets (3). Indeed, so good were her illustrations that later in life the eminent
illustrator John Everett Millais would say to Beatrix, "plenty of people can
draw, but you…have observation" (3).

As was typical for wealthy families at the time, the Potters took extended
family vacations, first to Scotland and later to the Lake District. These
holidays were an escape from the boredom of the London family home for
Beatrix and her brother, and they had free range to explore the local coun-
tryside, which was actively encouraged by their parents. During one of these
vacations, Beatrix met Charles McIntosh, an eminent naturalist and amateur
mycologist who also happened to be their mailman. McIntosh was to inspire
a love of fungi within Beatrix, helping her to improve the accuracy of her
illustrations while teaching her taxonomy. It is possible to see this influence in
her drawings and eventual watercolors (4, 5). For the next decade, McIntosh
continued to send Beatrix live fungal specimens, and she, in turn, sent him
originals of her paintings. In total, Beatrix painted 350 illustrations of fungi,
which were of such quality that many are still used to teach fungal taxonomy
today, with some of her original art representing the first descriptions of some
species within the British Isles (Fig. 2) (5).

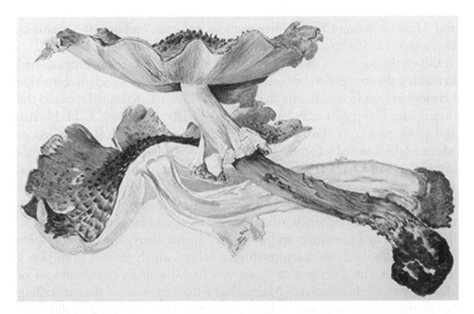

Figure 2 An illustration of the freckled dapperling *Lepiota friesii* (now more commonly known as *Echinoderma asperum* or *Lepiota aspera*) by Beatrix Potter. Courtesy of the Armitt Museum and Library, with permission.

Beatrix was also fortunate in that her aunt, Lucy Potter, was married to Henry Roscoe, who was Chair of the Chemistry Department at Owens College, Manchester. Roscoe studied chemistry in Germany under Robert Bunsen; the two were the first to demonstrate that it was possible to burn magnesium as a light source, allowing the eventual development of flash photography. A prominent photochemist in his own right, Roscoe was one of the first people to use Bunsen's famous burner in his research, and he went on to serve as president of the British Royal Chemical Society. Roscoe was also influential in Beatrix's life and encouraged both her illustration and her budding scientific studies. Through her continuing attention to detail in her painting and encouragement by McIntosh, Beatrix began using a microscope to describe her fungi. In doing so, she became one of the first mycologists to successfully grow fungi directly from their spores, while accurately logging their timed generation curves, particularly in the *Basidiomycetes*. She found and reported a rare new fungus (*Strobilomyces decastes*), anticipated the close association between certain fungi (*Lachnellula wilkommii*) and trees, and even suggested that the dry rot and fungal growth on trees could possibly help to cure cancer. Beatrix even found and studied a fossil fungus, *Boletus luvidus*. Her sense of humor is apparent as she incorporated brachial fungi and *Myxomycetes* into her children's stories and even had one of her animal characters reacting to hallucinogenic effects of "toadstool tartlets" (1).

Given the success of Beatrix's scientific studies, Roscoe was instrumental in making the connection between her and mycologists at Kew Royal Botanic Gardens in London, including the director William Thistleton-Dyer and the mycologists George Murray (famous for his collaboration with T. H. Huxley and participation in the Discovery expedition to Antarctica) and George Massee (also a noted painter). There is some controversy as to how Beatrix's mycological research was viewed by Murray, as much of the scientific material she presented was not preserved and we rely on her personal diary for details. Much weight has been given to the idea that Murray dismissed Beatrix's work on the symbiotic relationship between algae and fungi in lichens due to her position as an amateur and a woman. In her diary, she describes presenting her research on the growth of a lichen, which elicited from him a response that she described as "…so very high-handedly contemptuous of old-fashioned lichenologists." Murray had a different view of Beatrix, calling her so tenacious in persistence during visits that he "…fled, and so did Miss Smith the librarian" (6).

Other mycologists have suggested (from the conversations she described in her diary and likely contamination) that Beatrix was actually arguing that lichen could be grown from a single fungal spore, supporting the idea of lichen as a novel form of fungi that contained chloroplasts (2). This photosynthetic-fungus idea had already been dismissed by Simon Schwendener almost 40 years earlier, in favor of a symbiotic relationship between algae and fungi, which would have already been known to Murray and colleagues. What was never in doubt, however, was the quality of Beatrix's work on fungal sporulation. One modern analysis suggests that in order to record some of her observations, Beatrix would have needed to record microscopic observations at least once every 20 minutes. It appears that while Massee was skeptical of the amateur lichen research carried out by Beatrix, he was more sympathetic to her sporulation work (even though Beatrix in her journal remained less sympathetic to Massee) (2).

With encouragement and editorial advice from her uncle, Beatrix submitted her research to the Linnean Society of London in a paper titled "On the Germination of the Spores of *Agaricineae*," which featured the first image of a germinating basidiospore. The contents of this paper have been lost, although there is evidence that the spore work led Beatrix to suggest that many molds were in fact simply a life cycle stage for mushroom-forming fungi, rather than a unique species (a generally accepted fact within mycology today). She even appears to have described the jelly fungi (the *Tremella*) as mycoparasites, which was not determined by modern mycologists until the 1950s. As only fellows of the society were allowed to present papers, and women were not allowed to join the society until 1905, it was Massee who presented the work on Beatrix's behalf on 1 April 1897. It was common at the time for such papers to be withdrawn after they were presented (for modification based on the comments of society members); however, it was unclear as to whether sexism and Beatrix's amateur status played a role in the society's reaction to the paper. Beatrix appeared to have been undeterred, continuing her sporulation studies until the following September. Nonetheless, at some point later in the year, at the age of 31, Beatrix turned to another outlet that was more acceptable for women at the time—writing.

Beatrix had always remained close to one of her governesses, Annie Moore, who was only a few years older than she. She had remained an active correspondent, often sending small drawings of playful animal characters to Annie's children. One of them, Noel, was frequently ill, and Beatrix (having run out of ideas for a letter) wrote him the story that would eventually turn into *The Tale of Peter Rabbit* (3). In 1901 Beatrix produced her own limited printing of this story, which was eventually published as a book with the help of Norman Warne, an editor at Frederick Warne and Company. Beatrix and Norman fell in love and became secretly engaged, against the wishes of her family, although the engagement was short-lived when Norman died of cancer. Heartbroken, it was almost a decade later before she fell in love with another man, William Heelis, who managed her farm in the Lake District. Again, her parents objected, but she finally married him. After her father died, her mother even moved in with them. Beatrix continued to write children's books, the proceeds of which allowed her to accumulate 4,000 acres of additional land around her farm, which she donated to the British National Trust conservation society upon her death (Fig. 3).

While Beatrix's fame was born out of her children's books (*The Tale of Peter Rabbit* has been translated into 45 languages and sold 450 million copies worldwide), it was many years before her scientific contributions and fungal work were recognized. In 1958 her diaries were translated by Leslie Linder to reveal the detail of her scientific investigations (2), while in 1967, W. Phillip Findlay used 59 of her fungal drawings in his book *Wayside and Woodland Fungi*. In 1997, the Linnean Society published a posthumous apology for the way in which her original paper had been handled, while they produced a special volume of *The Linnean* in 2000 to highlight her contributions to science (6). In 2007, historian Linda Lear published a biography

Figure 3 A photo of Beatrix Potter from 1913.

of Beatrix that highlighted her scientific work to modern readers, particularly on lichen and fungal spores (3). Finally, in May 2002, the Smithsonian's Natural History Museum in Washington, DC, displayed an exhibit of her life and multiple works, including her science in a variety of other fields.

CITATION

Goldschmidt ME. 2018. Beatrix Potter: an early mycologist, p 237–242. *In* Whitaker RJ, Barton HA (ed), *Women in Microbiology*. American Society for Microbiology, Washington, DC.

References

1. **Taylor J, Whally J, Hobbs AS, Battrick EM.** 1987. *Beatrix Potter 1866–1943. The Artist and Her World.* Frederick Warne and Co Publishers, London, United Kingdom.
2. **Linder L.** 1966. *The Journal of Beatrix Potter from 1881–1897.* Frederick Warne and Co Publishers, London, United Kingdom.
3. **Lear L.** 2014. A scientist's eye. *Nature* 508:454–455.
4. **Lear L.** 2007. *Beatrix Potter: A Life in Nature.* St. Martin's Press, New York, NY.
5. **Jay E, Nobel M, Stevenson Hobbs A, Potter B (illustrator).** 1992. *A Victorian Naturalist: Beatrix Potter's Drawings from the Armitt Collection.* Frederick Warne and Co Publishers, London, United Kingdom.
6. **Watling R.** 2000. Helen Beatrix Potter. *Linnean* 16:24–31.

Women in Microbiology
Edited by Rachel J. Whitaker and Hazel A. Barton
© 2018 American Society for Microbiology. All rights reserved.
doi:10.1128/9781555819545.ch27

Abigail Salyers: An Almost Unbeatable Force   27

Rachel J. Whitaker[1]

Abigail Salyers is arguably the mother of microbiome research, the revolutionary science that explores the impact of the microbial ecosystem on human health. She dedicated years of her life to developing genetic tools for a nonmodel bacterium that makes up one-quarter of the cells in the human colon. She fought to make microbiology relevant and accessible in the medical curriculum and "to promote the cause of microbiology as THE biological discipline of the coming decades." By her own facetious account, her inspiration for this dedication was the flowers blooming from an unlikely place: in the lower right of *The Garden of Earthly Delights* by Hieronymus Bosch (Fig. 1). This irreverent, self-deprecating, and bawdy revelation represents Abigail Salyers well. With the brilliant mind of a physicist, and a very strong sense of humor, she ventured with indefatigable drive into the dark and inhospitable anaerobic world of the human gut. Through this pursuit, she became a giant in microbiology, publishing over 220 scientific articles and winning numerous awards for teaching and research, as well as becoming the first woman granted tenure in the Microbiology Department of University of Illinois, a codirector of the Microbial Diversity course in Woods Hole, MA, and the president of the American Society for Microbiology (ASM) during the U.S. anthrax crisis.

Abigail was born in 1942 in Louisville, KY. In spite of being pregnant in her senior year (and facing threats of expulsion for this), she graduated from Wakefield High School in Arlington, VA, in 1959. She then went directly to George Washington University, where she earned her B.A. with honors in

---

[1]Department of Microbiology, University of Illinois Urbana-Champaign, Urbana, IL 61801

Figure 1 (Left) Center panel from *The Garden of Earthly Delights*, by Hieronymus Bosch. (Right) Zoomed-in view of an unlikely bloom.

mathematics and her Ph.D. in physics. By the age of 27, Abigail was a single mother to her 10-year-old daughter, a published author of two papers on nuclear physics in *Physical Reviews Letters*, and an assistant professor in physics at St. Mary's College in Maryland.

Abigail's early career at St. Mary's focused on her teaching and on counseling physics students. She did not feel that she fit in in the all-male department, but was dedicated to her students. In addition to this focus, and raising her daughter, while she lived in Virginia, she authored a column in the Lexington Park newspaper entitled "A Radical's Eye View." The articles were humorous but challenging pieces, every week bringing new perspectives to the conservative Lexington Park and surrounding communities. This was the basis of what would be her superpower: her ability to write quick, witty, and compelling arguments, which she would use to great success throughout her career. Abigail's daughter, Georgia, remembers how importance was placed on writing with clarity, a lesson for which she will always be grateful.

After achieving tenure in 3 years, Abigail decided to move on. According to her daughter, "She didn't see a challenging, interesting, or appealing future in teaching physics. She had the drive and capability for research. She decided to retrain herself for a fresh start in a new career." Abigail took a lectureship in physics at Virginia Polytechnic Institute and State University in Blacksburg, VA, and there took classes and worked as a research associate in the anaerobic microbiology lab of Tracy Wilkins. This was when Abigail turned to what she called the "dark side": anaerobic microbiology. This is where she found her love for the organism that would define her career—the anaerobic bacterium *Bacteroides*.

Within 3 years, Abigail had transformed herself and her research and became an assistant professor now working in the area of anaerobic microbiology at Virginia Polytechnic Institute and State University. This gave her the background she needed to apply for a job in the Microbiology Department at the University of Illinois. Emeritus faculty member Ralph Wolfe recalled how he and others in the department were floored by her application. She had a Ph.D. in physics and B.A. in math. Wolfe remembers thinking, "We could really use someone like that around here."

Abigail set up her lab at Illinois in 1978, the same year as professor Jeff Gardner, who would become her long-term partner and unabashed biggest fan. They met sorting through equipment from the laboratory of Sol Spiegelman, who had just moved to Columbia University. Abigail took "the old stuff," much of which still exists today. This reflected another of her superpowers: she saw opportunity in what others overlooked. Their labs eventually merged, including some of that rediscovered "old stuff." Jeff and Abigail never formally married, and it did not matter to either of them. She gleefully armed Jeff against the marriage pleas from her brother with suggestions about the size of the dowry he could expect.

From the start, to the chagrin of some in the Microbiology Department, Abigail was not interested in *Escherichia coli* and focused instead on the physiology of the (then) nonmodel organism *Bacteroides*. Wolfe encouraged her work, saying, "Funny bugs needed some people working on them." Explaining her focus on *Bacteroides* years later, Abigail said, "From the 1950s to the late 1980s, the field of microbiology was dominated by the paradigm organism mentality: the idea that a few microbes could serve as representatives of the entire microbial world." Her passion was to expand the appreciation for what we now know is a vast and diverse microbial world and its fundamental impact on all ecosystems.

Abigail and her new lab began developing genetic tools in *Bacteroides* so that she could extend work begun with Wilkins on polysaccharide transport and fermentation in this bacterium (Fig. 2). The development of a new genetic system was painstaking work. No individual step stood on its own, but the steps together as a set of tools had a transformative impact,

Figure 2 Abigail Salyers, pictured here in her lab in the late 1980s.

opening a new understanding of a group of bacteria that play a vital role in human, animal, plant, and natural ecosystems. As these tools developed, Salyers's work would identify the processes through which complex carbohydrates are sequestered from other bacteria inside the individual cell and then degraded into small subunits anaerobically for growth. This metabolism of bacteria in the human colon produces by-products that are available to their human hosts and are vital to other bacteria inhabiting the human ecosystem. The physiology of *Bacteroides* was the start and continued focus of Abigail's research throughout her career and the basis for the microbiome research that has accelerated with increased impact on health and disease over the past 5 years. Perhaps this was the flowering from within that she referred to in the Bosch painting.

Along the way to developing genetic tools, the Salyers lab discovered a set of mobile genetic elements, called conjugative transposons. The lab tamed these elements and used them for genetic analysis, but they realized that outside of the lab, in the human gut, conjugative transposons move important genes carrying antibiotic resistance between bacterial cells. Adding a new component to her research, Salyers used the *Bacteroides* system to elucidate the molecular mechanisms through which these mobile genetic elements moved genetic material. The Salyers lab discovered and dissected the impact of *Bacteroides* as a reservoir of resistance genes. They uncovered the shocking and sobering fact that these elements are tightly regulated to spread resistance by mobilizing resistance genes to other bacteria in response to antibiotic treatment. Realizing the importance of this process to the distribution of antibiotic resistance in human pathogens, Salyers became one of the first to sound alarm bells about antimicrobial resistance. Today this is widely recognized as a global challenge to human health on the scale of climate change. Pushing the boundaries of microbiology into the dark world of anaerobic commensal organisms transformed our understanding of their impact (both positive and negative) on human and ecosystem health.

Key to her success in developing this model system was a long-term collaboration with Nadja Shoemaker. Nadja joined her lab 3 years after Abigail moved to Illinois, and together they changed microbiology. Nadja played a pivotal scientific role in guiding the progress of the lab. She was also the social glue. Nadja was the ombudsman of the Salyers lab. She kept students on track and communicated directly with Abigail. In her words, "You need the dreamers but they don't run things." Abigail "was the global thinker out in outer space," while Nadja would "bring it down so human beings could do those projects." Abigail loved to talk and write, and Nadja loved to run experiments. Their partnership propelled the lab forward. Nadja joked, "You didn't want her in the lab with you—she would mouth pipette acrylamide!" Yet she credits Abigail for "always pushing forward; she wouldn't luxuriate in the success of the moment." Abigail also kept the lab focused on important contributions to the field of microbiology. Nadja noted that "she couldn't stand to do things that were not relevant." Abigail would ask, "Where is this going? You have to have a good rationale!" The legion of students she mentored through the years admired this virtue. A recounting of stories from the pair showed a dynamic that was fun and supportive but always on the cutting edge. See Boxes 1, 2, and 3 for a few firsthand reflections about Abigail from her students.

In addition to her research, Abigail was also hired at Illinois to teach clinical microbiology to first-year medical students. At the start of this endeavor she was appalled at the boring and inaccessible curriculum, which was based on memorizing the many organisms that cause disease as "bugs on parade." She wanted to make the subject real for the students and would not be satisfied until they appreciated the importance of microbiology. As she revamped the curriculum to meet these objectives, Abigail decided to write a textbook to go with it. Dixie Whitt worked with her on the class and the textbook, *Bacterial Pathogenesis: A Molecular Approach*. On his first read, then-ASM Press Director Patrick Fitzgerald was stunned—and thrilled. He fought for its acceptance against what Whitt described as "dyed-in-the-wool microbiologists who thought it was not sophisticated enough, and too much fun." Abigail didn't want the book "full of jargon and details

Box 1 Kyung Moon, Ph.D., Senior Scientist, NIH, NIDDK

Abigail was a giant, yet she stood up for the weak and powerless. She accepted our failures and mistakes but did not hesitate to go out of her way to encourage us, because she believed in the second chance. Without her belief in us, many of us, including myself, would never have had the chance to be where we are today. She had the uncanny ability to find each student's particular talent when no one else could or was willing, and she nurtured those talents to full success in each of us. When needed, she raised her voice and fought for us with tremendous compassion. Abigail saw potential even among students at the bottom of the pit and brought them up to the top of the mountain. She was like Ms. Anne Sullivan was to Helen Keller, and like Mrs. Baker[a] was to Abigail Salyers. Today, living in this so self-absorbed society, I miss Abigail, who saw things differently and taught us to have a hope even in the darkest time of life.

[a]Mrs. Baker was Abigail's high school English teacher who inspired her in writing, helped prevent her from being kicked out of school when she became pregnant, and helped her get into college despite the fact that admitting pregnant teenagers was not typical at the time.

that students don't care about and could not possibly remember." Instead, calling on her writing skills, she made it "easy and succinct in a way that they enjoy reading." In the book and in the class, Abigail went to great lengths to get her message across, sometimes even surprising her close colleagues. She mercilessly teased the medical students and even insulted them. At the same time, when she heard about their parties, she was sure to attend. On the day they discussed sexually transmitted diseases, Abigail dressed in fishnet stockings and stilettos just to maintain their attention. It worked. She even incorporated "book bombs"—small asides to engage the reader.

Building on her success to revitalize the medical school curriculum, Salyers was invited to be codirector of the Microbial Diversity course at the Marine Biological Laboratory in Woods Hole, MA, one of her greatest

Box 2 Matthew Parsek, Ph.D., Professor, Department of Microbiology, University of Washington

We can all probably identify pivotal moments in our early career that shaped us. My single most important moment involved Abigail. I was an undergraduate at the University of Illinois in my junior year, and my major was engineering. I took Abigail's Bacterial Pathogenesis class as an elective to check off one of the boxes I needed to graduate. This was my first serious introduction to microbiology. She was a compelling, electric teacher. She crystallized for me the notion that our understanding of microbiology was in its infancy and that there were many unexplored vistas of research. I eventually graduated with a biology degree, and she helped me through the process of choosing a graduate school. The time she gave me through this whole process is amazing to me, particularly as I get older and have the professor's perspective of shrinking time and mounting responsibilities.

Thus, I can confidently say that if it weren't for Abigail, I'd be in an office somewhere with a slide rule sticking out of my shirt pocket. She made all the difference.

Box 3 Ann Stevens, Ph.D., Professor, Department of Biological Sciences, Virginia Tech

When I joined Abigail's lab, she was the only female faculty member in the department and the only to earn tenure. When a second female faculty member joined the department, Abigail was surprised to hear the male faculty talking about how great it was to have a woman in the department. I think she was surprised by the realization that she hadn't been viewed as a woman faculty member but was just "one of the guys," an equal in their eyes. Interestingly, shortly after this Abigail began making a more concerted effort to express her feminine side by a change in hairstyle and wardrobe. As a graduate student in her lab, I never saw my gender as an issue to my ability to succeed as a scientist. She and Nadja [Shoemaker] were great role models.

undertakings and most dearly held successes. As described by Ralph Wolfe, she was the obvious choice to modernize and reorganize the course and "bring it into the 21st century," including integrating the molecular revolution into the curriculum of the course. Abigail took on the challenge as she did everything: with full force. She was not satisfied until it was reinvigorated and introduced students to the diversity of the microbial world and the up-to-date tools for investigating it.

After 5 years codirecting the Microbial Diversity course, Abigail's engagement with the scientific world increased in scope. Her passionate defense of microbiology turned to arguing that microbiology would be the focus of biology moving forward. She spelled out this belief using her well-honed writing skills again to craft a compelling vision for the future of microbiology as a candidate for ASM president.

> I believe that the time has come to make a more concerted effort to promote the cause of microbiology as THE biological discipline of the coming decades...I would work to minimize the fragmentation that has occurred within microbiology itself, especially the rift between those who characterize themselves as "environmental microbiologists" and those who characterize themselves as "clinical microbiologists." These two groups need to interact more with each other. I believe that if we could forge these two areas into a single cohesive unit, we could become an almost unbeatable force in biology.

Against a more established candidate, this vision won Abigail the presidency. As president, Abigail served critical functions in the anthrax scare, the rise of antibiotic resistance, and promoting her vision for microbiology and its future. Her focus was on microbiology and not everyday matters (the team at ASM had to take her out shopping for respectable clothes to wear in Washington, DC as ASM president).

The road from there was long and not easy. Abigail encountered resistance to her expansive view of microbiology from within her department at Illinois and the ASM, those who practiced clinical microbiology, and colleagues and friends. She met each challenge head on. Abigail was famous for being the one in the room who would say what needed to be said "because the others were too chicken to say it." Armed with insight, a vivid sense of humor, and an uncanny ability see the truth, Abigail fought for what she called "prokaryotic pride."

Abigail Salyers was the inspiration for this book. She does not have an archive of her work or writings. There is one box with some of her drawings and pictures of her dressed up as a microbe, posing in the lab. When she died in 2013, there was a simple obituary in the *News-Gazette*, in *ASM News*, and on the Marine Biological Laboratory website. It could not possibly have represented the impact Abigail had on so many lives and the science of microbiology. Nor can I. I had the pleasure of attending the Salyers Symposium, in which many who were inspired by her came to tell stories. Everyone there agreed on her impact, particularly as the pioneer on the human microbiome. They spoke of her with humor, appreciation, and love. To hear an interview with Abigail, visit http://bit.ly/ WIM_Salyers.

I did not know Abigail well personally. I first met her at a party after she spoke at the first graduate student-organized Microbiology Symposium at the University of California, Berkeley. She had just been diagnosed with breast cancer, but you would never know it. She was in the middle of a room full of students, cracking jokes and egging people on to enjoy themselves and bask in the success of an event born of our own burgeoning "prokaryotic pride." I followed her thereafter from afar and was lucky enough to have her as a senior colleague when I started my lab in the Microbiology Department at the University of Illinois. Having her there, as a strong woman unafraid to be herself and to ask the hard questions; full of integrity, passion, and humor and with the ability to disarm others with sarcasm, promoted bravery in me as well as the students and scientists around her. I remember her irreverently poking fun at senior colleagues and giving them nicknames that they had to accept because they accepted her. I remember her standing up at seminars of students and senior researchers alike and demanding a big-picture explanation of why their research was important. I attended a semester of her teaching General Microbiology to hear her spontaneous stories about what she had told Congress during testimony, or how she had insulted seemingly untouchable senior scientists with the truth. Abigail, like so many others featured in this book, came from an unlikely place and inspired many other women scientists to do the same.

ACKNOWLEDGMENTS

I extend many thanks to Jeff Gardner, Dixie Whitt, Nadja Shoemaker, Ralph Wolfe, Brenda Wilson, and Georgia Will. Through laughter and enduring admiration, they related more stories than I can share. I also thank Jeff Karr for archival material from ASM and Diane Tszevelos and Deborah LeBaugh for support and access to Salyers's files.

CITATION

Whitaker RJ. 2018. Abigail Salyers: an almost unbeatable force, p 243–251. *In* Whitaker RJ, Barton HA (ed), *Women in Microbiology*. American Society for Microbiology, Washington, DC.

Women in Microbiology
Edited by Rachel J. Whitaker and Hazel A. Barton
© 2018 American Society for Microbiology. All rights reserved.
doi:10.1128/9781555819545.ch28

Christa Schleper: Enthusiasm and Insight in the World of *Archaea*

28

Sonja-Verena Albers[1]

Christa Schleper is an esteemed microbiologist and leader of the Archaeal Biology and Ecogenomics Division at the University of Vienna. Her research focuses on the molecular biology of *Archaea* and the search for and characterization of new archaeal species. She grew up in the German microbiology tradition, with a diploma in biology from the University of Konstanz in Southern Germany, and performed her thesis work in 1989 in the lab of Professor Wolfgang Zillig at the Max Planck Institute for Biochemistry in Munich. At the time, her mentor, Zillig, was one of the leading biologists in Germany to establish *Archaea* as a separate domain in the tree of life. His research focus was to find and characterize viruses and plasmids in archaea, mainly *Sulfolobus* species, and Christa was an integral participant in these research efforts. After completing her thesis in Zillig's lab, Christa considered changing her research focus to immunology for her Ph.D. She was intrigued that immunology would be a "hot" topic and would fit her ambition to become a research leader. However, after some consideration at this key juncture in her career, she realized that the perceived "hotness" of the topic was less important than finding a home lab with supportive conditions for the demanding work of attaining a Ph.D (Fig. 1).

For her Ph.D. thesis, she stayed in the lab of Wolfgang Zillig to work on genetic elements of *Sulfolobales*, with the goal to one day develop a genetic system for a model organism in the crenarchaea, a highly divergent division of the archaeal domain. She established transformation protocols for *Sulfolobus*, studied insertion sequence elements, and finally was able to

---

[1]Molecular Biology of Archaea, University of Freiburg, Institute of Biology II, 79104 Freiburg, Germany

Figure 1 Photo of Dr. Christa Schleper.

take part in one of Zillig's famous expeditions to Japan to hunt for new archaea and their viruses and plasmids. In Japan, they discovered *Picrophilus oshimae*, an archaeon that grows at 60°C and an amazing low pH of 0.7. In the Japanese *Sulfolobus* strains they also found the first archaeal conjugative plasmid, pNOB8. Christa said that this experience really inspired her to continue to look for new organisms and hunt for new archaeal strains, which would drive her to many revolutionary discoveries of groups of archaea with large impacts in the understanding of nitrogen cycling of soil and to a restructuring of the tree of life.

Christa finished her Ph.D. in 1995 and took a postdoctoral position in Edward DeLong's lab at the Monterey Bay Aquarium Research Institute in California. There she pursued work characterizing a low-temperature crenarchaeal lineage that could not be cultured but could be enriched through its close symbiotic association with a marine sponge. She pioneered work in the field that would become metagenomics to clearly identify low-temperature crenarchaea and its unique DNA polymerase that looked more like that of eukaryotes than bacteria.

She brought this knowledge back to Germany, where she met her husband and established her first group as an assistant professor at the Technical University of Darmstadt, where she stayed from 1998 to 2004. She later moved to the University of Bergen in Norway, where both she and her husband secured research positions. In Bergen Christa grew her personal and scientific family with two daughters and many scientific friends. She continued to recognize the importance of having people around to relate to personally and scientifically. In 2007, Christa was recruited for a professor appointment at the University of Vienna and was also able to negotiate a position for her husband. There, she established a very successful group that developed cultures of new archaeal lineages related to the crenarchaea in a group now known as the thaumarchaea. This novel group occupies an important niche and filled a gap in our understanding of the nitrogen cycle of the soil and raised the *Archaea* to environmental relevance in global nutrient cycling. Studies of this novel group, additional new lineages, and in the biology of *Sulfolobus* viruses and their interactions with the CRISPR-Cas system are ongoing in her active and dynamic lab.

Throughout her career, Christa's influence has been far reaching. As one of the few women in the archaeal world, her large personality and unrelenting insight precede her and leave a strong impression on everyone she meets. The minute she walks into a room she raises the intensity and excitement of the scientific discussion. She does not work behind the scenes but has pushed the frontier with incisive questions and confrontational insights for every audience. In doing so she has reshaped the tree of life and the structure of its growth from the last universal common ancestor. She put *Archaea* on the map for ecological relevance in all environments, not just the extreme.

I took over Christa's work on the conjugative plasmid of *Sulfolobus*, pNOB8, for my own diploma work in the Zillig lab. When I joined the lab in late 1995, she was no longer there, but it was clear that the few people left in his lab (Dr. Zillig was 75 years old and close to retirement) felt Christa's absence every day. A few weeks after I joined, Dr. Zillig started to shout at me in the corridor, and I was at first a little surprised and then realized he actually wanted to have a scientific discussion. Ingelore, who had been Zillig's technician for many years and whom he eventually married a few years later, told me that was how Zillig used to have discussions with Christa —shouting through the corridors while exchanging scientific ideas, and enjoying the battle. This built a lasting impression of the woman in whose footsteps I was following.

In March 1996 I planned to go to the German Society for General and Applied Microbiology meeting in Jena. Just a day before I had to leave for the meeting, Christa appeared in the lab in Munich without warning but to great fanfare. It turned out that she was also going to Jena to give a presentation. We agreed to travel together to Würzburg to stay with a friend of mine and then continue to Jena the next day.

It is the train trip from Munich to Würzburg with Christa that I remember most fondly. We were talking science, only science, and it was fantastic. She had just left behind unfinished experiments on pNOB8, which I was now pursuing. I had the feeling that this was the first time that I was talking to someone who was as interested in science as I was. This is one of Christa's most outstanding qualities: she is an incredibly good listener and takes the time to show whomever she is speaking with that she is interested in them and thinks what they are doing is a worthwhile pursuit. I believe it is this interest and enthusiasm that make her so inspiring, especially to young scientists. During the train journey we were so busy talking about science that we nearly missed our stop in Würzburg. That night we continued our talk, to the point that Christa did not get much sleep and still had to prepare her presentation for the next day.

It was this meeting with Christa and my time with Dr. Zillig that seeded a great interest in archaea for me and vindicated my passion for science. As Zillig's lab was closing down, I completed my Ph.D. work on sugar transport of *Sulfolobus* in Wil Konings's lab in Groningen, the Netherlands. My path crossed with Christa's again several times as independently and through collaboration we built a genetic system for *Sulfolobus*. Ten years after Christa had started this work in the genetics of *Archaea*, Melanie Jonuscheit, a Ph.D. student in Christa's lab, and I managed to clone the ara inducible promoter into the virus vector and used it for expression of homologous and heterologous proteins, which was a key step towards bringing the *Archaea* into a tractable experimental system.

I still see Christa regularly at *Archaea*-centered meetings. It is always difficult to catch her, as many people want to talk to her, but if you get a chance to spend even five full minutes with her to discuss new ideas and embrace science, you will experience the great joy of her unyielding enthusiasm and inspiring personality.

CITATION

Albers S-V. 2018. Christa Schleper: enthusiasm and insight in the world of *Archaea*, p 253–256. *In* Whitaker RJ, Barton HA (ed), *Women in Microbiology*. American Society for Microbiology, Washington, DC.

Women in Microbiology
Edited by Rachel J. Whitaker and Hazel A. Barton
© 2018 American Society for Microbiology. All rights reserved.
doi:10.1128/9781555819545.ch29

Marjory Stephenson: An Early Voice for Bacterial Biochemical Experimenters

29

Stephen H. Zinder[1]

The place of bacteria in evolution is a question very hard of approach; we have, for example, no idea whether the forms familiar to us resemble primitive bacterial types or whether, like modern animals and plants, they are the successful competitors of the ages. **Perhaps bacteria may tentatively be regarded as biochemical experimenters; owing to their relatively small size and rapid growth variations must arise very much more frequently than in more differentiated forms of life, and they can afford to occupy more precarious positions in natural economy than larger organisms with more exacting requirements.**

MARJORY STEPHENSON, 1930,
in the preface of the first edition of
her monograph *Bacterial Metabolism* (1)

I have used the bold portion of the above quote for over 30 years at the beginning of an undergraduate course as one of the rationales for studying microbial diversity. I can no longer remember where I first saw the quote, but it struck me with its prescience and clear thinking, having been formulated when we knew so little about bacteria and few even thought they evolved. Stephenson's name was familiar to me because I had read a classic 1955 review by C. B. van Niel (2) for a graduate course and the review's subheading was "The Second Marjory Stephenson Memorial Lecture"; her

---

[1]Department of Microbiology, Cornell University, Ithaca, NY 14853

name stuck in my head, perhaps because it was unusual for a lecture series to be named after a female scientist, since there were so few at that time.

Over the years I kept coming across her name, since part of my research and teaching involved anaerobic metabolism and therefore overlapped with her work. For example, I saw her name associated with the Stickland reaction and on a figure showing a time course for the acetone-butanol fermentation (3) that I used in teaching, taken from the first edition of Gottschalk's classic *Bacterial Metabolism* text (4). Finally, I found Stephenson's third edition of *Bacterial Metabolism* (5) while researching a book on microbial diversity that I attempted to write in the 1990s. Again, I was impressed and inspired by its clarity and breadth of coverage of the microbial world. My interest in this remarkable scientist, who was one of the first two women elected fellows of the Royal Society of London, was solidified. I was therefore intrigued to write this chapter. It gave me the opportunity to read several biographies/ obituaries (6–11), and I was fortunate that an excellent monograph on Stephenson's life and work recently became available (12), which I've used as a main source for biographical details.

EARLY LIFE

Marjory Stephenson was born 24 January 1885 in a village about 10 miles from Cambridge. Her father, who owned and operated a large successful farm, had a secondary school education and was a well-read autodidact in science, especially as it pertained to agriculture. When Stephenson was about 10, her father described symbiotic nitrogen fixation (being elucidated by Beijerinck at that time) to her while walking through a clover field. Her mother fostered interests in literature and art (12).

In 1903, Stephenson enrolled at Newnham College, one of two women's colleges affiliated with Cambridge University, concentrating in the natural sciences (chemistry, physiology, and zoology). Although women could attend some of the same classes as the men, they were not considered full members of the university and received certificates of proficiency rather than degrees. While there, she was influenced by Ida Freund, the first woman in the United Kingdom appointed as a chemistry lecturer, and by a guest lecturer in a physiology class, Frederick Gowland Hopkins, a Cambridge biochemist who would later become her mentor and colleague (12).

After graduating in 1906, she took various teaching jobs, landing in 1910 in the Home Science and Economics Department at Kings College, London, teaching material related to food chemistry and nutrition. In 1911, Robert H. A. Plimmer, a nutritionist/biochemist and a cofounder of the Biochemical Society, invited her to University College, London, to teach courses in biochemistry and nutrition and to join his research group (12), his quick action a testament to how impressive Stephenson must have been.

By 1912, Stephenson had already published a paper (13) on lactase (beta-galactosidase) in dog intestines, a subject of study in the Plimmer laboratory, showing that its activity was inhibited by glucose but not galactose. She also obtained a Beit memorial fellowship that provided her salary for a few years. However, after World War I began, Stephenson volunteered to join the Red Cross in 1914 and served as a director of cooking/dietitian at field hospitals, eventually managing kitchens for 17 hospitals in Greece (6), until late 1918. She received an MBE (Member of the Order of the British Empire) for her distinguished service and ingenuity (12).

In 1919 she was recruited by F. G. Hopkins, an outstanding biochemist who would win a Nobel Prize in 1929 for his discovery of water-soluble B vitamins. He had also discovered tryptophan and glutathione, was knighted in 1925, and founded the illustrious Laboratory (Institute) of Biochemistry at Cambridge. His philosophy was to recruit the best young scientists he could find, with little prejudice, and give them good resources and as much freedom as possible. The Institute counts among its alumni Nobelists Albert Szent-Györgyi, Hans Krebs, Ernst Chain, Rodney Porter, Frederick Sanger, and Peter Mitchell (12). At the time it was nearly impossible for a woman scientist to succeed on her own in England, where academia, even more than in the United States, was governed by hidebound traditions and prejudices. It is clear that Stephenson benefited from the support of Hopkins, a greatly admired scientist and true believer in meritocracy, of whom British Medical Research Council (MRC) head Sir Walter Fletcher, reflecting the prejudices of his time, once said, "His place bristles with clever young Jews and talkative women" (14). At least nine other women in Hopkins's group had sustained careers in research (14). Even with Hopkins's support, academic rank and honors commensurate with her accomplishments were greatly delayed. However, I suspect that she would say that the most important thing was that she was able to do her science and interact with great students and colleagues.

RESEARCH ON MICROBIAL METABOLISM

Stephenson, having resumed her Beit fellowship, began work with Hopkins on vitamin A, but he soon encouraged her to work on biochemistry of bacteria, which he believed would serve as tractable model systems for many biochemical reactions and pathways. Stephenson and graduate student Margaret Wetham began a collaboration with Juda H. Quastel, a just-minted Cambridge Ph.D. biochemist who had begun experiments using washed "resting" cells of *Bacillus* (*Escherichia*) *coli* grown in defined medium rather than in the complex media typically used at that time. These better-defined conditions allowed them to determine material balances of reactants and products, and versions of these methods are used to this day, as of course is *E. coli* as a model organism. Together they demonstrated that *E. coli* could use fumarate, reducing it to succinate, as an electron acceptor for growth with nonfermentable electron donors like lactate, and also reduce nitrate to nitrite (15, 16). Electron transport in these cells was examined using redox indicator dyes, especially methylene blue, in evacuated anaerobic (Thunberg) tubes, and oxygen consumption/gas production measurements used Barcroft and later Warburg manometers. Until a reliable method to rupture cells was developed (17), Stephenson mainly studied enzyme systems assayed in whole cells, many of which were really multienzyme complexes/pathways.

By the late 1920s, her research moved more towards anaerobic metabolism. Stephenson and her student Leonard Stickland showed that sulfate-reducing and methane-producing enrichments derived from polluted river sediments (18) reduced methylene blue upon addition of hydrogen (19), as indeed did *E. coli*, whereas a *Pseudomonas* strain and *Bacillus subtilis* did not. They named this enzyme hydrogenase, a name still in use today (as of October 2017 there were nearly 2,000 publications in the PubMed database with hydrogenase in their titles), and described it as reversibly catalyzing the uptake or evolution of hydrogen.

Subsequent experiments showed that *E. coli*, already shown to contain a formate dehydrogenase (20), formed hydrogen when incubated with formate, and she named the activity formate hydrogen lyase (21). We now know that this enzyme complex, only recently isolated in active form (22) because of stability problems, has subunits homologous to membrane-bound proton pumps and is believed to participate in energy conservation in anaerobically grown *E. coli*.

Stickland went on to study amino acid degradation in *Clostridium sporogenes*, showing that coupled oxidative and reduction deamination reactions of amino acids allow energy conservation (23). Stickland's eponymous reaction is one of the major routes of anaerobic protein degradation. Stephenson wasn't a coauthor on these papers (as was common at that time, she didn't put her name on unless she did a full portion of the research [10]), but she was acknowledged for her "advice and criticism."

During the World War II years, her lab studied acetone/butanol fermentation (3) since acetone was needed for munitions production, as it was in World War I. The approach was typical of earlier ones, designing a medium that minimized (but could not eliminate) the addition of complex nutrients and studying metabolism in washed cells harvested at various stages of the growth cycle.

Stephenson complained (1) that bacterial chemists (biochemists) considered bacteria "bearers of enzymes" rather than living organisms that physiologically adapt to their environments, and she became interested in these adaptations. Some work had already been done, and the terms "constitutive" and "adaptive" for enzymes were already in use. Stephenson and her colleagues used their washed-cell techniques to test whether appearance of an enzyme in response to an environmental stimulus was the result of "natural selection," i.e., the selection of variants/mutants in the population, or what they called chemical adaptation, which they described as "stimulating formation of the enzyme by the cells" (24). They showed that formate hydrogen lyase activity increased greatly on addition of formate to washed cells, with little or no increase in cell numbers, thus ruling out "natural selection" (24). A later experiment was done by student John Yudkin on "galactozymase" (enzymes needed for galactose catabolism, including galactokinase) adaptation in yeast (25). These studies showed that galactose utilization rates greatly increased in washed glucose-grown cells exposed to galactose in the absence of growth.

Once her Beit fellowship expired, Stephenson was supported at Cambridge by yearly fellowships from the MRC and made a favorable impression on its head, Sir Walter Fletcher. With Hopkins's supplications to Fletcher, she became a full-time salaried "external" member of the MRC staff in 1929 (9), which provided her with some security. However, she was never offered the directorship of a research unit in microbial chemistry, despite serving as one *de facto*. In 1936, Cambridge awarded her the degree of Doctor of Science, a degree that several of her students had attained before her. Although she was an associate and later a fellow at Newnham College, she had no formal appointment at Cambridge until she was appointed lecturer in 1943, despite having taught Cambridge students advanced biochemistry for 18 years (12).

One of Stephenson's most highly influential contributions, in addition to her research, was her book entitled *Bacterial Metabolism* (7, 11). Hopkins edited a series called *Monographs in Biochemistry* and encouraged Stephenson to write one on bacterial metabolism, a subject not covered previously. She undertook this task, and the first edition was published in 1930 (1). An expanded second edition was published in 1939, and a third edition was published posthumously in 1949 (5). Each version demonstrates the enormous progress made in understanding critical cellular processes, metabolic pathways, and electron transport, which she describes in biochemical and thermodynamic detail. The texts overall are built around experimental data, with many tables and figures. The results are described, followed by an almost conversational critical analysis of the strengths, weaknesses, and implications of the experiments. The chapters on metabolism particularly played to Stephenson's fortes, including chemistry, thermodynamics, and general microbiology, and were the definitive descriptions of these topics for years to come. There are extensive sections on anaerobic respiration, a topic never covered before in a cohesive fashion. Notably, despite all of the biochemical detail, and clear mastery of the topic, in the third edition of the book, there was no mention of how these components might be coupled with energy conservation in respiring cells. I believe that this reflects Stephenson's aversion to speculation without facts (7, 10–12), and of course chemiosmosis was not yet a gleam in Peter Mitchell's eye.

The final chapter in the third edition, introduced in the second edition (12), is entitled "Enzyme Variation and Adaptation." It attempted to parse the many contradictory studies in the field at a time when researchers were a long way from understanding the mechanisms involved. Like her research, it differentiated between permanent mutations in bacteria, such as sulfonamide resistance, and reversible adaptations.

This chapter, rather than the one on nucleic acids, includes Griffith's experiments on transformation of "pneumococcus" from rough to smooth form. More importantly, it was apparently the first textbook (12) to describe the findings of Avery, McLeod, and McCarty (26) that highly polymerized DNA was the "transforming agent," a finding only 3 years old when the book was prepared in 1947, and widely disbelieved (DNA was considered by many as just scaffolding for the proteins). Stephenson called these studies "irrefutable proof that nucleic acid controls enzyme production." She finished the section stating that "it is a piece of experimental evidence whose significance it would be difficult to overrate" (5).

Demonstrating the broad impact of Stephenson's book, a young Jacques Monod described some experiments to André Lwoff showing biphasic (diauxic) growth by *E. coli* on certain sugars; Lwoff immediately recognized it as a form of adaptation and recommended that he read Stephenson's recent (1939) text. Monod later said (27) that the "chapter summarized with great insight the still few studies concerning the phenomenon." Of course, Monod would go on to study the *lac* operon and its regulation for nearly the rest of his career, and more precise terms were coined for regulatory phenomena: induction, repression, catabolite repression, derepression, promoter, operon, and allosteric regulation.

In addition to her influential research, writings, and teaching, Stephenson played a pivotal role in founding the Society for General Microbiology (SGM; now called the Microbiology Society) in 1945. To some extent the Society of American Bacteriologists (SAB; now the American Society for Microbiology [ASM]), established in 1899, was a model.

Stephenson was a member of the founding committee in 1943 and presided over some of its meetings. She believed that the time was ripe for biochemistry and microbiology to interact and originally conceived it as a bacteriology society, but later others convinced her to include other microbes and have the society be as all-encompassing as possible. Rather than focus on medical aspects, then the dominant area of microbiology, an announcement in *Nature* (28) stated that the SGM would deal "predominantly with the more fundamental aspects of the study of these forms, including their physiology, nutrition, chemotherapy, systematics, and ecology." Alexander Fleming was the first president of the SGM, serving from 1945 to 1947, and the second was Stephenson, who served from 1947 until her death in 1948. According to its website, the SGM/Microbiology Society would not have another female president (they serve 2- to 3-year terms) until Hilary Lappin-Scott in 2009, 60 years later! Stephenson's students Ernst Gale and Sidney Elsden served terms in 1967 and 1969. In comparison, the SAB/ASM, which elects new presidents yearly, had two female presidents by 1947: Alice Evans in 1928, who worked on *Brucella* pathogenesis, and Rebecca Lancefield in 1943, who pioneered the serology of streptococci. Only two more women served as ASM president between 1948 and 1985; since then, 14 of 32 ASM presidents have been women. In recognition of her contributions a memorial fund was begun by the SGM, and in 1953, the first Marjory Stephenson Memorial Lecture (and Prize later on) was established to be given every other year. The first lecture was given by her former student Donald D. Woods (29), and the second was by C. B. van Niel (2), as mentioned previously. Since then, among the lecturers have been her students/colleagues G. S. Wilson, Bert Knight, Muriel Robertson, Sidney Elsden, and Ernst Gale, as well as André Lwoff, Jacques Monod, Renato Dulbecco, Patricia Clarke, Paul Nurse, Rudolf Thauer, and Stanley Falkow.

Clearly the most notable honor Stephenson received during her lifetime was being one of two women elected to the Royal Society of London in 1945 (the other was Kathleen Lonsdale, a crystallographer). In 1919, British Parliament passed the Sex Disqualification (Removal) Act, which allowed women to join chartered organizations, including learned societies (14). When the president of the Women's Engineering Society inquired whether women were eligible for membership in the Royal Society, the Secretary wrote back in 1925 that women were indeed eligible "provided that their scientific attainments were of the requisite standard (14)." Of course the measure of this was a vote by the all-male membership, and it took another 20 years before a woman was elected.

Stephenson's election was championed by the polymath and political radical J. B. S. Haldane, who knew Stephenson from his time working on his degree at the Cambridge Institute of Biochemistry. He stated in private correspondence that "I have no doubt at all that had (Dr. Stephenson) been a man she would have been elected to the Fellowship some time ago" (14). Of course there was stiff opposition by some members, calling it a "break from tradition," but Stephenson's qualifications were so clear that the vote was 339-37 in favor (14), and the Royal Society was no longer a men's club. In the following years a few more women were elected, including Stephenson's friend and colleague in Hopkins's lab Dorothy Needham and, later, crystallographer and Nobelist Dorothy Hodgkin (14).

Stephenson tended to pooh-pooh exclusive groups and the awards that scientists gave each other, once stating in this context, "These young men fuss about their reputations as if they were ageing virgins in a Victorian novel" (7). Nevertheless, she did accept membership to the Royal Society, perhaps realizing that it was an important step forward for women, regardless of her opinion.

In 1943–1944, Stephenson found out she had breast cancer and had a mastectomy (12). There are few details about this since she told almost no one. In 1947, she found out that the cancer had metastasized to her lungs, and she wrote to Elsden towards the end that her brain was becoming increasingly foggy (12). Elsden finished his obituary (7) saying that she had the "wisdom and courage to remain gay, argumentative, and active to within a few weeks of her death." Marjory Stephenson died on 12 December 1948 at the age of 63.

Obituaries and reminiscences of Stephenson paint a picture of someone lively, enthusiastic, forthright, and encouraging but with a critical streak (7, 10). Probably the most personal reminiscence was that of Ernst Gale, her student and eventually her successor in the microbial chemistry group at Cambridge, who later gave the 10th Marjory Stephenson Memorial Lecture (8). Its title, "Don't Talk to Me About Permeability," comes from a conversation he had with her concerning non-lactose-fermenting *E. coli* mutants that regained the ability to use lactose after treatment with membrane-disrupting agents, in which she stated the above followed by "(permeability) is the last resort of a biochemist who cannot find any better explanation." Gale noted the irony that he spent much of his later career studying membrane transport and imagines Stephenson changing her opinion in the face of hard data, as he had seen her do in the past, saying, "Now perhaps we can talk about permeability."

Gale also mentions that sometimes "MS's laughter rang through the building" (she was called MS by all who knew her). Indeed, Stephenson had an active sense of humor. *Brighter Biochemistry* was a satirical pamphlet occasionally published by members of the Hopkins lab, and one of her contributions was the parody "Through the Microscope and What Alice Found There" (30), with an illustration of Alice standing next to a giant rod-shaped bacterium, apparently drawn by Stephenson and closely resembling illustrations from the original Alice books. The story begins with Alice looking down a microscope, unable to see anything on a slide, and then wishing she could go down there to see for herself. She suddenly shrinks down to bacterial size ("Don't ask me to tell you how"), meets *Bacillus pyocyaneus* (*Pseudomonas aeruginosa*), "Pyo to real friends," and has several adventures culminating with her being tried for "treason against the Bacterial state." Just as she is convicted and about to be sent to the autoclave, she wakes up screaming and her professor suggests that she should go and have some tea. Stephenson also wrote a brief satirical review of her *Bacterial Metabolism* book in a 1930 issue of *Brighter Biochemistry*, describing it as a "racy and intimate account of the greatest bacterial metabolists" (12).

Stephenson is not well known today to most microbiologists, but few scientists of her era are, unless they made a groundbreaking discovery or perhaps had something named after them, as Stickland did. Moreover, metabolic biochemistry was displaced in the second half of the 20th century by molecular biology and genetics as the leading area of biology. However, in the 1980s, with improved anaerobic and other techniques, there was a surge of interest in the biochemistry of methanogens and other anaerobes and in syntrophic interactions that brought back her approaches of quantitative mass balances, pathway elucidation, and thermodynamic thinking. I believe the author who had written a paper 50 years earlier entitled "The Bacterial

Formation of Methane by the Reduction of One-Carbon Compounds by Molecular Hydrogen" (18) would have fit right in, made major contributions, and been enthusiastic and argumentative about the fine points of anaerobic metabolism of microbial "biochemical experimenters" at Gordon Conferences. It should be recognized that her life and work set the stage for the ongoing realization that microorganisms are indeed "the successful competitors of the ages" (1).

CITATION

Zinder SH. 2018. Marjory Stephenson: an early voice for bacterial biochemical experimenters, p 257–267. *In* Whitaker RJ, Barton HA (ed), *Women in Microbiology*. American Society for Microbiology, Washington, DC.

References

1. Stephenson M. 1930. *Bacterial Metabolism*. Longmans, Green and Co, London, United Kingdom.
2. van Niel CB. 1955. Natural selection in the microbial world. *J Gen Microbiol* 13:201–217.
3. Davies R, Stephenson M. 1941. Studies on the acetone-butyl alcohol fermentation: nutritional and other factors involved in the preparation of active suspensions of Cl. acetobutylicum (Weizmann). *Biochem J* 35:1320–1331.
4. Gottschalk G. 1979. *Bacterial Metabolism*. Springer-Verlag, New York, NY.
5. Stephenson M. 1949. *Bacterial Metabolism*, Longmans, Green and Co, London, United Kingdom.
6. Cope J. 9 December 2016. *Marjory Stephenson*. https://www.bioc.cam.ac.uk/about/history/members-of-the-department/marjory-stephenson-scd-frs-1885-1948/view. Accessed 16 November 2017.
7. Elsden SR, Pirie NW. 1949. Stephenson, Marjory—1885–1948. *J Gen Microbiol* 3:329–338.
8. Gale EF. 1971. 'Don't talk to me about permeability.' The Tenth Marjory Stephenson Memorial Lecture. *J Gen Microbiol* 68:1–14.
9. Kohler RE. 1985. Innovation in normal science. Bacterial physiology. *Isis* 76:162–181.
10. Robertson M. 1949. Marjory Stephenson. *Obituary Notices Fellows R Soc* 6:562–577.
11. Woods DD. 1950. Marjory Stephenson, 1885–1948. *Biochem J* 46:377–383.
12. Štrbáňová S. 2016. *Holding Hands with Bacteria: The Life and Work of Marjory Stephenson*. Springer Nature, London, United Kingdom.
13. Stephenson M. 1912. On the nature of animal lactase. *Biochem J* 6:250–254.
14. Mason J. 1992. The admission of the 1st women to the Royal Society of London. *Notes Rec R Soc Lond* 46:279–300.
15. Quastel JH, Stephenson M. 1925. Further observations on the anaerobic growth of bacteria. *Biochem J* 19:660–666.
16. Quastel JH, Stephenson M, Whetham MD. 1925. Some reactions of resting bacteria in relation to anaerobic growth. *Biochem J* 19:304–317.

17. Booth VH, Green DE. 1938. A wet-crushing mill for micro-organisms. *Biochem J* 32:855–861.
18. Stephenson M, Stickland LH. 1933. Hydrogenase: the bacterial formation of methane by the reduction of one-carbon compounds by molecular hydrogen. *Biochem J* 27:1517–1527.
19. Stephenson M, Stickland LH. 1931. Hydrogenase: a bacterial enzyme activating molecular hydrogen. The properties of the enzyme. *Biochem J* 25:205–214.
20. Stickland LH. 1929. The bacterial decomposition of formic acid. *Biochem J* 23:1187–1198.
21. Stephenson M, Stickland LH. 1932. Hydrogenlyases: bacterial enzymes liberating molecular hydrogen. *Biochem J* 26:712–724.
22. McDowall JS, Murphy BJ, Haumann M, Palmer T, Armstrong FA, Sargent F. 2014. Bacterial formate hydrogenlyase complex. *Proc Natl Acad Sci USA* 111:E3948–E3956.
23. Stickland LH. 1934. Studies in the metabolism of the strict anaerobes (genus *Clostridium*): the chemical reactions by which *Cl. sporogenes* obtains its energy. *Biochem J* 28:1746–1759.
24. Stephenson M, Stickland LH. 1933. Hydrogenlyases: further experiments on the formation of formic hydrogenlyase by *Bact. coli*. *Biochem J* 27:1528–1532.
25. Stephenson M, Yudkin J. 1936. Galactozymase considered as an adaptive enzyme. *Biochem J* 30:506–514.
26. Avery OT, Macleod CM, McCarty M. 1944. Studies on the chemical nature of the substance inducing transformation of pneumococcal types: induction of transformation by a desoxyribonucleic acid fraction isolated from pneumococcus type iii. *J Exp Med* 79:137–158.
27. Monod J. 1966. From enzymatic adaptation to allosteric transitions. *Science* 154:475–483.
28. Anonymous. 1945. Society for General Microbiology. *Nature* 155:340.
29. Woods DD. 1953. The integration of research on the nutrition and metabolism of microorganisms; the Inaugural Marjory Stephenson Memorial Lecture. *J Gen Microbiol* 9:151–173.
30. Stephenson M. 1930. *Brighter biochemistry: through the microscope and what Alice found there.* https://upload.wikimedia.org/wikipedia/commons/c/cb/Brighter_Biochemistry_Journal_Wellcome_L0070073.jpg. Accessed 16 November 2017.

Women in Microbiology
Edited by Rachel J. Whitaker and Hazel A. Barton
© 2018 American Society for Microbiology. All rights reserved.
doi:10.1128/9781555819545.ch30

Michele Swanson: A Rewarding Career and Life in Balance

30

Brian K. Hammer[1]

Passage of the Equal Opportunity in Education Act Title IX in 1972 aimed to ensure that "no person in the United States shall, on the basis of sex, be excluded from participation in, be denied the benefits of, or be subjected to discrimination under any education program or activity receiving federal financial assistance." Olympic athletes like Florence Griffith-Joyner, Mia Hamm, Kerri Walsh, and Misty May-Treanor are often mentioned as success stories of this legislation, which enabled female participation in intercollegiate athletics. However, the primary intent of the law was educational reform. Many premier universities began accepting female students for the first time in the late 1960s and early 1970s, with Yale University graduating its first female Yalies in 1973. Four years later, Ohio native and high school three-sport athlete Michele Somes attended a summer field hockey camp in southeast Michigan. She intended to apply to Michigan and one of her in-state universities in the fall. However, during that pivotal summer sports camp in 1977 she met the Yale field hockey coach who encouraged her to apply to his university's undergraduate program. Though Ivy League schools were not something Michele had considered prior, she applied and was accepted, arriving at Yale in 1978. To this day Michele Somes Swanson credits Title IX with giving her the educational opportunity the law was designed to allow.

Some may view Title IX as legislation benefitting women at the expense of men. Such a simplistic view is utter nonsense. In 2001, I defended my Ph.D. as Michele Swanson's first graduate student. I am a current associate

---

[1]School of Biological Sciences, Georgia Institute of Technology, Atlanta, GA 30332

professor at Georgia Tech and a beneficiary of Title IX's success. The same can be said for her other 20+ current and former Ph.D. students and postdocs, a near 50:50 mix of women and men. As I learned of Michele's own training while writing this chapter and reflecting on her career since I departed her lab, I have a deeper understanding of my mentor and the motivations for my adviser, my role model, and my friend who trained me. There are many different ways to conduct research, to train students, and to lead. Michele's way has made an indelible impact on me as a scientist and a person.

In 2018, 40 years after she began her biology training, Michele Swanson begins her term as the president of "the largest and oldest single life science membership organization in the world," the American Society for Microbiology (ASM). Michele succeeds a long line of successful male and female microbiologists who have held that esteemed position since 1899. Her journey is extraordinary not just because of what has been accomplished but also because of how it has been accomplished. Michele has not simply followed in others' footsteps but has charted a unique, rewarding, and balanced path. Michele has contributed in many significant ways to the field of microbiology, recognized and seized opportunities, and done so by making wise professional and personal choices. Michele's advice to her students and her own children has been, "You can't be good at everything." Certainly, Michele's accomplishments are far more than simply "good enough" for this recognition; they are remarkable. Her path is inspirational and serves as a reminder that there are many possible ways to reach our goals.

SPOTTING TALENT

In the 1970s, women in U.S. colleges and universities were vastly outnumbered by their male counterparts, a trend that has reversed in recent years. Even greater was the disparity in the ratio of male and female tenure track faculty members, an imbalance that sadly still persists. Consequently, prospects for women in academics at the time were modest, and a scientific career was hardly on the mind of Michele Somes during her undergraduate studies at Yale. Rather, Michele's attention was directed towards varsity field hockey and softball, counseling first-year students, her part-time job, and enjoying campus life. During her senior year, her interest in experimentation was kindled when she enrolled in a course taught by world-renowned developmental biologist John Trinkaus, legendary for his contributions to our understanding of *in vivo* cell biology. Trinkaus's engaging lectures were infused with his passion for scientific discovery. With graduation on the horizon and a burgeoning interest in experimentation, Michele sought opportunities to obtain a research experience. She eventually inquired about a lab technician position at Rockefeller University in New York City. There she met her first research mentor, cellular immunologist Sam Silverstein.

"DON'T SELL YOURSELF SHORT"

Current emeritus professor at Columbia University Sam Silverstein is known for his research successes uncovering macrophage structure and function. He also directs a summer research program for science teachers, which has provided hands-on teacher training for a quarter century. Remarkably, Silverstein not only is celebrated for his academic achievements but also is recognized as a mountaineer of distinction. Michele joined the lab of this adventurous spirit in the summer of 1982 and then moved with his group the following year, when Silverstein accepted the chair position at Columbia. Silverstein recalls vividly the first day he met young Michele, handed her a calculator, and challenged her with his famous "1 molar quiz," designed to sort exuberant students from those who also had reasonable quantitative skills. He still recounts Michele returning the calculator and confidently replying "I'd weigh out 58 grams of sodium chloride in a tared beaker, add distilled water, and bring the solution to one liter in a volumetric flask"; hired on the spot, Michele is remembered fondly as a valuable and contributive member of his group.

While in the Silverstein lab, Michele met postdoc Joel Swanson, whom she married the following year. A Renaissance man himself, Joel's research on macrophage biology complements his own passions for music, performing, and painting. Joel believes that the lab's zeitgeist had a profound impact on Michele, as Silverstein "encouraged everyone in his group to live up to their potential, both scientifically and personally." Throughout her nascent career, Michele had already watched other talented female students leave academia feeling inadequate, perhaps in part due to insufficient encouragement, guidance, and role models. After 2 years in her nurturing environment, Michele mustered up confidence to ask her mentor for a letter of recommendation to pursue a master's degree. To this day, both clearly remember the watershed moment...and Silverstein's categorical refusal to write that letter: "Why not apply to doctoral programs? You have the ability to become an outstanding scientist." Perhaps grudgingly, Michele reflected on these prescient comments and applied to, and was accepted into, Columbia's Ph.D. program in genetics.

GRADUATE SCHOOL—MICROBIAL GENETICS AND REGULATION

Michele's doctoral training began in 1984 under the guidance of yeast geneticist Marion Carlson at Columbia; however, soon after joining the Carlson lab, Michele made the difficult decision to move to Boston with her husband Joel, who had accepted an assistant professorship at Harvard. The choice of how to balance one's family life and career is a personal one, and Michele's resolve certainly led some to question whether this was wise. I recall Michele's own reflections of this pivotal moment and her conviction that it was her decision to make, and hers alone. So, as Joel launched his cell biology career, Michele joined Fred Winston's yeast genetics lab at Harvard, where she continued work on yeast transcription factor SSN20/SPT6, which she had initiated at Columbia.

Michele was "mature and broad in her thinking and ambitious in her experiments," says Winston, "serving as an intellectual leader whose ideas helped not only her own projects, but those of others." Winston also credits Michele with several notable firsts. The former Yale athlete earned the position of shortstop on the department's Terminators softball team, the lab's first starting player. Two years later, in 1988, Michele would give birth to daughter Hannah and became the first Winston lab member with a child. As a bench scientist and parent at Harvard, Michele experienced considerable challenges and social pressure during graduate school. But with support from Winston and other faculty, and soon with her Ph.D. defense in sight, Michele would begin to seek out a postdoctoral research position. Her goal was to engage in science that would make an impact ("Big Science," with a capital "S," as she liked to call it) and to do so in an environment where one's life choices were respected.

POSTDOCTORAL TRAINING—*LEGIONELLA PNEUMOPHILA*

In 1976, a major outbreak of fatal pneumonia occurred at a Philadelphia Legionnaires' convention. The following year, McDade et al. demonstrated that a previously unknown fastidious bacterium, *Legionella pneumophila*, was responsible for the outbreak (1), which Sam Silverstein and Marcus Horwitz would later discover replicated in macrophages. In the early 1990s, Howard Shuman and Ralph Isberg were developing cell biology and genetics tools to dissect the pathogenesis of this intracellular pathogen. In a friendly discussion with another colleague from Rockefeller days, Dan Portnoy, Michele learned about the burgeoning new field of cellular microbiology and contacted Isberg at Tufts Medical School in Boston, where his lab was focusing their efforts on the immune-evading bacterium *L. pneumophila*.

It was while 8 months pregnant with son Len that doctoral candidate Michele Swanson interviewed with Isberg. Her superb prior training in genetics with Winston and macrophage biology with Silverstein was a great scientific fit for the Isberg lab. Isberg recalls his excitement at the possibility that Michele might join his lab; Isberg also appreciated that a healthy blend of family and work could indeed be an asset, not a liability. In fact, years later Michele would remind her own students with children (including me) that "every second spent in lab is a second spent away from your kids, so you had better make it count." Michele's postdoc in the Isberg lab would become an invaluable experience as she transitioned from student to faculty.

Michele was entering into the field of *Legionella* pathogenesis at an opportune time. Pioneering work from the Isberg and Shuman groups in the early 1990s first identified transposon mutants of *L. pneumophila* with defects in interactions with macrophages. Both Vogel et al. and Segal et al. first identified avirulent mutants with disruptions in genes for what we now call the type IV secretion system (2, 3). Identifying the myriad of substrates secreted by this molecular apparatus and their modes of action remains an ongoing active research area. Carving her own path via her unique training in genetics, immunology, and microscopy, Michele focused on the unusual cell biology of this intracellular macrophage parasite. Isberg remembers that for Michele "browsing the new Molecular Probes catalogs for fluorescence tools became a hobby, in the same sort of way that kids used to wait in anticipation of toy catalogs" (Fig. 1). Michele eventually identified a distinct set of intracellular trafficking mutants obtained by chemical mutagenesis that were predicted to have regulatory defects. These *Legionella* mutants served as raw material for her transition into her tenure track faculty position at the University of Michigan, where she would launch her adventurous, playful new lab that she dubbed the "clubhouse."

Figure 1 Postdoc Michele Swanson at work with her "toys" in the Isberg lab.

THE UNIVERSITY OF MICHIGAN—THE EARLY DAYS OF RESEARCH IN THE CLUBHOUSE

Isberg recalls that during her 4-year postdoc, Michele developed a reputation for being "uncommonly mature, never hurried, fascinated by the biology of the system, and ready to think of a new idea or strategies to use." These attributes served her well during the job search, distinguishing her from a formidable set of peers entering the field at that time. Candidates applying for faculty positions at an R1 doctoral university require a substantial academic record with evidence of research productivity, and they must also demonstrate during the interview process their capacity to train students and interact productively with their peers. As Michele began considering faculty positions, and with the notoriously difficult Harvard promotion seeming unlikely for Joel, the two chose to launch a search together. In 1995 the University of Michigan Medical School's Department of Microbiology and Immunology advertised a microbial pathogenesis position, which ultimately led to a dual-career hire for the Swansons. Departmental Chair Mike Savageau was a progressive architect for his department and a master at establishing the connections often necessary for such hires, and the impressive cohort of women scientists he hired along with Michele now comprise a substantial portion of the department's senior leadership (Fig. 2).

Assistant professor Vic DiRita, now chair of Microbiology and Molecular Genetics at Michigan State University, served on that search committee and

Figure 2 Michele Swanson (seated right) with (from left to right): Denise Kirschner, Oveta Fuller, Malini Raghavan, Alice Telesnitsky, and Cheong-Hee Chang.

recalls that "the candidate pool was very impressive…[But] even as a young investigator just emerging from her postdoctoral training towards independence, Michele exhibited remarkable maturity, poise, and intelligence, and it seemed clear she would apply those traits not only as an investigator, but also as a colleague, as an adviser for our trainees, and as an academic leader. That impression was confirmed over and over in the ensuing two decades." The department was adding outstanding young faculty to its ranks, and each new hire afforded tremendous opportunities for incoming graduate students like me that were joining this community.

Michele arrived in Ann Arbor and quickly established a lab culture that developed out of her own conviction and prior experiences. She instilled in her group members a sense of enjoyment and wonder with the opportunity and the privilege to pursue research in balance with a healthy life outside the lab. Michele's unflappable, calming mentoring style developed at this time, along with her unwavering confidence in the potential and capability of each lab member.

Before meeting Michele, I had worked as an infectious disease lab technician for several years in Boston and then moved to Ann Arbor with my wife, Tracy, to pursue a M.S. in aquatic ecology. I was fortunate to receive an assistantship to teach the Introduction to Microbiology labs, which stimulated my interest in aquatic microbes. In 1995, I had been accepted into several out-of-state doctoral programs, although personal commitments caused me to consider the University of Michigan. After several stalled attempts to obtain a position, I convinced DiRita to hire me as a part-time technician; in this capacity I learned fundamental molecular biology working on *Vibrio cholerae*, and then I entered the doctoral program. During that first year Michele caught my attention with her "A Career and a Life, Too" panel discussion at a graduate career fair. It was DiRita that suggested I consider Michele's lab for my doctoral training. When I met Michele, I confessed that I was excited with the potential of *Legionella* genetics but intimidated by the eukaryotic cell biology. "Perhaps that IS why you should consider my lab," Michele replied. By the end of that third summer rotation, I had constructed my first functional gene fusion of the *flaA* promoter to the gene for green fluorescent protein, the lab's first paper was about to be published (4), and I had convinced both myself and Michele that I would be a good fit for her clubhouse.

Michele is a wordsmith. She can happily spend an hour debating the merits of a single word for use in a presentation or manuscript. I recall that after many cycles crafting the organization, content, and pace of an oral presentation for a particular audience, each and every slide received her undivided attention. "Let's start with slide 1, top line. Why that font style? Size? Color? Justification? ..." Manuscripts received the same level of scrutiny. Indeed, after countless rounds of revisions to the text of my first manuscript (5), I accidentally toggled on the "track changes" feature of Microsoft Word, which neither of us had used before. I still recall my shock when we discovered the scant words, mostly articles and prepositions, from my original text that had survived the successive rounds of now-fabled Swanson revisions! Later, a departmental colleague, realizing I was distraught, took me aside to show me her own heavily marked-up manuscript she had given a colleague to review. Her sage advice to me was "This is normal. Get used to this process. There is no such thing as a perfect document." Over time I would become comfortable with, and welcome, such attention to detail from my mentors, and I would in later years require fewer revisions to my manuscripts. The finished product, whether a manuscript or oral presentation, is always polished to a shine. Michele's current Microbiology and Immunology Department colleague Mary O'Riordan offers that, beyond formal instruction, "I learned more about the craft of scientific writing from reading the eloquent and precise language in Michele's manuscripts." Thanks to Michele's commitment to training me and many others' effective writing and presentation skills, my own students are the recipients of similar scrutiny. I am hopeful they too will come to appreciate my constructive criticism and my deft usage of Word's 'track changes' feature.

It was exhilarating as a young student to obtain my first substantial results that Michele and I were confident would become a component of a publication. However, I recall on several occasions being advised by some of our department's more reserved colleagues to curb my excitement over fresh results and to dampen my enthusiasm during my oral presentations to our department. Michele's advice was quite the opposite: be yourself and relish the process of research. A career in science is dominated by long stretches of delayed gratification, since the scientific method entails significant modification, replication, and validation. When we'd get an interesting, perhaps unexpected, result that was repeatable, and our controls had been performed properly, Michele would remind us, "Don't forget to pause and enjoy this moment; it's unclear when the next one will come." We took time to reflect and to savor the moment each paper was accepted for publication. The discoveries that emerged during that time period would serve as a solid foundation to launch the maturing Swanson lab into a period of notable creativity and discovery.

I had initiated my doctoral training at age 28, a bit older than most of my cohort. Soon after I passed my qualifying exams, my son Benjamin was born. I was initially concerned that faculty and peers might perceive that I was no longer a serious scientist. Of course, Michele had managed similar challenges and scrutiny as a graduate student and postdoc. She would assure me in times of doubt that "we'll show 'em," meaning that with her support, I would simply model how to balance a successful scientific career with the rewards and challenges of parenting. As my new atypical schedule emerged, our lab events became family friendly: Benjamin joined weekly lab meetings and our occasional lab outings, while Michele could often be found entertaining my young son in her office when I needed to complete additional DNA preps or start my overnight cultures for the next day. "Michele challenged the dogma that a successful scientific career had to come at the expense of family and a full life," says O'Riordan. The work/life balance fostered during my graduate training remains a core component of my own approach to science and to mentoring, and each Swanson lab member I corresponded with for this chapter embraces a conspicuously similar code.

Michele relishes questioning the status quo and enjoys sharing with her lab members some of her secrets of success. For example, many find the slow-growing nature of *Legionella* frustrating. The standard model bacterium *Escherichia coli* has a doubling time of about 20 minutes and will form a colony on an agar plate over the course of 1 day. In contrast, *L. pneumophila* has a doubling time of 3 hours! In graduate school, it is not uncommon to hear that a successful career in science requires 24/7/365 attention, sacrifice, and uncompromising dedication. So, to some, this organism's stubborn lifestyle might appear ill suited if one's goals are simply expediency and efficiency. Michele's simple suggestion was that we view *Legionella*'s slow growth not as a challenge but as an opportunity. As she told Mike Bachman during his rotation, "It takes 3 days to grow (a visible colony on a plate). You can do an experiment on a Friday, plate the bacteria, enjoy the weekend, and count colonies on Monday." This is not typical advice from a Ph.D. mentor, but it reflects Michele's appreciation for work time and play time. In a similar vein, when I'd sheepishly inquire about taking some time off after extended stretches of focused, time-consuming bench work, her common reply was, to quote Sam Silverstein, "you can get more done in 11 months than you can in 12." In Michele's view, people who can disconnect from work when needed often return with fresh perspective that allows greater productivity and creativity than if they'd not left. Now more than ever, it is harder for each of us to disconnect and get off the grid in our fast-paced society of tweeting, texting, and emailing around the clock. Michele's advice to occasionally look up from the bench is something many lab members, including me, took to heart and now share when mentoring our own students.

Michele's counsel and support have benefited students both within and outside her lab group. Not long after I had joined the lab, third-year doctoral student Amrita Joshi from a neighboring lab sought out Michele, feeling that her training to date had been less productive than anticipated. While Amrita was intelligent, well qualified, and eager to participate in our lab's science, it is also certain that Amrita's interview while 8 months pregnant evoked in Michele a recollection of her own first meeting with Ralph Isberg years prior! Amrita soon joined the lab and completed her Ph.D., with several published papers, and she remains a productive, active researcher on campus. Mairi Noverr was a member of my small cohort of incoming graduate students, a resident of another lab on our floor in Med Sci II, and another female scientist who often sought advice from Michele when encountering challenges in her own training. After 3 years in the program she too had experienced limited success in her project and was on the cusp of being one of the countless talented individuals who could abandon their career aspirations. Michele provided Mairi with much-needed encouragement to remain in our program and find a new lab, and she successfully petitioned the administration for modest funds that would allow Mairi to find a new mentor able to support her at this relatively late stage. Now a full professor at Louisiana State University, Mairi reflects that it is role models like Michele "who selflessly offer mentoring and support, that make the difference for the success of young scientists."

As early residents of Michele's clubhouse, each of us recalls Michele's intellect, along with her calm demeanor, reassuring tone, and quick wit. We were the first beneficiaries of many of her maxims, which I use now when my own students are frustrated with their progress and the constant troubleshooting required in this discipline. "What we do is called 're-search' for a reason." Science is not a sprint but a marathon, so pace yourself and enjoy the process if you want to be in it for the long haul. Michele's daughter Hannah learned the same lessons from her mother: "Prioritize wisely." "Do what you love. Say no to things you don't." "Allow yourself to be less than perfect." By my defense date, not only had I learned bacterial genetics but also I had an impressive collection of new aphorisms to live by.

Long after departing Michele's lab, I continue to benefit from her advice and encouragement. Indeed, I remember clearly the first day in my new office, feeling both excited and overwhelmed about what lay ahead. I set up my new computer, installed necessary software, and checked my new email system. There in my inbox was a message from my colleague Dr. Michele Swanson, inviting me to give an oral presentation of my research at a national meeting she was organizing the following year. One of Michele's many qualities that has made her an outstanding mentor is her belief that the driver of scientific progress is people. As DiRita eloquently attests, Michele is "peerless at creating opportunities for others to shine, whether in her own research group, or at any other level where she has influence." It is hard to imagine a better Ph.D. mentor.

LEADERSHIP IN TEACHING AND SERVICE

At the University of Michigan, Michele has chosen a variety of leadership roles that reflect her commitment to promoting biomedical research and the people who engage in that science. She began participating as an assistant professor in instruction at the medical school. While assuming the demands of launching her successful research program, she embraced the challenge of serving as a small-group leader for an infectious disease sequence at the medical school in 1997. Several years later Michele began codirecting this program with Carol Kauffman from the Department of Internal Medicine and Donna Shewach from the Department of Pharmacology. Michele served as codirector for nearly a decade, during which time the three women designed a program that would expose medical students to the excitement of microbiology and infectious disease research. Kauffman, a long-term member of the University of Michigan community, recalls that Michele was "the most exciting lecturer that I have worked with in my long tenure at the University of Michigan." Kauffman observed Michele has been "especially proactive in mentoring young women scientists." O'Riordan likewise notes that Michele has "always looked for ways to promote and highlight the work of up-and-coming female scientists." When several graduate students joined her lab as M.D./Ph.D. students, Michele became an active member of the University of Michigan Medical Scientist Training Program (MSTP, i.e., the M.D./Ph.D. program). Reflecting on her roles in the MSTP governing body and admissions committee, Director Ron Koenig singled out Michele's commitment to serve as a "strong advocate for graduate students, and especially for women in science." Currently, Michele directs the Office of Postdoctoral Studies at Michigan. In this capacity, Michele developed a "postdoc preview" program for senior graduate students to visit the Ann Abor campus and identify research labs for carrying out their training. Her tireless work on behalf of postdoctoral researchers at Michigan led the medical school to implement a minimum salary for postdoctoral trainees. Michele's work on behalf of young scientists, both graduate students and postdocs, male and female, has enriched the Michigan experience for many entering the field of microbiology.

Throughout the last 20 years, Michele has also chosen to give back to the broader community by becoming an active participant and leader in the ASM, serving as a journal editor for several years and more recently as a member of its Board of Directors. Last year, she joined Gemma Reguera to coauthor the second edition of ASM's *Microbe* textbook along with the first edition's authors Elio Schaechter and Michigan's Fred Neidhardt, who sadly passed away during completion of the text. She also enjoys cohosting ASM's bimonthly podcast *This Week in Microbiology* (*TWiM*), where cohost Schaechter delights in boasting that listeners can hear Michele's "voice of imagination and wisdom, both delivered with a soft touch." Michele's voice on any given week on *TWiM* is still music to my ears.

For some in academic positions, one's research productivity is the sole delineator of scholarship, with mentoring and service viewed as more ancillary activities. Michele Swanson's group continues to contribute to our understanding of *L. pneumophila* and its exploitation of our immune systems and, more recently, the public health challenges this organism presents to communities at risk. By her varied leadership roles with the University of Michigan system and at the ASM, Michele models a particular example of a successful and rewarding academic career that balances these research activities with mentoring and service. As Vic DiRita eloquently states, "I have been privileged to serve in many capacities alongside Michele, and she has taught me, as she teaches everyone fortunate enough to work with her. Her most impressive lesson, and which she routinely puts into practice, is that leadership—in its highest and best form, the type at which she excels —is service."

ONWARD AND UPWARD

In 2018 Michele Swanson assumes the leadership role of president of the ASM. Like Fred Neidhardt, Elio Schaechter, and other peers of Michele who once held the position, she will direct the organization's ongoing mission to advance the microbial sciences. Michele brings to the position her wisdom, her open-mindedness, and her own confident, thoughtful approach to the post. Title IX's ratification sought to provide women equal opportunities in education. Perhaps, as some believe, its success is mixed, with gender imbalance persisting and female faculty often called upon to perform more service than male peers. Michele Swanson is undeniably a Title IX success story. We can all derive inspiration from Michele, who will continue to lead by performing impactful research, imparting wise counsel to trainees, and providing leadership in our scientific and broader communities. She is indeed a remarkable woman in microbiology.

ACKNOWLEDGMENTS

I thank many people whose assistance was invaluable in writing this reflection. It was a joy corresponding with Michele, who clarified many details. I received helpful advice and suggestions from Tracy Hammer and Mirjana Milosevic-Brockett, who edited the text. I am grateful for the comments, stories, and quotations provided by Mike Bachman, Brenda Byrne, Zach Dalebroux, Vic DiRita, Rachel Edwards, Ralph Isberg, Amrita Joshi, Carol Kauffman, Ron Koenig, Mairi Noverr, Mary O'Riordan, J. D. Sauer, Elio Schaechter, Sam Silverstein, Hannah Swanson, Joel Swanson, Len Swanson, and Fred Winston.

CITATION

Hammer BK. 2018. Michele Swanson: a rewarding career and life in balance, p 269–281. *In* Whitaker RJ, Barton HA (ed), *Women in Microbiology*. American Society for Microbiology, Washington, DC.

References

1. McDade JE, Shepard CC, Fraser DW, Tsai TR, Redus MA, Dowdle WR. 1977. Legionnaires' disease: isolation of a bacterium and demonstration of its role in other respiratory disease. *N Engl J Med* **297**:1197–1203.
2. Vogel JP, Andrews HL, Wong SK, Isberg RR. 1998. Conjugative transfer by the virulence system of *Legionella pneumophila*. *Science* **279**:873–876.
3. Segal G, Purcell M, Shuman HA. 1998. Host cell killing and bacterial conjugation require overlapping sets of genes within a 22-kb region of the *Legionella pneumophila* genome. *Proc Natl Acad Sci USA* **95**:1669–1674.
4. Byrne B, Swanson MS. 1998. Expression of *Legionella pneumophila* virulence traits in response to growth conditions. *Infect Immun* **66**:3029–3034.
5. Hammer BK, Swanson MS. 1999. Co-ordination of *Legionella pneumophila* virulence with entry into stationary phase by ppGpp. *Mol Microbiol* **33**:721–731.

Women in Microbiology
Edited by Rachel J. Whitaker and Hazel A. Barton
© 2018 American Society for Microbiology. All rights reserved.
doi:10.1128/9781555819545.ch31

The Legacy of Patricia Ann Webb: Broken Vials and Urgency

31

May C. Chu[1]

It was 11 October 1976 when an aliquot of blood from a fatal case of unknown hemorrhagic fever in Zaire arrived at the Centers for Disease Control (CDC) Special Pathogens Branch. Patricia Ann Webb, a medical virologist, opened the package and found a mess of broken glass, blood, and cotton packing. Knowing the urgency of what the contents might reveal, Patricia, with little regard to personal protection, retrieved the soaked cotton packing from the box, squeezed out what she could, and inoculated cultured Vero cells in hopes of recovering viable virus. Two days later the Vero cells began to show signs of cytopathology, and Patricia collected drops of the culture fluid for Frederick Murphy to examine by negative-contrast electron microscopy. Fred says that what he saw "raised the hairs on the back of his neck." This is the story of a distinguished public health medical virologist working with some of the most consequential infectious diseases of our time (Fig. 1).

EARLY YEARS

Patricia was born in 1925 in Cambridge, England, where her father was a professor of pathology at the University of Oxford. She and her siblings were evacuated to the United States in 1940 to escape the Blitz. After graduating from Agnes Scott College at age 20, she received her medical doctor degree in 1950 from Tulane University. She completed her pediatric residency in 1953 at the Kern General Hospital in Bakersfield, CA (1). There, Patricia investigated an extensive Western equine encephalitis (WEE) outbreak that occurred in

---

[1]Colorado School of Public Health, Anschutz Medical Center, Aurora, CO 80045

Figure 1 Patricia A. Webb at the CDC, 1976. Courtesy of Frederick A. Murphy.

Kern County and San Joaquin Valley during the summer of 1952 (2). The WEE study was the beginning of a remarkable lifetime of contributions to the body of knowledge on arboviruses and hemorrhagic fever viruses.

Between 1955 and 1961, she was with the U.S. Army Medical Research Unit in Kuala Lumpur, Malaysia, which was a leading study site for tropical disease ecology, known for its scrub typhus, typhoid fever, and respiratory disease studies. She was there with Bennett Elisberg, her first husband, a respected rickettsiologist, and their family. They crossed paths with many of the recognized tropical medicine pioneers, such as Robert Traub, Robert Shope, and C. E. Gordon Smith. Patricia became interested in studying viral fevers of unknown origin and infections of the respiratory tract; the first isolate of the 1957 H2N2 pandemic influenza virus was recovered as part of her studies. Patricia returned to the United States in 1961 and began her career with the U.S. Public Health Service, where she joined Robert Chanock's infectious disease laboratory at the National Institutes of Health (NIH). This is where she met her second husband, Karl Johnson. Over the next 15 years, the two of them went on to make significant discoveries while describing and investigating viral hemorrhagic fevers.

HEMORRHAGIC FEVERS IN SOUTH AMERICA

Patricia joined the NIH's Middle America Research Unit (MARU) in the Panama Canal Zone from 1963 to 1975. Karl had taken a position at MARU and had moved there ahead of Patricia in 1962. This was an exceptional and productive period of virological exploration beginning with the investigation of an outbreak of hemorrhagic fever in northeastern Bolivia. Machupo virus was discovered as the cause of Bolivian hemorrhagic fever (BHF). In May of 1963, Karl, Ronald MacKenzie, and Merl Kuns chartered an old bomber to transport laboratory and field equipment to the remote village of Magdalena near San Joaquin, Bolivia. They had laid out a plan to study the natural history of Machupo virus. By July 4, Karl had to organize the return to Panama of his ill companions, Ron and Merl, at which point Karl also fell ill; they all had come down with BHF (and later recovered). Patricia, who was still at NIH, flew from Bethesda to care for Karl and also contracted BHF (3). Thus, this series of studies was not without personal sacrifice, but the team succeeded in describing how Machupo virus was transmitted and identified *Calomys callosus*, a peridomestic mouse, as the reservoir and vector (4). At MARU, Patricia ran a strictly controlled laboratory operation, as working with Machupo virus was risky; only persons immune to the virus could work alongside her in the special suite. She devised experimental techniques to refine the antigenic grouping of Machupo and related viruses, evaluated and characterized typing sera for serological tests, and defined the morphological and viral replication of the viruses in tissue culture. A critical success was the breeding of a captive colony of *C. callosus*, which then allowed the team to study closely the interactions of the virus with its rodent host, laying out the pathogenic mechanisms of chronic infection and excretion of Machupo virus by the rodents as well as ascertaining the effect of tolerance to the virus on reproduction of the rodents (5). Their findings and study methods attracted other researchers to join them in the effort. This collaboration led to definitive studies that mapped the antigenic relationship of the related Amapari, Junin, and Tacaribe complex viruses and their rodent and bat hosts (6–8). With the help of Fred Murphy, who provided the electron micrographs, and other collaborators who provided evidence of the antigenic relatedness of the Central and South American viruses to lymphocytic choriomeningitis virus and Lassa virus, the group identified a new genus, *Arenavirus* (9). Arenaviruses now comprise more than 30 members and are mostly identified with their host rodents or bats. Some of these infect humans, including in laboratory-acquired infections (http://www.cdc.gov/vhf/virus-families/arenaviridae.html).

Living in the Panama Canal area was a special experience. Patricia and Karl welcomed staff, locals, and visitors to their home; their parties were, as recounted by those who were there, memorable. Patricia had a deep love for animals, especially Labrador retrievers, which were always with her. Thomas Walton recounts how his family was gifted with a puppy. In the evening the dogs would have a group run in the community, which was surely very rambunctious and lively. Tom and Patricia's friendship and trust were tested when Patricia asked Tom, as the MARU veterinarian and assisted by C. J. Peters, a physician, to spay her beloved dog Cricket, a procedure that turned out to be more difficult than expected. The pressure with Patricia on the other side of the surgical room door made it quite tense. Fortunately, Cricket recovered from the ordeal, as did Tom and C. J. MARU was closed in 1975 and Patricia and Karl moved to Atlanta, GA, to join the CDC Special Pathogens Branch.

EBOLA HEMORRHAGIC FEVER

Fred saw that the cell culture sample that Patricia had given him contained a virus that was nearly identical to the 1967 Marburg virus. Karl and Patricia were called in to have a look, and they were all spellbound and electrified (Fig. 2). At this point, the Tropical Diseases Institute in Antwerp, Belgium, and the Microbiological Research Establishment in Porton Down, United Kingdom, already were involved in analyzing samples from the Zaire outbreak and from a concurrent outbreak in Sudan. The Belgian and the British scientists knew that both the Zaire and Sudan samples contained a Marburg-like agent, but it was Patricia's immunofluorescence test results that confirmed that the virus was a new discovery (10). The naming rights to this new virus fell to the CDC scientists and, with the blessing of the Belgian and British colleagues, it was named Ebola, for a river near the origin of the Zaire outbreak (10). The CDC laboratory tests showed the close relatedness of the Zaire and Sudan isolates, which were different from Marburg virus; up until then, Marburg virus had no known virus relatives. As of 2016, *Marburgvirus* and *Ebolavirus* are classified in *Filoviridae*, with a recent addition of a bat virus genus, *Cuevavirus*.

The natural habitat and ecology of Ebola virus remain elusive 40 years after its discovery. This is crucial information if we are to understand the factors that allowed it to spill over into humans, as it did explosively in 2014 in West Africa. The Special Pathogens Branch laboratory developed reagents and tools to detect Ebola virus and antibodies in wild animals, insects, and birds. This was done in collaboration with scientists from all

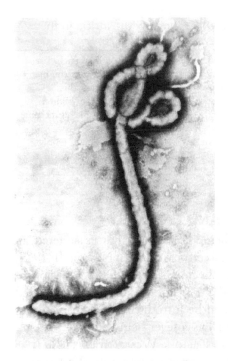

Figure 2 Ebola virus as it appears by electron micrograph, 1976. Courtesy of Frederick A. Murphy.

over the world who would come to Atlanta to learn the techniques. Oyewale Tomori, a visiting professor then but now the president of the Nigeria Academy of Sciences, shares his account of what it was like to work with Patricia in Box 1.

When samples from the Cameroon pygmy population showed 17% antibody reactivity to Ebola virus (from unpublished notes by David Heymann), Patricia mounted a campaign to visit the source site. David organized a field investigation in February of 1980 in the Mbatika-Sesse village in southeast Cameroon, with the goals of identifying what animals the pygmies hunted and ate and studying their lifestyle for any clues to Ebola virus's natural habitat. John Krebs, a mammologist on the team, remembers that Patricia was not amused when the tribal chief prohibited women to hunt. She gamely, as noted in her own words, stayed behind to observe tribal social interactions and attempted to catch fish, which was women's work, while fighting off loa-loa flies and fleas. According to Patricia's trip report, the team examined 32 species, mostly rodents, with the only positive Ebola virus antibody found in a flying squirrel and confirmation of 4 pairs of human serum samples showing conversion to Ebola virus antigens. This is illustrative of the difficulty of many exploratory efforts to identify the natural life cycle of this virus (11–13).

Box 1 Recollection from Oyewale Tomori, who trained with Patricia in the CDC Special Pathogens Laboratory

I remember Patricia Webb as a stickler for order in the lab, and neatness on the bench. I see her now, checking my table as I round up. "Wally, clear the table! Wally, leave it as you met it!" Indeed that was what many remember, but outside the lab, Patricia would ask this Nigerian man, "Are you hearing from your family back in Nigeria? How are your children doing?" Patricia was one for safety and orderliness in the laboratory, reminding me all the time to be careful—it is easy to get infected in the lab.

Patricia was special; she cared about people in her own way, a tough "disciplinarian"—only tough because she does not want you to make the last and fatal mistake. She has escaped many: "You may not be as lucky as I am, and we still want you around." This is Patricia for you.

LASSA FEVER EPIDEMIOLOGY AND ECOLOGY

Lassa fever was described in 1969, when 3 missionary nurses suffered an acute and severe febrile illness in a hospital setting in Lassa, Nigeria. In 1972, Thomas Monath, later to become the director of the CDC's Division of Vector-Borne Infectious Diseases (DVBID) in Fort Collins, CO, described Lassa fever as a serious disease that could be introduced into hospital settings, but he was more concerned about Lassa fever being an endemic and recurring illness in affected communities across West Africa (14). It is estimated that up to 500,000 cases of Lassa fever occur across West Africa, with about 5,000 deaths each year (http://www.cdc.gov/vhf/lassa/index.html). Like its New World arenavirus relatives, Lassa virus is carried and shed by *Mastomys natalensis*, a ubiquitous peridomestic rodent found in Southern and Western Africa. These findings led the CDC to establish a long-term Lassa fever project (1976 to 2000) in Kenema, Sierra Leone (15). Joseph McCormick was its first director, and he set up an ambitious program to elucidate the pathological, virological, and ecological aspects of this widely spread disease in West Africa. Patricia joined the Lassa fever project first as a medical virologist and took over as the director from 1978 to 1981. During that period, Patricia worked with many colleagues, including Joe, Robert Craven, David Morens, John Krebs, and a roster of short-term experts and Sierra Leonean staff (16–19).

The most poignant tributes about Patricia come from the Sierra Leonean staff, a number of whom returned to Kenema to work at the hospital after the cessation of the internal strife in 2004 and are still engaged in Lassa fever surveillance and ecological studies there. They are part of the team that diagnosed the first case of Ebola virus infection in Sierra Leone in 2014 (12). James Massally, Kandeh Kargbo, and James Koninga, who worked with John Krebs, give the account of Patricia with her 3 dogs (Limpopo, Radcliff,

Figure 3 Patricia A. Webb collecting *Mastomys natalensis* samples in Sierra Leone, ca. 1978. Courtesy of Joseph B. McCormick.

and Beans) and cat (Tigger) living in Kenema in 1978. She would visit them at the field sites where they had mapped out grids to capture and recapture rodents (Fig. 3) in Panguma, Konia Kpindima, and Yengema. They remember that Patricia would have a staff dinner party every 3 months without fail. Austin Demby (Box 2) pays a special tribute to Patricia, for she mentored him and created opportunities for him to develop his career. John Krebs recounts how Patricia would play golf wherever they were, and in Sierra Leone, that meant mostly playing on sand-based greens. At the end of her assignment in 1981, Patricia, menagerie and all, moved to Fort Collins to join Tom Monath at DVBID. The fact that Lassa fever cases were being reported to the World Health Organization (WHO) in 2016 (http://www.who.int/mediacentre/factsheets/fs179/en) is indicative that the virus's prevalence and transmission to humans remain a public health issue.

EPIZOOTIC VSV

The Colorado Front Range was struck in 1982 by a major epizootic of vesicular stomatitis virus (VSV) disease, affecting livestock, horses, wild animals, and humans. This outbreak provided an opportunity to conduct

Box 2 Recollections from Austin Demby, who started his career with Patricia

I first met Dr. Patricia Webb in July of 1980. I had just graduated from the University of Sierra Leone with an honors degree in zoology and was wondering what I should do next. I went back home to Kenema with my undergraduate thesis in hand, looking for a job. My first and only stop was at the CDC Lassa fever field station, and without an appointment I walked into the office and interviewed with her. A week later, she offered me a job with CDC. For the next 25 years we were never more than a phone call away. Thirty-five years later, I am still employed by the same institution, now as the director of the PEPFAR program for the Department of Health and Human Services.

Patricia taught me the fundamentals of critical thinking and problem solving. She taught me work ethics and gave me confidence to pursue knowledge that would help humanity. She would tell me that my place of birth in the small and dusty little town of Kenema should be a source of pride and not an impediment for standing tall among international peers. I recruited and led several Sierra Leonean budding scientists like Augustine Goba, James Massally, and Paul Kamara. She introduced me to famous virologists including Karl Johnson, Joe McCormick, Tom Monath, Fred Murphy, and others.

I spent many a Christmas holiday with Patricia and her dogs on three different continents. On every occasion we started the holidays by working a complex puzzle; by the end of the holiday, we never had an incomplete puzzle.

I always think of Patricia and her counsel to be "diligent and steadfast in the service of humanity in any and every way you can."

studies to understand where and how such epizootics emerge every 10 to 15 years. This investigation reunited Tom Walton with Tom Monath and Patricia, along with arbovirus ecologists, wildlife specialists, and veterinarians from the DVBID, Colorado State University, and the U.S. Department of Agriculture to pool their knowledge and skills to pursue the answers. Patricia and her graduate student Suzanne Vernon developed an improved serological test, a much-needed method to distinguish VSV antibodies in livestock (20). This was a multifaceted effort that required energetic and tenacious fieldwork to pull together the many threads of the VSV epizootic picture (21–23). Most remember Patricia as a tough—even relentless—taskmaster, but fun to be with. Suzanne would comment that working with her was not for the "faint of heart," but all appreciated the experience and remember it fondly.

HUMAN IMMUNODEFICIENCY VIRUS (HIV) TRANSMISSION BY ARTHROPODS

Between 1982 and 1987, Belle Glade, FL, an agricultural community located about 45 miles west of Palm Beach, reported a particularly high incidence of AIDS (24). The higher rate of cases (8%, compared to the 3% reported nationally) with no explainable source of infection led to public conjecture that mosquitoes and bed bugs might have been transmitting the virus, despite evidence that AIDS was not infecting all age groups. In the fall of 1985, hysteria reigned in the press and Belle Glade was labeled the "AIDS capital of the world." Decades later, the stigma still remains (http://www. palmbeachpost.com/news/news/belle-glade-the-town-that-fear-could-not-destroy/nLnjh/); people recall vividly that sports teams coming to Belle Glade would avoid touching residents, especially shaking hands or taking offered snacks.

Harold Jaffe, Kenneth Castro, and Tom Monath at the CDC felt that a definitive, rigorous vector competency study was needed. Tom turned to Patricia, and she volunteered to assemble a skilled team to address this at the Fort Collins laboratory and insectary. This was not an easy undertaking, since it combined propagating and purifying high doses of infective HIV with the risky processes of preparing HIV-spiked blood feedings and injecting abdomens of *Cimex hemipterus* (bed bug) and *Toxorhynchites amboinensis* (mosquito), followed by harvesting, dissecting, pooling, and testing of the respective samples for evidence of residual and replicating HIV in insect tissue culture and by molecular detection methods. This study in Fort Collins was conducted with trepidation and with some DVBID staff disquiet, so the team had to take extra control measures for safety and security in the basement laboratory and insectary rooms. The study team members, Christine Sykes, Gary Maupin, Barbara Johnson, and Chin Yin Ou, understood the importance of the study and knew of the risks and expectations. It was extremely stressful work. They volunteered because with their skills and Patricia's leadership, they had confidence that this could be done. The results of the study convinced scientists and the public that the risk of insect transmission of HIV was extremely low or nonexistent (25). Since its publication, vector-borne AIDS has not been an issue or of "headlines" public concern.

PERSONAL REMEMBRANCE OF PATRICIA

Patricia was a mentor and a friend to me. At first she intimidated me: she had doubts about someone who wanted to study dengue virus molecular epidemiology without the benefit of field experience. We would take walks with the dogs, for example, with the famous Cricket (Fig. 4), and I got to enjoy her hospitality (at her curry parties). I learned from her the nuances and the fine art of serological cross-typing and virus plaque assays. She came around to appreciate the results derived from virus sequences that showed genetic relatedness as well as the PCR's utility to detect virus in field samples. She awarded Barb Johnson, Vincent Deubel, and me buttons that had

Figure 4 Patricia A. Webb and Cricket in Rist Canyon, Colorado, ca. 1985. Courtesy of Christine Sykes.

3-prime ends (of sows), and she gave a 5-prime-end button to Dennis Trent, our laboratory chief. I really felt her presence when I was assigned by the WHO to restart the reference Lassa fever laboratory on the Kenema General Hospital grounds after a devastating nosocomial Lassa fever outbreak in 2004. Some of the field surveillance team that had been hired to work for the CDC Lassa project in 1978 offered to work in the new Lassa laboratory; the first thing they asked was how Patricia, Bob Craven, and John Krebs were. I was touched by this gesture; such was the impact she had on those she worked with, decades later. Beyond these personal interactions, Patricia's legacy of studying hemorrhagic fevers will last for generations to come. I took a picture of the team in front of the new Lassa laboratory and sent that to Patricia in January 2005 in California, where she had moved to be with family. She passed away only a few weeks later.

ACKNOWLEDGMENTS
It took a village to pull all the pieces of this picture of Patricia together, and it was possible only because so many kindly took the time to provide me with their recollections, photographs, and unpublished memoirs. I especially thank Peter Johnson and Michael Webb for sharing their mother's letters, notes, and

trip reports. Using Patricia's own words allowed me to connect all the stories. I extend much appreciation and gratitude to Joel Breman, Dick Bowen, Austin Demby, Charlie Calisher, Cookie Cloninger, Augustine Goba, David Heymann, Harold Jaffe, Barbara Johnson, Karl Johnson, Kandeh Kargbo, James Koninga, John Krebs, Jimmy Massally, Joe McCormick, Tom Monath, Fred Murphy, Stuart Nichol, Christine Sykes, Oyewale Tomori, Suzanne Vernon, and Tom Walton; you all made this happen.

CITATION

Chu MC. 2018. The legacy of Patricia Ann Webb: broken vials and urgency, p 283–294. *In* Whitaker RJ, Barton HA (ed), *Women in Microbiology*. American Society for Microbiology, Washington, DC.

References

1. Monath TP, Johnson KM. 2005. In memoriam. Patricia Ann Webb (1925–2005). *Arch Virol* 150:1268–1270.
2. Cohen R, O'Connor R, Townsend TE, Webb PA, McKey RW. 1953. Western equine encephalomyelitis; clinical observations in infants and children. *J Pediatr* 43: 26–34.
3. Douglas RG Jr, Wiebenga NH, Couch RB. 1965. Bolivian hemorrhagic fever probably transmitted by personal contact. *Am J Epidemiol* 82:85–91.
4. Johnson KM, Kuns ML, Mackenzie RB, Webb PA, Yunker CE. 1966. Isolation of Machupo virus from wild rodent *Calomys callosus*. *Am J Trop Med Hyg* 15:103–106.
5. Justines G, Johnson KM. 1970. Observations on the laboratory breeding of the cricetine rodent *Calomys callosus*. *Lab Anim Care* 20:57–60.
6. Webb PA, Johnson KM, Mackenzie RB. 1969. The measurement of specific antibodies in Bolivian hemorrhagic fever by neutralization of virus plaques. *Proc Soc Exp Biol Med* 130:1013–1019.
7. Murphy FA, Webb PA, Johnson KM, Whitfield SG. 1969. Morphological comparison of Machupo with lymphocytic choriomeningitis virus: basis for a new taxonomic group. *J Virol* 4:535–541.
8. Rowe WP, Pugh WE, Webb PA, Peters CJ. 1970. Serological relationship of the Tacaribe complex of viruses to lymphocytic choriomeningitis virus. *J Virol* 5:289–292.
9. Rowe WP, Murphy FA, Bergold GH, Casals J, Hotchin J, Johnson KM, Lehmann-Grube F, Mims CA, Traub E, Webb PA. 1970. Arenoviruses: proposed name for a newly defined virus group. *J Virol* 5:651–652.
10. Johnson KM, Lange JV, Webb PA, Murphy FA. 1977. Isolation and partial characterisation of a new virus causing acute haemorrhagic fever in Zaire. *Lancet* i:569–571.
11. Breman JG, Johnson KM, van der Groen G, Robbins CB, Szczeniowski MV, Ruti K, Webb PA, Meier F, Heymann DL, Ebola Virus Study Teams. 1999. A search for Ebola virus in animals in the Democratic Republic of the Congo and Cameroon: ecologic, virologic, and serologic surveys, 1979–1980. *J Infect Dis* 179(Suppl 1):S139–S147.

12. Wauquier N, Bangura J, Moses L, Humarr Khan S, Coomber M, Lungay V, Gbakie M, Sesay MS, Gassama IA, Massally JL, Gbakima A, Squire J, Lamin M, Kanneh L, Yillah M, Kargbo K, Roberts W, Vandi M, Kargbo D, Vincent T, Jambai A, Guttieri M, Fair J, Souris M, Gonzalez JP. 2015. Understanding the emergence of Ebola virus disease in Sierra Leone: stalking the virus in the threatening wake of emergence. *PLoS Curr* 2015(7):ecurrents.outbreaks.9a6530ab7bb9096b34143230ab01cdef.

13. Marí Saéz A, Weiss S, Nowak K, Lapeyre V, Zimmermann F, Düx A, Kühl HS, Kaba M, Regnaut S, Merkel K, Sachse A, Thiesen U, Villányi L, Boesch C, Dabrowski PW, Radonić A, Nitsche A, Leendertz SA, Petterson S, Becker S, Krähling V, Couacy-Hymann E, Akoua-Koffi C, Weber N, Schaade L, Fahr J, Borchert M, Gogarten JF, Calvignac-Spencer S, Leendertz FH. 2015. Investigating the zoonotic origin of the West African Ebola epidemic. *EMBO Mol Med* 7:17–23.

14. Monath TP. 1973. Lassa fever. *Trop Doct* 3:155–161.

15. McCormick JB. 1996. *Level 4: Virus Hunters of the CDC*, p 32–39. Turner Publishing, Inc, Atlanta, GA.

16. Keenlyside RA, McCormick JB, Webb PA, Smith E, Elliott L, Johnson KM. 1983. Case-control study of *Mastomys natalensis* and humans in Lassa virus-infected households in Sierra Leone. *Am J Trop Med Hyg* 32:829–837.

17. Webb PA, McCormick JB, King IJ, Bosman I, Johnson KM, Elliott LH, Kono GK, O'Sullivan R. 1986. Lassa fever in children in Sierra Leone, West Africa. *Trans R Soc Trop Med Hyg* 80:577–582.

18. Helmick CG, Webb PA, Scribner CL, Krebs JW, McCormick JB. 1986. No evidence for increased risk of Lassa fever infection in hospital staff. *Lancet* ii:1202–1205.

19. McCormick JB, Webb PA, Krebs JW, Johnson KM, Smith ES. 1987. A prospective study of the epidemiology and ecology of Lassa fever. *J Infect Dis* 155:437–444.

20. Vernon SD, Webb PA. 1985. Recent vesicular stomatitis virus infection detected by immunoglobulin M antibody capture enzyme-linked immunosorbent assay. *J Clin Microbiol* 22:582–586.

21. Walton TE, Webb PA, Kramer WL, Smith GC, Davis T, Holbrook FR, Moore CG, Schiefer TJ, Jones RH, Janney GC, Janney GC. 1987. Epizootic vesicular stomatitis in Colorado, 1982: epidemiologic and entomologic studies. *Am J Trop Med Hyg* 36: 166–176.

22. Webb PA, Monath TP, Reif JS, Smith GC, Kemp GE, Lazuick JS, Walton TE. 1987. Epizootic vesicular stomatitis in Colorado, 1982: epidemiologic studies along the northern Colorado front range. *Am J Trop Med Hyg* 36:183–188.

23. Webb PA, McLean RG, Smith GC, Ellenberger JH, Francy DB, Walton TE, Monath TP. 1987. Epizootic vesicular stomatitis in Colorado, 1982: some observations on the possible role of wildlife populations in an enzootic maintenance cycle. *J Wildl Dis* 23:192–198.

24. Castro KG, Lieb S, Jaffe HW, Narkunas JP, Calisher CH, Bush TJ, Witte JJ. 1988. Transmission of HIV in Belle Glade, Florida: lessons for other communities in the United States. *Science* 239:193–197.

25. Webb PA, Happ CM, Maupin GO, Johnson BJB, Ou CY, Monath TP. 1989. Potential for insect transmission of HIV: experimental exposure of *Cimex hemipterus* and *Toxorhynchites amboinensis* to human immunodeficiency virus. *J Infect Dis* 160:970–977.

Women in Microbiology
Edited by Rachel J. Whitaker and Hazel A. Barton
© 2018 American Society for Microbiology. All rights reserved.
doi:10.1128/9781555819545.ch32

Donna M. Wolk: It's Never Too Late To Bloom

32

Natalie N. Whitfield[1]

As I sat in the hotel lobby waiting to meet with Dr. Donna Wolk, I could not fathom how my life was about to change. I had inquired about potential postdoctoral opportunities in her research laboratory when she suggested that I meet her for lunch. Donna was giving a "Careers in Clinical Microbiology" lecture at the National Institutes of Health (NIH)-funded Kadner Institute held at Michigan State University. Little did I know that during that lunch I would receive my very own personal workshop session on the topic, having little prior knowledge about this career path. I quickly learned that this exemplified Donna's generosity—to give of her personal time, especially when it's an opportunity to advance the career of another woman in science. Though we sat and discussed the focus of her research laboratory and why a traditional research postdoctoral fellowship may not be applicable to her laboratory, I was more intrigued with her journey in clinical microbiology and its relationship to her research emphasis.

Donna did not set out to be a "rock star" clinical microbiologist, her humble beginnings likely not predicting the successes she would have in her future. She grew up in a rural town of less than 500 people in northeastern Pennsylvania. A first-generation college student, she will tell you that "[she] didn't know what broccoli was…" until setting foot on the Penn State campus, where she began her career in the sciences, graduating with a bachelor's in science in microbiology and clinical laboratory sciences. She obtained her medical technology certification, becoming a medical technologist at Geisinger Medical Center in Danville, PA, where she had

---

[1]Clinical Services Laboratory, OpGen, Inc., Gaithersburg, MD 20878

performed her clinical internship. Donna quickly moved up the ladder, becoming the manager of clinical microbiology. Rather content, she had successfully completed her degree and had a respectable, satisfying career. However, one day, she was told by her superior that though she was the best of all the supervisors in the laboratory and, in fact, the only female supervisor, her salary was significantly less than those of her male counterparts. He told her, "...if you go back to school and get your master's degree, I'll correct your wages." Though she had vowed that she would never go back to school ("I could not understand why anyone would ever want to go back...!"), she decided it was her only option to rectify the wage disparity, and so she pursued her master's degree in health administration at Wilkes University. Her successful completion of this degree bumped up her salary and provided her with further knowledge and skills for her management position. She persisted in her role as manager of the clinical microbiology laboratory until another path presented itself almost 14 years later.

Though not planned for, Donna became the single mother of a 4-year-old. Knowing that microbiology was her core skill set, she made the choice to once again do that which she said she would never do, go back to school to pursue her Ph.D. In a clinical laboratory, the next attainable management level from a manager is either the laboratory's administrative director or the specialized technical directors, which are medical doctors (M.D.) or Ph.D. scientists. Having a specialty in microbiology and over 10 years' experience as a bench technologist, the path to become the technical director seemed like the next logical step. Although the motivation was partially financial, her hope to provide a better life for her daughter was the true motivation. She subsequently applied for doctoral degree programs and was accepted to the University of Arizona in Tucson. Donna knew she faced two major challenges: (i) to return to school at the age of 36 and (ii) to uproot her daughter and move so far away from her family. However, she pushed on and packed up her life in a U-Haul, and they made their way across the country to Tucson.

At the University of Arizona, Donna pursued her doctoral degree with a focus in medical parasitology, establishing chlorine treatment as an effective water treatment for *Encephalitozoon* syn. *Septata intestinalis*, a group of pathogens with the potential for waterborne transmission, while also showing how spectrophotometric methods can be substituted for labor-intensive hemacytometer counting methods in laboratory-based chlorine disinfection studies (1, 2). During this time, Donna developed lifelong friendships and a love for Arizona while raising her daughter, but as she grew close to the completion of her doctorate, her passion for medical microbiology swayed her towards applying for a clinical microbiology fellowship. This did not come without naysayers, faculty members who couldn't understand why she would want to waste her time with applied sciences when she was "way too smart for that!" Fortunately, Donna is not one to let the opinions of others influence her decisions, so upon completing her doctoral degree, she once again packed up to head north to complete a clinical microbiology fellowship at Mayo Clinic, Rochester, MN, in the Division of Clinical and Molecular Microbiology. Here, Donna would begin to hone her skills as a director-level clinical microbiologist, increasing her understanding of the importance of diagnostic microbiology and the role that clinical microbiologists play in the full spectrum of patient care. During her fellowship, she published journal articles on bacterial sepsis and viral hepatitis while also developing a laboratory real-time PCR assay for the detection of *E. intestinalis* from stool specimens, essentially bringing her graduate work full circle with her clinical microbiology fellowship (2–4).

Upon completion of the fellowship, she took a position back in Arizona as the Director of the Molecular Diagnostics & Research Laboratory of the Southern Arizona VA Health Care System (SAVAHCS); while in this position, she established the Molecular Genetics Pathology Fellowship Program, the 12th accredited in the United States by the Accreditation Council for Graduate Medical Education. As Donna expanded the clinical microbiology services to include molecular diagnostics at the SAVAHCS, she also served as a consultant jointly providing clinical microbiology services as an assistant professor in the Pathology Department of the University of Arizona. Shortly thereafter, Donna became the full-time Division Chief of Clinical and Molecular Microbiology at the University Medical Center.

Donna's desire to perform translational research and her vision of a Clinical Laboratory Improvement Amendments of 1988 (CLIA)-compliant research laboratory to conduct clinical trials for novel infectious disease molecular diagnostics seeking FDA approval were not accomplished without hard work and sacrifices. She built a translational clinical microbiology research program from the ground up, receiving minimal financial or laboratory infrastructure support from her pathologist-dominated, primary department, which did not share her vision. However, she persisted, obtaining affiliations in other departments in the university system and securing research laboratory space to create the Infectious Diseases Research Core (IDRC) in the state-of-the-art, interdisciplinary Bio5 Institute. The aim of the IDRC was translational research with a focus on molecular diagnosis, epidemiology, and pathogenesis of infectious disease/public health threats, specifically sepsis and drug-resistant microbes. While in operation, the IDRC contributed to the advancement and availability of several FDA-cleared molecular diagnostics for infectious diseases that are widely utilized in clinical microbiology laboratories today (5–11).

In 2013, Donna was offered the opportunity to return to her roots at Geisinger Medical Center as the Director of Clinical Microbiology. She faced a difficult dilemma, once again deciding to do something else she said she would never do, which was move back to Pennsylvania. There were many significant benefits to accepting the position, both professionally and personally, but she felt a deep responsibility to those in her Arizona laboratory, so she labored over the decision. Ultimately, the opportunity won: Donna returned to Geisinger and reestablished an infectious diseases translational research laboratory with the vision of "advancing diagnostics…saving lives" by using a bench-to-bedside approach to assess the outcomes of implementing novel diagnostics in the clinical microbiology laboratory.

Donna's contributions to the field of clinical microbiology are numerous. Her work has helped change the understanding of broad-range antimicrobial therapy through the use of advanced rapid diagnostics for cases of bacteremia in emergency departments (12). She has also played a leading role in introducing mass spectrometry as a diagnostic tool into clinical laboratories for the identification of bacteria and yeasts and the differentiation of antimicrobial resistance markers (6, 8, 13), setting the foundation for our clinical laboratory to become one of the first laboratories in the country to use matrix-assisted laser desorption ionization–time of flight mass spectrometry for routine identification. As one of seven national leaders to introduce evidence-based outcome studies and systematic reviews, Donna endorses the importance of outcomes studies to clinical microbiology laboratories as part of the Centers for Disease Control and Prevention's (CDC's) Laboratory Medicine Best Practices Initiative (LMBP). The first LMBP comprehensive systematic review specifically evaluated the evidence for the effectiveness of three rapid diagnostic practices in decreasing the time to targeted therapy for hospitalized patients with bloodstream infections (14). Donna's belief is that through these outcomes studies the true impact of the work that clinical laboratory scientists do daily and the vital role they play in health care can be recognized.

Donna's work that helped change the face of diagnostic testing for hospital-acquired infections in laboratories was recognized by the 1885 Society at the University of Arizona with the Distinguished Scholar Award for Excellence in Research, Service, and Education. The award, created in 2012, acknowledges outstanding mid-career faculty whose leadership, research, scholarship, and creative contributions promise to catapult their disciplines to new levels of innovation. In addition, she was recognized with the Becton Dickinson Award for Research in Clinical Microbiology. The American Society for Microbiology (ASM) awarded this national honor to acknowledge Donna's translational research, which complies with the FDA's good laboratory practice regulations and serves as a model for quality practices in molecular diagnostic microbiology and clinical/translational research. Donna refers to her work as "diagnostic intervention," or the development, implementation, integration, and evidence-based evaluation of infectious disease diagnostics that are aimed to improve patient outcomes, reduce health care costs, and improve patient satisfaction. In a recent ASM Speakers Bureau, Donna spoke of the importance of this approach: "...There's nothing more satisfying...in the workplace...than seeing that an intervention you have done saves someone's life."

Donna truly is a force of nature and one of the hardest-working people I know, her successes all well deserved. However, to not include the impact she has had on the success of others, besides me, would be leaving out a huge part of Donna's story. At the University of Arizona, she revived the Diagnostic Laboratory track of the Professional Science Master's Program for students interested in careers in the diagnostic industry. She was instrumental in establishing relationships with industry partners to advise the program on the employment needs of diagnostic companies and encouraged these partners to provide internship opportunities. She contributed to the success of many students throughout her tenure as the director of the program, establishing a strong foundation that continues to make it one of the most popular tracks in the program today (Fig. 1) (15). Donna also maintained a welcoming clinical laboratory environment for clinical laboratory science interns from the University of North Dakota to address the shortage of medical technologists in microbiology, while creating a readily trained and available workforce for our own laboratory. Donna encouraged the medical technologists in the laboratory to seek certification, making textbooks available and holding study sessions on her personal time for those preparing for the examination. Laurel Burnham, a medical technologist mentee, said of Donna, "Dr. Wolk has a perspective on laboratory operations that is unique in my experience as a microbiologist. She finds a way to balance staying on the forefront of technology while still keeping the patient first. Dr. Wolk always found ways to make the system work for her and her department.

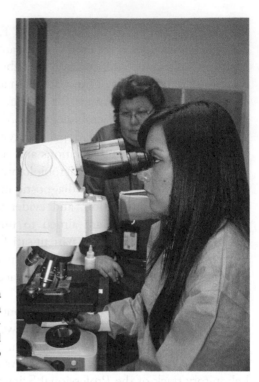

FIGURE 1 Donna Wolk observes as a laboratory scientist, Candi Grado, uses a microscope to read a fluorescent peptide nucleic acid FISH test. Reprinted with permission from UANews. Photo credit: Beatriz Verdugo/UANews.

Getting to be a part of that has helped me tremendously being the supervisor of a microbiology laboratory to bring in new technology and offer better care to the patients we serve. She always encouraged me to be a better microbiologist: to always ask the hard questions and think about the impact our decisions make. Her insight on interpersonal relationships and searching out the people who can help you grow was so important in my career decisions. Her own desire to keep growing in her own career inspired me to see that change is a good thing and expanding your horizons is an important part of 'growing up.' Having her as a mentor so young in my microbiology career helped me learn how to be challenged, to try harder, to think smarter, and find a career path that would fit with my own personal goals."

The level of respect, guidance, and promotion of her colleagues really speaks volumes about Donna and her genuine commitment to the success of others. In working with Donna, I was *always* her associate, never her subordinate, which empowered me in my role and in our working relationship. Donna strategically collaborates with women microbiologists on research projects, clinical trials, committees, and publications. If you happen to be walking through a conference with Donna and she is stopped by another colleague, she will make sure to introduce you, ensuring that she extends her professional network to her mentees and colleagues. Raquel Martinez, Donna's associate microbiologist at Geisinger, says, "Donna is one of my favorite people and I will always be grateful to her for all that she has done for me, both professionally and personally. She is not only my mentor and colleague; she is my friend and role model. I often ask myself, 'WWDWD: what would Donna Wolk do?' I am honored to have been taught by such a smart, knowledgeable, driven, thoughtful, good-natured microbiologist who has a passion for providing the best in quality patient care and for advancing women in clinical microbiology."

When I had lunch with Donna in 2010, I was somewhat desperate, working 7 days a week as a single mother. I needed a change. Though I had been strongly opposed to a postdoctoral fellowship, I would reluctantly take one if it meant that I had some balance in my life. When Donna and I met, she quickly dismissed the idea of a research postdoctoral or research specialist position, not because she questioned my scientific abilities but because she was able to offer me something better. At the time, the starting annual NIH postdoctoral salary was $37,740, which most institutions follow in setting their postdoctoral salaries (16). She refused to offer me a job that paid a postdoctoral salary because "we are worth more than that." Instead, she presented an alternative. The manager position of her clinical microbiology laboratory was vacant, and she knew that I had managed a research laboratory's operations. So she suggested that I apply for that vacant position. "If you get the manager position, I can teach and mentor you for the boards in medical microbiology, and by the way the manager salary is better than a post-doc!" It was unbelievable that after sitting with me for an hour, she was willing to invest her time and effort for the required three year mentorship.

In the weeks following our meeting, Donna made herself available for questions and interview prep. I grew to know Donna, where she came from and who she is, and I am not surprised by her willingness and availability to help. After my interview, Donna drove me around Tucson on a housing tour, providing data on which schools would be best for my child. Upon my arrival in Tucson, Donna supported me both professionally and personally. Knowing the difficulties I faced being in a new place, venturing on a new career path and with a child, she truly reminded me of the value of having a community by becoming an instrumental part of my community. Whatever the emergency, Donna was there to help. When I needed to travel, Donna opened her home to my child. My interactions with Donna have taught me that she lives by two very simple principles: (i) do the right thing and (ii) treat people the way you want to be treated. Putting it simplistically, during our first meeting and always, Donna held to these principles.

But mine isn't a remarkable story; instead, it involves a remarkable woman. Beth Marlowe, a fellow clinical microbiologist, tells of her first "Donna encounter" and friendship.

> Donna Wolk is one of the bravest women I have ever known.... We met the day I walked in late to our graduate microbial genetics class. I sat next to her. She handed me her notes and told me to just give them back to her next class. I almost fell out of my chair. This class was full of medical school wannabes who would undermine you to get ahead. The result of that day was a friendship that has resulted in a sisterhood. Four years later there were three CPEP fellowships available in the U.S. and two of the matches to those available programs came out of the University of Arizona. If not for Donna's advice and guidance in those years I would likely have ended up on a very different career path.
>
> I have learned many things from Donna and most of it has nothing to do with microbiology. Donna always has this way of putting things in perspective. To this day I call her when I need a friend who can honestly tell me what I need to hear. One day driving into work at the very large clinical laboratory I directed...feeling completely inadequate [to] the amount of work and responsibility, I called Donna. She let me know she has days like that too and it was OK. She also reminded me to never forget the golden rule, "Never take yourself too seriously. And on days when you feel really overwhelmed, always get some sleep, because things just don't seem as big after you have had some sleep." Much of this advice I find myself quoting to others and myself when a reminder is needed.
>
> Recently I was talking with a friend who, in their early forties, was beginning to ask...in the midst of a very successful career what they really want to do. I reminded them about my dear friend Donna, whose 40th birthday we celebrated in graduate school, and it wasn't until after her two-year fellowship at Mayo Clinic that she really made a name for herself in her career. I look at her and still don't know how she did it while raising a young daughter, who I might add has grown into one of the most impressive young women you could meet.
>
> I look back at Donna and have to smile. Sometimes in life we get fortunate enough to be surrounded by quality people who continue to add value to our lives. My only regret is I don't get to spend enough time with her. As Donna would say, find your work family, treat them well and keep them close.

Through her comedic stories of her car catching fire on her way to give a talk, or of toting her laptop along to basketball tournaments to get work done while cheering her daughter on, I got to know Donna and at the same time gained an excellent role model. Her stories of how she balanced her work with her responsibilities as a parent provided the hope and security I needed to see my challenges through. Her stories are the ones that let women know that it can be difficult at times, everyone is not always nice, you will make sacrifices to have this career, and giving up may sometimes feel like the only option, but if we do it together, it can be done. Today, I can think back to that initial meeting and easily name many of the countless little things that Donna has done to guide me since then and that have impacted my career in clinical microbiology. But above all, the true benefit is that I not only gained a mentor but also gained a family and a friend.

ACKNOWLEDGMENTS
I extend my sincerest thanks to Laurel Burnham, Elizabeth Marlowe, and Raquel Martinez for providing their personal stories and quotes and a special thank-you to Linda Hilbert for reviewing and editing this chapter. I also thank Daniel Stolte and UANews for permission to use the photograph in Fig. 1.

CITATION
Whitfield NN. 2018. Donna M. Wolk: it's never too late to bloom, p 295–304. *In* Whitaker RJ, Barton HA (ed), *Women in Microbiology*. American Society for Microbiology, Washington, DC.

References
1. Marshall MM, Naumovitz D, Ortega Y, Sterling CR. 1997. Waterborne protozoan pathogens. *Clin Microbiol Rev* 10:67–85.
2. Wolk DM, Johnson CH, Rice EW, Marshall MM, Grahn KF, Plummer CB, Sterling CR. 2000. A spore counting method and cell culture model for chlorine disinfection studies of *Encephalitozoon* syn. *Septata intestinalis*. *Appl Environ Microbiol* 66:1266–1273.
3. Wolk DM, Jones MF, Rosenblatt JE. 2001. Laboratory diagnosis of viral hepatitis. *Infect Dis Clin North Am* 15:1109–1126.
4. Baghai M, Osmon DR, Wolk DM, Wold LE, Haidukewych GJ, Matteson EL. 2001. Fatal sepsis in a patient with rheumatoid arthritis treated with etanercept. *Mayo Clin Proc* 76:653–656.
5. Wolk DM, Picton E, Johnson D, Davis T, Pancholi P, Ginocchio CC, Finegold S, Welch DF, de Boer M, Fuller D, Solomon MC, Rogers B, Mehta MS, Peterson LR. 2009. Multicenter evaluation of the Cepheid Xpert methicillin-resistant *Staphylococcus aureus* (MRSA) test as a rapid screening method for detection of MRSA in nares. *J Clin Microbiol* 47:758–764.

6. Wolk DM, Struelens MJ, Pancholi P, Davis T, Della-Latta P, Fuller D, Picton E, Dickenson R, Denis O, Johnson D, Chapin K. 2009. Rapid detection of *Staphylococcus aureus* and methicillin-resistant *S. aureus* (MRSA) in wound specimens and blood cultures: multicenter preclinical evaluation of the Cepheid Xpert MRSA/SA skin and soft tissue and blood culture assays. *J Clin Microbiol* **47:**823–826.

7. Wolk DM, Marx JL, Dominguez L, Driscoll D, Schifman RB. 2009. Comparison of MRSASelect agar, CHROMagar methicillin-resistant *Staphylococcus aureus* (MRSA) medium, and Xpert MRSA PCR for detection of MRSA in nares: diagnostic accuracy for surveillance samples with various bacterial densities. *J Clin Microbiol* **47:**3933–3936.

8. Kaleta EJ, Clark AE, Johnson DR, Gamage DC, Wysocki VH, Cherkaoui A, Schrenzel J, Wolk DM. 2011. Use of PCR coupled with electrospray ionization mass spectrometry for rapid identification of bacterial and yeast bloodstream pathogens from blood culture bottles. *J Clin Microbiol* **49:**345–353.

9. Marner ES, Wolk DM, Carr J, Hewitt C, Dominguez LL, Kovacs T, Johnson DR, Hayden RT. 2011. Diagnostic accuracy of the Cepheid GeneXpert vanA/vanB assay ver. 1.0 to detect the vanA and vanB vancomycin resistance genes in *Enterococcus* from perianal specimens. *Diagn Microbiol Infect Dis* **69:**382–389.

10. Kaleta EJ, Clark AE, Cherkaoui A, Wysocki VH, Ingram EL, Schrenzel J, Wolk DM. 2011. Comparative analysis of PCR-electrospray ionization/mass spectrometry (MS) and MALDI-TOF/MS for the identification of bacteria and yeast from positive blood culture bottles. *Clin Chem* **57:**1057–1067.

11. Schieffer KM, Tan KE, Stamper PD, Somogyi A, Andrea SB, Wakefield T, Romagnoli M, Chapin KC, Wolk DM, Carroll KC. 2014. Multicenter evaluation of the Sepsityper™ extraction kit and MALDI-TOF MS for direct identification of positive blood culture isolates using the BD BACTEC™ FX and VersaTREK® diagnostic blood culture systems. *J Appl Microbiol* **116:**934–941.

12. Stoneking LR, Patanwala AE, Winkler JP, Fiorello AB, Lee ES, Olson DP, Wolk DM. 2013. Would earlier microbe identification alter antibiotic therapy in bacteremic emergency department patients? *J Emerg Med* **44:**1–8.

13. Wolk DM, Kaleta EJ, Wysocki VH. 2012. PCR-electrospray ionization mass spectrometry. *J Mol Diagn* **14:**295–304.

14. Buehler SS, Madison B, Snyder SR, Derzon JH, Cornish NE, Saubolle MA, Weissfeld AS, Weinstein MP, Liebow EB, Wolk DM. 2016. Effectiveness of practices to increase timeliness of providing targeted therapy for inpatients with bloodstream infections: a laboratory medicine best practices systematic review and meta-analysis. *Clin Microbiol Rev* **29:**59–103.

15. Stolte D. 2011. Making germs glow: new test helps save lives and cuts costs. *UANews*, Tucson, AZ.

16. National Institutes of Health. 2010. *NOT-OD-10-047: Ruth L. Kirschstein National Research Service Award (NRSA) stipends, tuition/fees and other budgetary levels effective for fiscal year 2010*. National Institutes of Health, Bethesda, MD.

Women in Microbiology
Edited by Rachel J. Whitaker and Hazel A. Barton
© 2018 American Society for Microbiology. All rights reserved.
doi:10.1128/9781555819545.ch33

Esther Miriam Zimmer Lederberg: Pioneer in Microbial Genetics

33

Rebecca V. Ferrell[1]

In the 1940 U.S. Census, Esther Miriam Zimmer was 17 years old, had completed 2 years of college, and lived with her parents and brother in the Bronx, NY. Her 1939 earnings were $60 for the equivalent of 3 weeks' work, an entry that looks like it was changed from $0 (1). Perhaps the census taker assumed a young college girl would have no earnings and someone, probably Esther, asked him to correct the data. She would have asked gently, say those who knew her (M. O. Martin, personal communication), but she would have asked. To watch the author and Mark O. Martin discuss Esther further, visit http://bit.ly/WIM_Lederberg.

Esther's father, David Zimmer, was an immigrant from Romania who ran a print shop. Seeing her father's press at work may have helped Esther invent replica plating later when she needed an efficient way to inoculate bacteria onto different plates of media to test their traits. Esther used sterile velvet, pressed onto a plate of bacterial colonies and then stamped onto plates of media with different ingredients that allow traits to be observed (2), a powerful bacterial printing technique that has been useful in many discoveries.

David, born in Romania in the Austro-Hungarian Monarchy, was Jewish and considered "illegitimate" because non-Christian marriages were not recognized (3). His family immigrated to America and by 1910 lived in a crowded Manhattan building occupied mostly by Yiddish-speaking immigrants from Austro-Hungary. Nearly everyone in the building was a garment worker, including David's father and siblings (4). Esther's mother, Pauline Geller, was born in New York City to parents from Romania and Galatia.

---

[1]Department of Biology, Metropolitan State University of Denver, Denver, CO 80217

David and Pauline married in 1920, and Esther Miriam Zimmer, their first child, was born 18 December 1922 at Bronx Maternity Hospital, followed by her brother Benjamin in 1925 (3).

Esther's family was poor; she shared a memory with her second husband, Matthew Simon, that her lunch was often "a piece of bread, upon which her mother squeezed juice from a tomato." Esther's zayde (grandfather) taught her Hebrew, and she was a more enthusiastic pupil than her male cousins. Despite disappointment that the boys couldn't read Hebrew during Passover, says Simon (3), "Esther made her grandfather happy by doing all the reading."

Esther Zimmer graduated from New York's Evander Childs High School in June 1938 (3) and won a scholarship to enter Hunter College that fall (Fig. 1). Expected to study language or literature, Esther chose instead to focus on biochemistry (5). While a student at Hunter, Esther worked on the mold *Neurospora* under Bernard Ogilvie Dodge at the New York Botanical Garden. In a 1944 letter, Dodge recommended Esther to George Beadle as "a very bright girl" who "would do very well" for a fellowship (3). Graduating *cum laude* from Hunter College in June 1942 (6), she went to work in the U.S. Public Health Service's Industrial Hygiene Research Laboratory, exposing *Neurospora* spores to UV radiation and analyzing the resulting mutants. Reference letters in 1944 from associates at the Industrial Hygiene lab describe Esther as "a quiet and reserved young lady who gets along very well with her fellow workers" and "very cooperative and industrious" with an

Figure 1 Esther Zimmer, age 15, shortly before her high school graduation in 1938. ©Esther M. Zimmer Lederberg memorial website (http://www.estherlederberg.com/home.html).

APRIL 1938

"intelligent approach" that meant "Miss Zimmer did not accept opinions uncritically," a "great asset in a research assistant" (3).

So Miss Zimmer went to Stanford, where Beadle and Edward Tatum were working with mutations in *Neurospora* (7). Three years before, they had published experiments using X rays to mutate single *Neurospora* spores. Each was grown on complete medium and then screened for inability to grow on minimal media, indicating a mutation. Then each mutant was individually tested for growth with various supplements, labor-intensive before Esther invented replica plating. Beadle and Tatum identified three mutants in *Neurospora* but characterized only one (8). It was Tatum who introduced Esther to genetics with a sink-or-swim approach.

Esther crossed the United States by train, surviving lost luggage and a seedy hotel her first night in Palo Alto. The next day she met Tatum, who helped her find a dorm room and asked what she wanted to study. Esther said, "Genetics." He said it wasn't offered until winter quarter. She persisted. The next day he left for her, without comment, a bottle of *Drosophila* fruit flies in which one fly had eyes of a different color. Esther worked out why, learning genetics as she went, so successfully that Tatum asked her to be his teaching assistant (TA) for the course (3). Esther's research was to analyze one of Beadle and Tatum's *Neurospora* mutants.

It was not easy for Esther as a master's student. In exchange for a room in a home she did the household laundry. She later told Matthew Simon (3) that she was so poor, "she and another TA ate the frog's legs after student dissections were over." Supporting herself as a TA, Esther completed her research and in August 1946 submitted her M.A. thesis at Stanford University, entitled "Mutant Strains of *Neurospora* Deficient in Para-Aminobenzoic Acid" (9).

Four months later, she married Joshua Lederberg. Their first known letter was 2 July 1946, when Joshua, Tatum's student at Yale University, wrote to Esther at Beadle's suggestion to ask about a *Neurospora* strain (3). The same week, Joshua announced "bacterial sex" at the Cold Spring Harbor Symposium, causing "lively" discussion (10). Tatum, using methods pioneered with *Neurospora*, had made *Escherichia coli* nutritional mutants (11). When the mutants were mixed, recombinant colonies grew on minimal medium, allowing the discovery of conjugation (12). After completing her M.A., Esther moved to Yale's Osborn Botanical Laboratory, where Joshua was, and worked with Norman Giles on mechanisms of reversion in *Neurospora* (13). On 13 December, Esther Zimmer, age 23, and Joshua Lederberg, age 21, were married by a justice of the peace in New Haven (3). Joshua later said they were on their honeymoon at the American Association for the Advancement of Science meeting in Boston that December (14).

Joshua, "descended from a long line of rabbis" (14), was born in 1925 to parents recently arrived from Israel, his father an orthodox rabbi at a New York synagogue (15). Interested in science from an early age, Joshua graduated from male-only Stuyvesant High School at age 15 and entered Columbia University at age 16 (15). During World War II, Joshua was in a program combining medical training with scientific research. Scheduled to return to New York to finish medical school in fall 1947, he completed his Yale Ph.D. that summer. Days before the move to New York, at the age of 22, he was offered a job as assistant professor to work on bacterial genetics at the University of Wisconsin (10). Esther, who had continued at the Osborn Lab, had a tuition grant to study botany at Columbia University that fall but instead resigned the grant (3) and moved to Madison with Joshua.

Joshua wrote later of considerable discussion at Madison about hiring one of "the first Jewish professors in a midwestern college of agriculture," noting that he found out later that support from a few senior faculty members, including R. A. Brink, had made the difference (10). At a time when anti-Semitism and anti-immigrant sentiments were widespread in the United States, Esther would have been conscious of these issues. She did her doctoral work under Brink's sponsorship (3), and his unbiased support made her next steps in science possible. Esther received fellowships from University of Wisconsin and the National Cancer Institute, and in 1950 she completed her dissertation, "Genetic Control of Mutability in the Bacterium *Escherichia coli*" (16).

Esther wrote later that Joshua advised her to delay work on bacteriophage λ, which she had recently discovered, and on F (conjugation's fertility factor) because "he thought that finishing my thesis work was a priority," adding that, "the λ discovery was written up in MGB (microbial genetics bulletin)... an informal round robin publication" (3, 17). Esther published a short abstract on λ in 1951 (18). In May 1952 she and Joshua submitted a detailed account of crosses between her *E. coli* K-12 strains lysogenic for the phage "now referred to as λ" and mutant strains from Joshua and Tatum's work that were sensitive to λ lysis. They said of Esther's discovery of λ only, "our interest in lysogenicity was provoked by the discovery that *E. coli* strain K-12 was lysogenic," defined as having the ability to transmit "virus potentiality" without lysis when bacteria multiply. Neither of her earlier λ papers is cited (19). They report on complex cross-streaking experiments, likely carried out by Esther, testing about 2,000 strains of *E. coli* for λ lysogeny and sensitivity, and conclude that a "chromosomal factor is altered when a cell becomes lysogenic." They offer two possibilities, either that a gene mutates to create lysogeny or that "the virus or provirus occupies a definite niche on the chromosome." They are close here to the modern understanding of a prophage integrated into the *E. coli* genome, but DNA is not mentioned; Watson and Crick's description of the double helix had not yet been published (20).

Esther and Joshua attended the 1951 Cold Spring Harbor Symposium, where his talk included some of Esther's doctoral work on *E. coli*, acknowledging her as second author (21). That summer they submitted a paper on indirect selection of mutants, introducing replica plating to pick up bacterial colonies with sterile velvet and stamp them onto plates of different media. Esther tried it first with her powder puff and refined it by testing various fabrics (5). They selected *E. coli* strains resistant to bacteriophage T-1 and, separately, strains resistant to the antibiotic streptomycin, in both cases with no contact between the resistant strain and the agent. Strong evidence for "the participation of spontaneous mutation and populational selection in the heritable adaptation of bacteria to new environments," these mutants were part of normal variation (preadaptation) in the *E. coli* population even without selection for their traits (2). The Lederberg experiment refuted Lamarck's idea that genetic traits change in response to the environment and was important in modern genetics' understanding of mutation as a random process rather than occurring due to need. Replica plating has been widely used to analyze mutants, but Esther is not always credited (22, 23).

Esther was sole author of a 1952 paper reporting experiments with lactose metabolism genes, analyzing Lac^+ recombinants from Lac^- × Lac^- crosses for stability of the Lac^+ trait (24). Examining over 31,000 recombinants, she found only one exception, concluding that Lac^+ reverse mutants are usually stable but identifying another genetic category, an unstable suppressor mutant able to "mimic wild type by suppressing the effect of Lac_1^- mutation."

In another 1952 paper, Joshua, Esther, and Luigi Cavalli described how an F^- conjugation recipient receives the F^+ agent and reported the inability of Cavalli's Hfr (high frequency of recombination) strains to transfer the F^+ agent (25). They developed this further in a 1953 paper, "An Infective Factor Controlling Sex Compatibility in *Bacterium coli*" (26).

The Lederberg lab's other important 1952 discovery, generalized transduction in *Salmonella*, was published by Joshua and Norton Zinder (27). Experiments designed to detect conjugation showed that genetic information was also transferred without cell-to-cell contact, via a filterable agent identified as temperate phage P22. Some genetics texts mention Esther's participation in this discovery (22).

In 1956, Larry Morse, Esther, and Joshua reported transduction by λ phage in "the sexually fertile K-12 strain" of *E. coli* previously described by Esther (17, 18) and further characterized by Esther and Joshua (19). Concentrated λ lysates were used for transduction, and K-12's F factor allowed "recombinational analysis of the sexual cycle." Analyzing thousands of recombinants, they saw stable linkage in transfer of λ prophage and *Gal* (galactose fermentation) genes, but not genes for eight other metabolic markers, proposing correctly that the prophage and *Gal* genes are near each other in the "genetic material" (DNA is not mentioned). Much was still not understood in bacterial genetics; they postulate that the prophage might be near a centromere or itself a centromere, proposing that segregation of traits could result from unaltered nuclei separating from the prophage-associated *E. coli* nucleus (28). Their follow-up paper on λ transduction offered diagrams of a possible mechanism including integration into the recipient *E. coli* chromosome, which would segregate by mitosis (29).

In 1957, Esther and Joshua traveled to Canberra, Australia, for a Symposium on Bacterial and Viral Genetics, and in 1958 Esther gave a talk on fine-structure mapping of the *Gal* gene region in *E. coli* at the 10th International Congress of Genetics in Montreal, Canada (4).

These years in Wisconsin were productive for the Lederbergs. Joshua was not a noted experimentalist, but Esther was. Much of their collaborative productivity was due to her research skills (3), giving Joshua time to publish numerous single-author papers. But Joshua's autobiographical account of the discovery of genetic recombination in bacteria does not mention her at all, either in his personal history or as a contributor to his lab's discoveries (10).

The 1958 Nobel Prize in Physiology or Medicine was awarded to Joshua, Beadle, and Tatum. Esther's scientific work was interwoven with the winners', but in Stockholm she was a laureate's wife. At his first press conference Joshua mentioned her briefly, saying that of numerous associates, "First among these is my wife, who is my close associate in the laboratory" (4). In his Nobel Lecture (30), Joshua said that his past studies, "in which I have enjoyed the companionship of many colleagues, above all my wife," were regularly reviewed so he focused instead on present and future directions of bacterial genetics. Citing the λ lysogeny work he said, "Dr. Esther Lederberg's first crosses were quite startling in their implication that the prophage segregated as a typical chromosomal marker." Esther's role as "Nobel wife" required three different pairs of gloves and an evening gown for the ceremony. Unwilling to spend time and money on this, Matthew Simon wrote, "little Esther (barely 5'3" tall) purchased a teenager's 'prom' dress for the ceremony" and then hand dyed slippers to match (3).

At this time Joshua became chair of the new Department of Genetics at Stanford University. The Lederbergs moved to California, and Esther joined the Department of Medical Microbiology (5). Noting that Stanford had no female professors, Esther and two other women asked the dean to use funds allocated for that purpose, arguing that some women did meet the criteria. All three were highly qualified but the dean offered only one position, which Esther received because she was willing to accept the rank of "research professor" without tenure (3).

A history of Joshua's Stanford years (1958–1978) does not mention Esther. Joshua had "only a small laboratory," limiting permanent faculty to himself and one other, a "deliberate strategy" to bring in a series of young scientists to work with Joshua (31). Esther and Joshua published their last paper together in 1964, describing suppression of a Gal operon mutation in *E. coli* by the antibiotic streptomycin (32). They divorced in 1966.

Esther continued at Stanford but was never tenured. Former collaborator Luigi Cavalli-Sforza wrote in 1974 supporting Esther's retention, noting her renewed interest in work after "the shock of a wrecked marriage," her studies on *Salmonella* phase variation, and interest in teaching. Citing her achievements, he said, "Dr. Esther Lederberg has enjoyed the privilege of working with a very famous husband. This has been at times also a setback, because inevitably she has not been credited with as much of the credit as she really deserved" (3). Stanford retained her, "coterminous with research support," meaning that her salary depended on grants. Esther continued collaborations, including transforming *Salmonella* with plasmid DNA with Stanley Cohen (33). Becoming director in 1976 of the Plasmid Reference Center (Fig. 2), she "named, organized, and distributed plasmids" (34), expanding to also allocate transposon and insertion sequence numbers (35). Esther retired in 1985 but stayed on at the plasmid center for a time.

Esther enjoyed performing medieval, renaissance, and baroque music on original instruments and was a founder in 1962 of the Mid Peninsula Recorder Orchestra. In 1989, Matthew Simon, an engineer newly arrived at Stanford, asked at a meeting if anyone knew about early music. Overhearing him, Esther began a conversation that led to marriage in 1993. They were together for the rest of her life, sharing a love of music, botany, and literature (3).

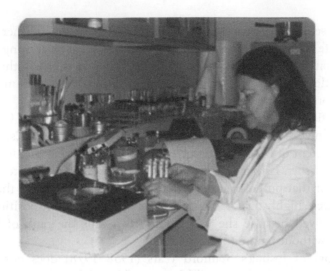

Figure 2 Esther Lederberg, Director of the Plasmid Reference Center at Stanford University, at work in the laboratory in 1977. ©Esther M. Zimmer Lederberg memorial website (http://www.estherlederberg.com/home.html).

The Stanford Memorial Resolution by Stanley Falkow and Lucy Tompkins remembers Esther as a distinguished scientist who made important contributions to genetic science, was entrusted with organizing the world's plasmid discoveries, brought music and snacks to meetings, and was "helpful to young investigators" and international scientists (3). Citing the hurdles in Esther's way, they say she faced obstacles with "extraordinary grace, gentleness, and with a respect and love for science that is important to remember and emulate."

Obituaries recognized Esther's contributions, and some raised questions about credit for discoveries. "There was also a female Lederberg," said Mercé Piqueras, who noted that Joshua's National Library of Medicine website did not mention Esther (36). *The Lancet* (37) called her a "pioneering bacterial geneticist and molecular biologist." "She did pioneering work in genetics, but it was her husband who won a Nobel prize," said *The Guardian* (5).

Matthew Simon (3) tells of Esther reflecting upon Joshua's Nobel Prize. "'How silly Josh was (to view himself in such inflated self-importance and grandeur),' she said with regret. 'Soon both of us will be forgotten.'" As the science of molecular biology began, Esther made key discoveries that helped change our understanding of life itself. That time and her role in it will not be soon forgotten. As Stanley Falkow said at her Stanford memorial (3), "Esther Zimmer Lederberg has an assured place in the history of science."

ACKNOWLEDGMENTS

I am grateful to Matthew L. Simon and the curators of the Esther M. Zimmer Lederberg Memorial Website (http://www.estherlederberg.com/home.html), who with the Stanford University Library have created an invaluable online resource of information about Esther's life and work. Mark O. Martin, who knew Esther when he was a graduate student at Stanford, kindly provided insight into her keen intelligence and generosity of spirit.

CITATION

Ferrell RV. 2018. Esther Miriam Zimmer Lederberg: pioneer in microbial genetics, p 305–315. *In* Whitaker RJ, Barton HA (ed), *Women in Microbiology*. American Society for Microbiology, Washington, DC.

References

1. **Department of Commerce, Bureau of the Census.** 1940. *Sixteenth census of the United States: 1940.* New York, New York City, Ward 2AD, Bronx, Block H sheet 2B. Viewed 28 September 2016, Denver Public Library, Denver, CO.
2. **Lederberg J, Lederberg EM.** 1952. Replica plating and indirect selection of bacterial mutants. *J Bacteriol* **63**:399–406.
3. **Simon M (ed).** Esther M. Zimmer Lederberg Memorial Website. Online repository of materials from the Esther Lederberg archives at Stanford University Library. http://www.estherlederberg.com. Accessed 30 October 2016 through 27 August 2017.
4. **Department of Commerce and Labor, Bureau of the Census.** 1910. Thirteenth census of the United States: 1910. New York, New York City, Borough of Manhattan, Enumeration District 764, sheet 20B. Viewed 28 September 2016, Denver Public Library, Denver, CO.
5. **Richmond C.** 12 December 2006. Esther Lederberg. *The Guardian*, London, United Kingdom. https://www.theguardian.com/science/2006/dec/13/obituaries.guardianobituaries.
6. **Scientific Legacies.** Esther Lederberg's diplomas and other awards. Scientific Legacies, Belmont, MA. http://www.scientificlegacies.org/esther-lederberg/diplomas.html. Accessed 23 August 2017.
7. **Judson HF.** 1979. *The Eighth Day of Creation*, p 368–369. Touchstone, Simon and Schuster, New York, NY.
8. **Beadle GW, Tatum EL.** 1941. Genetic control of biochemical reactions in *Neurospora*. *Proc Natl Acad Sci USA* **27**:499–506.
9. **Zimmer EM.** 1946. Mutant strains of *Neurospora* deficient in para-aminobenzoic acid. MA thesis, Stanford University, Stanford, CA.
10. **Lederberg J.** 1987. Genetic recombination in bacteria: a discovery account. *Annu Rev Genet* **21**:23–46.
11. **Tatum EL.** 1945. X-ray induced mutant strains of *Escherichia coli*. *Proc Natl Acad Sci USA* **31**:215–219.
12. **Lederberg J, Tatum EL.** 1946. Gene recombination in *Escherichia coli*. *Nature* **158**:558.

13. Giles NH Jr, Lederberg EZ. 1948. Induced reversions of biochemical mutants in *Neurospora crassa. Am J Bot* **35**:150–157.

14. Sarkar S. 2014. Lederberg on bacterial recombination, Haldane, and cold war genetics: an interview. *Hist Philos Life Sci* **36**:280–288.

15. Pevzner L. 1996. Interview with Prof. Lederberg, winner of the 1958 Nobel Prize in physiology and medicine. http://www.almaz.com/nobel/medicine/lederberg-interview. html. Accessed 20 August 2017.

16. Lederberg EZ. 1950. Genetic control of mutability in the bacterium *Escherichia coli.* Ph.D. dissertation. University of Wisconsin, Madison, WI.

17. Lederberg EM. 1950. Lysogenicity in *Escherichia coli* strain K-12. *Univ Wis Microb Genet Bull* **1**:5–9.

18. Lederberg EM. 1951. Lysogenicity in *E. coli* K-12. *Genetics* **36**:560. (Abstract.)

19. Lederberg EM, Lederberg J. 1953. Genetic studies of lysogeny in *Escherichia coli. Genetics* **38**:51–64.

20. Watson JD, Crick FH. 1953. Molecular structure of nucleic acids; a structure for deoxyribose nucleic acid. *Nature* **171**:737–738.

21. Lederberg J, Lederberg EM, Zinder ND, Lively ER. 1951. Recombination analysis of bacterial heredity. *Cold Spring Harb Symp Quant Biol* **16**:413–443.

22. Russell PJ. 2002. *iGenetics*, p 289, 564–565. Benjamin Cummings, Pearson Education, San Francisco, CA.

23. Willey JM, Sherwood LM, Woolverton CJ. 2014. *Prescott's Microbiology*, 9th ed. McGraw-Hill, New York, NY.

24. Lederberg EM. 1952. Allelic relationships and reverse mutation in *Escherichia coli. Genetics* **37**:469–483.

25. Lederberg J, Cavalli LL, Lederberg EM. 1952. Sex compatibility in *Escherichia coli. Genetics* **37**:720–730.

26. Cavalli LL, Lederberg J, Lederberg EM. 1953. An infective factor controlling sex compatibility in *Bacterium coli. J Gen Microbiol* **8**:89–103.

27. Zinder ND, Lederberg J. 1952. Genetic exchange in *Salmonella. J Bacteriol* **64**:679–699.

28. Morse ML, Lederberg EM, Lederberg J. 1956. Transduction in *Escherichia coli* K-12. *Genetics* **41**:142–156.

29. Morse ML, Lederberg EM, Lederberg J. 1956. Transductional heterogenotes in *Escherichia coli. Genetics* **41**:758–779.

30. Lederberg J. 1958. Nobel Lecture: A view of genetics. http://www.nobelprize. org/nobel_prizes/medicine/laureates/1958/lederberg-lecture.html. Accessed 23 January 2018.

31. Herzenberg L, Rindfleisch T, Herzenberg L. 2008. Joshua Lederberg: the Stanford years (1958–1978). *Annu Rev Genet* **42**:19–25.

32. Lederberg EM, Cavalli-Sforza L, Lederberg J. 1964. Interaction of streptomycin and a suppressor for galactose fermentation in *E. coli* K-12. *Proc Natl Acad Sci USA* **51**:678–682.

33. Lederberg EM, Cohen SN. 1974. Transformation of *Salmonella typhimurium* by plasmid deoxyribonucleic acid. *J Bacteriol* **119**:1072–1074.

34. Tran C. 7 December 2006. Esther Lederberg dies; pioneering microbial geneticist was 83. *The Scientist*, Ontario, Canada.

35. **Lederberg EM.** 1987. Plasmid Reference Center Registry of transposon (Tn) and insertion sequence (IS) allocations through December 1986. *Gene* **51:**115–118.
36. **Piqueras M.** 28 July 2014. *Esther Lederberg, pioneer of bacterial genetics. Small Things Considered blog.* http://schaechter.asmblog.org/schaechter/2014/07/esther-lederberg-pioneer-of-bacterial-genetics.html. Accessed 17 April 2017.
37. **Oransky I.** 2006. Esther Miriam Lederberg. Obituary. *Lancet* **368:**2204.

Women in Microbiology
Edited by Rachel J. Whitaker and Hazel A. Barton
© 2018 American Society for Microbiology. All rights reserved.
doi:10.1128/9781555819545.ch34

Women Microbiologists at Rutgers in the Early Golden Age of Antibiotics

34

Douglas E. Eveleigh[1] and Joan W. Bennett[2]

INTRODUCTION

The development of antibiotics, a transformative medical advance, occurred principally during World War II. The discovery of penicillin in England spurred the search for similar antimicrobial drugs from microbes. Under the leadership of Selman Waksman at Rutgers University in the United States, streptomycin was isolated, purified, and then tested in animal and human trials at the Mayo Clinic in collaboration with Merck Corporation, which scaled up production for clinical applications. This broad-spectrum antibiotic was active against Gram-negative bacteria, such as those that caused cholera, typhoid, bubonic plague, and dysentery, as well as tuberculosis. Since penicillin was inactive towards these pathogens, streptomycin was a major therapeutic breakthrough, particularly with respect to tuberculosis. The development of both drugs was accelerated by World War II and the need for drugs to treat wound infections and the many contagious diseases associated with wartime disruptions and crowding. Selman Waksman won the 1952 Nobel Prize in Medicine or Physiology for his hypothesis that soil microbes made useful antimicrobial substances, for coordinating the search for new antibiotics, and for the discovery of streptomycin. Most historical treatments about this era give emphasis to the disposition of the patent monies and recognition of credit at the Nobel Prize, especially with reference to a male graduate student, Albert Schatz, who isolated the first strepto-

[1]Department of Biochemistry and Microbiology, School of Environmental and Biological Sciences, Rutgers—The State University of New Jersey, New Brunswick, NJ 08901
[2]Department of Plant Biology and Pathology, School of Environmental and Biological Sciences, Rutgers—The State University of New Jersey, New Brunswick, NJ 08901

Figure 1 The Rutgers Microbiology Department, New Jersey College of Agriculture and Experiment Station, 1946. Top Row (L to R): Donald B. Johnstone (gs), Aldrage B. Cooper (lab tech), Dr. Merritt C. Fernald (asst. res. specialist), Lloyd R. Frederick (gs). Second Row (L to R): John D. Schenone (res. assoc.), Christine R. Frazier (lab asst.), Dorothy G. Smith (gs), Doris I. Jones (gs), [not recorded?], Kenneth L. Temple (gs). Third Row (L to R): John Q. Adams (lab asst.), Vivian Gerber (gs), Viola A. Battista (secretary), Dorothy Nycz (secretary), Clara H. Wark (lab asst.), Claire B. Landers (gs). Fourth and Fifth Rows (L to R): Donald M. Reynolds (gs), Dr. Robert Starkey, Doris W. Wilson (gs), Dr. Selman Waksman, Dorothy J. Randolph (lab asst.), Dr. Walton B. Geiger (assoc. biochemist), H. Christine Reilly (gs). gs, graduate student. Courtesy of Special Collections and University Archives, Rutgers University Libraries.

mycin-producing strains from an actinomycete named *Streptomyces griseus* (1–8). Here we present short biographies of six women who were pivotal in the research carried out in the Waksman laboratory during this remarkable time in antibiotic discovery (Fig. 1).

ELIZABETH SCHWEBEL HORNING, Ph.D., 1942 (1904–2000)

Elizabeth Horning was in her thirties by the time she began her graduate research in the Waksman laboratory. As part of her dissertation research, she developed a new screening protocol for isolation of soil microbes to find antagonistic cultures producing candidate antibiotics (9). Horning was on a short-term release from work to complete her degree, and time was of the essence. She changed the approach that had been developed by an earlier graduate student named Boyd Woodruff, who had discovered the first major antibiotic, which they named actinomycin. He used the classic enrichment protocol, which involved adding bacteria to the soil and then waiting for the growth of a soil population that was antagonistic towards these added species, which often involved a 1- to 2-month incubation period. Horning decided to simplify the procedure: she simply took diluted soil samples and plated them directly on selective media. Within a few days, she selectively recovered numerous microbial candidates with antagonistic activities. Most of her isolates were fungi or actinomycetes (which in those days were sometimes called "ray fungi") (9). In particular, her isolation of numerous antagonistic actinomycetes showed that these filamentous bacteria not only were abundant in soils but also commonly possessed the metabolic capacity to inhibit the growth of other microbes (9). Perhaps because penicillin had been isolated from a fungus, she selected two fungi for further study, *Aspergillus fumigatus* and *Aspergillus clavatus*, and found that they produced fumigacin and clavacin, respectively, the second and third antimicrobial substances isolated in the Waksman laboratory (10, 11, 12). Unfortunately, as was the case for most of these early, promising antibiotic products, both fumigacin and clavacin later proved too toxic for drug use; they were capable of killing bacteria but also mammalian cells.

Horning initiated the chemical characterization of these antibiotics with two associates, but her short leave of absence unfortunately precluded her from completing the analysis of the antagonistic activities she had found (13). Nevertheless, Horning had dramatically enhanced and sped up the screening protocol through use of soil dilution and direct plating. Her work lent credence to the hypothesis that soils were teeming with antibiotic-producing microorganisms.

ELIZABETH JANE BUGIE, M.Sc., 1944 (1920–2001)

Elizabeth ("Betty") Bugie (later Gregory) graduated with a degree in microbiology from the New Jersey College for Women and then began graduate studies working on fungi (14), specifically on two recently found new fungal metabolites, flavicin and chaetomin (15, 16). Departing from the usual emphasis on animal diseases, she examined antibiotics that are active against plant pathogens, including the fungus that causes Dutch elm disease (17, 18). On graduating, Bugie initially stayed at Rutgers to work on a new metabolite, micromonosporin (19), and later took a job at the nearby Merck evaluating the activity of several agents against *Mycobacterium tuberculosis* (20, 21).

The scientific contribution for which she is best remembered was made during her master's degree in collaboration with Albert Schatz,

Elizabeth Bugie was the middle author on the famous paper in which the discovery of streptomycin was announced (22); however, later when Waksman and Schatz patented the discovery, her name was not included (23). Several years passed, during which time Schatz and Waksman engaged in an ugly lawsuit about who deserved the glory and profits from the discovery of streptomycin. In this battle for priority and recognition, Bugie's contributions were barely discussed. Schatz later said that her role was limited to performing experiments that independently confirmed his data and insisted that he alone had discovered and isolated streptomycin: "The research which resulted in the discovery of streptomycin was done by a single individual, working alone in a basement laboratory" (6). Yet, it is manifest that Waksman had great faith in Bugie's abilities when he requested her to work with Schatz on the discovery and characterization study of this exceptional new antibiotic. Unfortunately, to our knowledge, no one at Rutgers thought to record her side of the story.

If Schatz later insisted that "he alone had discovered and isolated streptomycin," why was he willing to share authorship not only with his mentor but also with a fellow graduate student? We do know that when the Rutgers Foundation settled Schatz's lawsuit about patent profits out of court, Waksman received 10% of the royalties, Schatz 3%, and Bugie only 0.2%. Nevertheless, at minimum, this settlement suggests that Schatz's claim that he had "no help from anyone" was an exaggeration.

After her death in Pittsburgh, PA, in 2001, the reporter writing Bugie's obituary for the *Pittsburgh Post-Gazette* (24) quoted one of Bugie's daughters, Patricia Camp, who remembered her mother saying more than once, "They approached me privately and said, 'Someday you'll get married and have a family and it's not important that your name be on the patent.'" Camp said that her mother would smile and add, "If women's lib had been around, my name would have been on the patent." Her other daughter, Eileen Gregory, who became a microbiologist on the faculty at Rollins College, said, "She was a woman in a field of men and she was pressured. With graduate students, it's a tough line—who do you give credit to and where does it stop?" Eileen Gregory went on, "To be perfectly honest, she had a blast doing it. I think she thought she never needed to be rewarded for it. She got her reward" (24).

DORIS IRASIMUS JONES RALSTON, M.Sc., 1945 (1921–2011)

Doris Jones graduated with a degree in chemistry from the New Jersey College for Women in 1943. As an undergraduate, she had worked with

Professor H. J. Metzger in poultry husbandry, so it was a natural move for her to continue as a graduate student jointly mentored by Frederick Beaudette (poultry sciences) and Selman Waksman (microbiology). Her focus was to look for antiviral antibiotics (25). One approach included screening of the larynges of healthy chickens, with the idea of recovering microbes present that helped maintain the chicken's robust state. She swabbed chicken tracheas, streaked out microbes onto nutrient agar plates, and recovered the resultant culturable microorganisms. The recovered microbes were mixed populations, many of which showed intercolony antagonisms. Perspicaciously and magnanimously, she decided to bring these antagonistic strains to the attention of cograduate student Albert Schatz; she walked over to the next building, where he toiled in his basement laboratory, and gave him some of her discard plates. Schatz recovered a number of isolates from Doris's plates, including one that proved to be a streptomycin-producing strain of *Streptomyces griseus* (7). He named the strain D-1 ("Doris One") in her honor.

Later, Jones's research included assessing the effectiveness of streptomycin towards *Shigella gallinarum*. Her results were dramatic, as she showed *in vivo* (26) that streptomycin killed this common Gram-negative pathogen. Thus, streptomycin became the second antibiotic after penicillin to be effective *in vivo*. Equally important was that streptomycin was not toxic to chicks, unlike the first four potential "antibiotics" discovered by Waksman's group, which proved toxic. The work was also noteworthy in being the first study of the Waksman program with *in vivo* experimentation to address animal disease. Jones's work merited publication in *Science* (26, 27). However, her main thesis project was to search for antiviral antibiotics; her review earned Jones her second *Science* paper (27). She also addressed antiphage agents (28) and continued to publish with Schatz on the search for antiphage agents from microbes (29). They demonstrated that 28% of 176 actinomycetes tested produced apparent antiphage activity, although none demonstrated such activity in subsequent evaluations (30).

With the benefit of hindsight, we can understand why Jones could not discover an effective antibiotic that inhibited viruses: despite more than a half century of intensive research, there still are few known antiviral agents and viral diseases are primarily tackled with the use of vaccines. Despite the scientifically challenging nature of Jones's search for antiviral antibiotics, her sterling record helped her to receive financial support from a Yardley Fellowship of the N.J. Federation of Women's Clubs and as an Abraham Rosenberg Research Fellow. She carried on her research at the University of California, Berkeley, receiving a Ph.D. for her studies of bacteriophages

(31). She later wrote, "In those days women hardly ever got independent grants, and I would have been tickled to get peanuts for a salary, but through the encouragement of Dr. Joshua Lederberg, who served on the National Institutes of Health grant review study section, I ended up with a full-time salary to go with all my wonderful new equipment." Jones-Ralston worked with Sanford Elberg at the University of California, focusing on immuno-logic aspects of brucellosis. She went on to a successful career and retired in 1973 after 30 years in research. In a transcript of a lecture she gave at Rutgers in 1977 (preserved in the archives of the Associate Alumnae Douglass College), she wrote about her personal experiences during the antibiotic era, emphasizing how differently Waksman and Schatz dealt with it in their own minds (32). She was a good sport about the fact that during her years at Rutgers she had been given the unflattering nickname of "Moose." She modestly downplayed her own participation, was supportive of Albert Schatz, and with respect to recognition for the discovery of streptomycin and the Nobel Prize, she felt that there "could have been room at the top for two."

VIVIAN ROSENFELD SCHATZ, B.Sc., 1946 (1925–)

When Albert Schatz joined the Waksman team in 1942 (6), he was soon drafted into the Army and then unexpectedly invalided out due to a medical issue. He returned to the Waksman team and began his search for a soil microbe that would produce an antibiotic to be used in combatting infections of wounds and of diseases like tuberculosis. Using the Horning protocol and screening for antimicrobial agents against Gram-negative pathogens and also *M. tuberculosis*, Schatz found two *Streptomyces griseus* strains that were later shown to produce streptomycin. While the first strain was given to him by Doris Jones, he discovered the second through long hours of work (6, 22). While Schatz worked on the isolation and analysis of streptomycin, he met Vivian Rosenfeld, an undergraduate studying biology at the New Jersey College for Women. During their courtship, Vivian often helped him with projects in his basement laboratory. They married on 23 March 1945, a year and a half after his discovery of streptomycin and a year before she graduated from college. Perhaps to demonstrate their devotion to science, during their honeymoon, Schatz wrote several letters to Waksman, inform-ing him that they had traveled with four test tubes with white cotton stoppers containing *Actinomyces lavendulae*, the bacterium that produced strepto-thricin. "A week was too long to leave them unobserved," he explained, and "Vivian wondered whether there had ever been another man who took test tubes of multiplying microbes on his honeymoon." Albert continued, "Each morning and night Vivian and I examine the four agar slants of the different colonies of *A. lavendulae*. Vivian says it's strange to have 'business' with us now, but she is as interested in the cultures as I am" (3).

After Albert received his Ph.D., he went on to other jobs, including becoming a major figure in the antifluoridation movement, and Vivian continued to work closely with her husband (6, 7). Moreover, outside of science, Vivian Schatz accrued an admirable track record in working for international peace and justice. During the 1970s, she was an antiwar activist and also worked with the Chilean Emergency Committee. Her dedication to helping make the world a more humanitarian place was recognized on 27 May 2017 by the Women's International League for Peace and Freedom (WILPF) when The Greater Philadelphia Branch, WILPF, awarded her their Peace and Justice Dove Award. At the time of the award, Vivian was 92 (33).

As part of an oral history project about the history of antibiotics discovery for the American Society for Microbiology, one of us (J.W.B.) interviewed Albert and Vivian Schatz in 1998. It was apparent that Vivian had been an utterly loyal source of support for her husband during his long battle to receive credit for the discovery of streptomycin. Her idealism and formidable strength of character no doubt helped him find the courage to mount his David v. Goliath lawsuit against Waksman in 1950 and to continue his life-long quest for recognition. Like many strong women married to brilliant men, Vivian was a major participant in her husband's professional trajectory.

HILDA CHRISTINE REILLY, Ph.D., 1946 (1920–1989)

Christine Reilly studied at the New Jersey College for Women, where she majored in chemistry, receiving her bachelor's degree in 1941. She immediately joined Waksman's group, and her Ph.D. dissertation research focused on the antibacterial properties of streptomycin and streptothricin (34). However, during her years as a graduate student and subsequently as a research associate, Reilly was also involved to some extent with all the major projects in the Waksman laboratory. Thus, Reilly was instrumental in development of the breadth of the antibiotic program, including the isolation and identification of new streptomycin-producing strains, agar streak methods to rapidly assess their antibiotic production, and consideration of the resistance of bacteria towards antibiotics, including streptomycin. She also introduced consideration of their use in plant pathology against plant pathogens. With the continuing major international development of penicillin, which included the studies at nearby Merck and Co., Reilly evaluated *Penicillium notatum* strains for their productivity, and improved tests to assay penicillin production. Her broad expertise made her a "go-to" authority of the group. Reilly worked closely with Schatz on the characterization of streptomycin, which had emerged as the most promising antibiotic in the laboratory. Together they expanded the understanding of the metabolism, plus the chemical structure, of streptomycin (35). Significantly, they gave an early demonstration of antibiotic resistance among different strains of a supposedly susceptible bacterial species (36). The development of antibiotics had been progressing rapidly, and when it was found that pathogens could develop resistance to an antibiotic, this threw the whole concept of the application of antibiotics into a spin.

On receiving her Ph.D. in 1946, Reilly continued briefly as a research associate with the Waksman group, conducting a search for actinophage; it had become apparent that these viruses could be a major detriment in the large-scale fermenter production of antibiotics from actinobacteria (37). Despite all these efforts, of the 10 papers Reilly published during her highly productive years with the Waksman group, she was listed as first author only on her last actinophage paper. We infer that she not only was industrious and a good team player, but also was willing to defer to Selman Waksman's robust ego.

Reilly became a faculty member of the Sloan-Kettering Division of the Graduate School of Medical Sciences at Cornell in 1952 but then returned to her *alma mater*, Douglass College, to become chair of the Department of Bacteriology and guide a new generation of young women into the microbial sciences. In 1966, she was awarded an honorary D.Sc. degree as research scientist and teacher and in experimental chemotherapy, and in 1973, Dr. Reilly was one of the first recipients of The Associate Alumnae of Douglass College Society of Excellence Award for distinguished achievement.

DORRIS JEANETTE HUTCHISON, Ph.D., 1949 (1918–2007)

Dorris Jeanette Hutchison was educated at Western Kentucky University (B.Sc., 1940) and obtained her M.S. in 1943 from the University of Kentucky, studying the potential of microbial antagonists, including actinomycetes, to reduce the numbers of coliforms in drinking waters. Following a succession of teaching jobs at women's colleges, including Russell Sage College (Troy, NY), Vassar College (Poughkeepsie, NY), and Wellesley College (Wellesley, MA), she recognized that in order to have a successful academic career, she needed a Ph.D. With her interest in antagonistic microbes, Hutchison went to Rutgers University and the Waksman laboratory. By the late 1940s, it had become clear that streptomycin was therapeutically active against *M. tuberculosis*, the causative agent of tuberculosis, and therefore had enormous clinical and commercial potential. Streptothricin, long a focus of the Waksman laboratory, also showed activity against *M. tuberculosis* in a petri plate assay. Despite the fact that it was toxic for systemic use in mammals, it remained an antibiotic of intense interest. Variants of *S. lavendulae*, the streptothricin-producing species, included strains that could inhibit a streptomycin-resistant isolate of virulent *M. tuberculosis*. Hutchison's focus was to study these tantalizing new variants of *S. lavendulae* to clarify their diversity, to optimize their antibiotic production, and perhaps to fortuitously discover less toxic forms of streptothricin that could be exploited for therapeutic use. Hutchison detailed optimal cultural conditions and also demonstrated potential synergism between combinations of different antibiotics (38, 39).

On completing her Ph.D. in 1949, Hutchison, like Reilly before her, continued at Rutgers as a research associate, turning her attention towards neomycin, a compound recently discovered by Hubert Lechevalier. Her work provided a firm foundation for its potential as a new drug (40, 41); neomycin was later found to cause hearing and kidney problems, rendering it unsuitable for systemic use, but it was accepted as highly effective for external applications. Today, over half a century later, neomycin is formulated in a triple combination with polymyxin and bacitracin, which is sold widely as a topical salve under the brand name Neosporin.

After she left Rutgers, Hutchison went on to an exemplary career in cancer research as a professor and a medical researcher at the Cornell Sloan-Kettering Institute for Cancer Research. She addressed treatment of tuberculosis and leukemia as well as uses of chemotherapy in cancer-related studies, and she published more than 100 research papers. Along the way, she won a Philippe Foundation Fellowship in 1959 and was awarded a Bronze Medal Award by the Westchester Division of the American Cancer Society in 1984. She rose through the academic ranks, becoming associate dean from 1978 to 1987, and retired as professor emeritus in 1990, after nearly 40 years as a pioneer in cancer research.

Hutchison was a successful, self-confident woman who was not above teasing a former colleague. During a Rutgers symposium in 1993 to honor the 50th anniversary of the discovery of streptomycin, Hutchison commented on stage to Albert Schatz, the discoverer of streptomycin: "You know, Albert, if I had not been committed to my University of Kentucky scholarship in 1940, I would have accepted the Rutgers scholarship that you subsequently received, and I would have discovered streptomycin." With her quip, Hutchison spoke openly about the fact that hitting the antibiotic jackpot had involved luck: screening for new antibiotics was a soil sample isolation lottery, and all members of the group had been playing the same high-stakes scientific game. The fact that only one of them would win the top prize did not detract from the thrill of having been part of the action.

SUMMARY

Too often, we rely on the "great man" version of scientific advancement wherein a single brilliant leader who inspires and directs the work gets most of the acclaim for advances made by others. Anyone who has ever worked in a laboratory knows that most scientific developments are based on sustained cooperative work, not the work of a lone genius. In his autobiography, Selman Waksman compared himself to the conductor of an orchestra. Few accounts give much, if any, credit to the women who also were active participants in the Waksman laboratory or mention the way in which World War II had opened new doors for the careers of women in microbiology. The war not only created a huge impetus for finding novel anti-infective drugs but also meant that most young men were directly engaged in military activities, serving as troops, which thinned the number of male students available to conduct research in university laboratories. Women were welcomed to the search for new antibiotics partly because their presence kept graduate programs from withering through lack of "manpower." As always in history, women took advantage of their opportunity.

These profiles are part of our microbiological heritage. We hope our accounts of the lives of these female pioneers will help preserve their legacies. We also hope this book will inspire readers to look more closely at the history of other microbiological innovations and seek out the unsung women— and men—who were participants. Almost without exception, every laboratory group has members (*"Hidden Figures"*) whose contributions are minimized or overlooked. Too often, the overlooked members have been women and minorities, yet synergistic collaborations, based on the unique strengths and contributions of different people, create the climate that leads to breakthroughs in microbiology and all of science. It is time we celebrate the women who were the hidden figures during the golden age of antibiotic discovery.

ACKNOWLEDGMENTS

The assistance from the Rutgers Special Collections and University Archives, Rutgers University (Erika Gorder, Helen Hoffman, Thomas Frusiano, Tim Corlis, and Natalie Borisovets [Newark]), Jerry Kukor (Dean, School of Graduate Studies), the Associate Alumnae of Douglass College (Marie Siewierski, Flora Cowen, and Carol Hamlin), Merck and Co. (Ian McConnell) and Archives, and the American Society for Microbiology Archives (Jeff Karr), plus input from family members and coworkers, is gratefully acknowledged.

CITATION

Eveleigh DE, Bennett JW. 2018. Women microbiologists at Rutgers in the early golden age of antibiotics, p 317–329. *In* Whitaker RJ, Barton HA (ed), *Women in Microbiology*. American Society for Microbiology, Washington, DC.

References

1. Ryan F. 1992. *The Forgotten Plague. How the Battle against Tuberculosis Was Won—and Lost.* Little, Brown and Co, Boston, MA.
2. Meyers MA. 2012. *Prize Fight. The Race and the Rivalry to be the First in Science.* Palgrave Macmillan Ltd, London, UK.
3. Pringle P. 2012. *Experiment Eleven. Dark Secrets behind the Discovery of a Wonder Drug.* Walker and Co, New York, NY.
4. Rosen W. 2017. *Miracle Cure. The Creation of Antibiotics and the Birth of Modern Medicine.* Viking Press, New York, NY.
5. Lechevalier HA. 1980. The search for antibiotics at Rutgers University, p 120. *In* Parascandola J (ed), *The History of Antibiotics*. American Institute of the History of Pharmacy, Madison, WI.
6. Schatz A. 1965. Antibiotics and dentistry. Part 1. Some personal reflections on the discovery of streptomycin. *Pak Dent Rev* 15:125–134.
7. Schatz A. 1993. The true story of the discovery of streptomycin. *Actinomycetes* 4:27–39.
8. Woodruff HB. 2014. Selman A. Waksman, winner of the 1952 Nobel Prize for Physiology or Medicine. *Appl Environ Microbiol* 80:2–8.
9. Horning ES. 1942. *The distribution and properties of antagonistic fungi and actinomycetes in nature. Ph.D. thesis.* Rutgers, The State University of New Jersey, New Brunswick, NJ.
10. Waksman SA, Horning ES. 1943. The distribution of antagonistic fungi in nature and their antibiotic action. *Mycologia* 35:47–65.
11. Waksman SA, Horning ES, Spencer EL. 1943. Two antagonistic fungi, *Aspergillus fumigatus* and *Aspergillus clavatus*, and their antibiotic substances. *J Bacteriol* 45:233–248.
12. Waksman SA, Horning ES, Spencer EL. 1942. The production of two antibacterial substances, fumigacin and clavicin. *Science* 96:202–203.
13. Waksman SA, Horning ES, Welsch M, Woodruff HB. 1942. Distribution of antagonistic actinomycetes in nature. *Soil Sci* 54:281–296.

14. **Bugie EJ.** 1944. *Production of antibiotic substances by* Aspergillus flavus *and* Chaetomium cochliodes. *M.Sc. thesis.* Rutgers, The State University of New Jersey, New Brunswick, NJ.

15. **Waksman SA, Bugie E.** 1943. Strain specificity and production of antibiotic substances. II. *Aspergillus flavus-oryzae* group. *Proc Natl Acad Sci U S A* **29**:282–288.

16. **Waksman SA, Bugie E.** 1944. Chaetomin, a new antibiotic substance produced by *Chaetomium cochliodes*: I. Formation and properties. *J Bacteriol* **48**:527–530.

17. **Waksman SA, Bugie E.** 1943. Action of antibiotic substances upon *Ceratostomella ulmi*. *Proc Soc Exp Biol Med* **54**:79–82.

18. **Waksman SA, Bugie E, Reilly HC.** 1944. Bacteriostatic and bactericidal properties of antibiotic substances, with special reference to plant pathogenic bacteria. *Bull Torrey Bot Club* **71**:197–121.

19. **Waksman SA, Geiger WB, Bugie E.** 1947. Micromonosporin, an antibiotic substance from a little known group of microorganisms. *J Bacteriol* **53**:355–357.

20. **Solotorovsky M, Bugie EJ, Frost BM.** 1948. The effect of penicillin on the growth of *Mycobacterium tuberculosis* in Dubos' medium. *J Bacteriol* **55**:555–559.

21. **Solotorovsky M, Gregory FJ, Ironson EJ, Bugie EJ, O'Neill RC, Pfister K III.** 1952. Pyrazinoic acid amide—an agent active against experimental murine tuberculosis. *Proc Soc Exp Biol Med* **79**:563–565.

22. **Schatz A, Bugie E, Waksman SA.** 1944. Streptomycin, a substance exhibiting antibiotic activity against gram positive and gram negative bacteria. *Proc Soc Exp Biol Med* **55**:66–69.

23. **Waksman SA, Schatz A.** 21 September 1948. *Streptomycin and process of preparation.* US patent 2449866.

24. **Snowbeck C.** 14 April 2001. *Obituary: Elizabeth Gregory: did McCandless woman get fair shake for role in discovery of streptomycin?* Pittsburgh Post-Gazette, Pittsburgh, PA.

25. **Jones D.** 1945. *The effect of micro-organisms and antibiotic substances upon viruses. M.Sc. thesis.* Rutgers, The State University of New Jersey, New Brunswick, NJ.

26. **Jones D, Metzger HJ, Schatz A, Waksman SA.** 1944. Control of gram-negative bacteria in experimental animals. *Science* **100**:103–105.

27. **Jones D, Beaudette FR, Geiger WB, Waksman SA.** 1945. A search for virus-inactivating substances among microorganisms. *Science* **101**:665–668.

28. **Jones D.** 1945. The effect of antibiotic substances on bacteriophages. *J Bacteriol* **50**:341–346.

29. **Jones D, Schatz AL.** 1946. Methods of study of antiphage agents produced by microorganisms. *J Bacteriol* **52**:327–335.

30. **Schatz AI, Jones D.** 1947. The production of antiphage agents by actinomycetes. *Bull Torrey Bot Club* **74**:9–19.

31. **Jones D.** 1953. *The isolation and study of variants on* Staphylococcus *phage P1 and its host* Staphylococcus aureus. *Ph.D. thesis.* University of California, Berkeley, Berkeley, CA.

32. **Jones D.** 1977. A personal glimpse at the discovery of streptomycin. Unpublished talk presented at the Douglass College Alumnae Annual Reunion. Cited in the Douglass Associate Alumnae bulletin, Spring, p 6.

33. **Shelton T.** 17 July 2017. *Greater Philadelphia honors Vivian Schatz.* Women's International League for Peace and Freedom, Philadelphia, PA. http://wilpfus.org/news/updates/greater-philadelphia-honors-vivian-schatz. Accessed 6 February 2018.

34. Reilly HC. 1946. *Antibacterial properties of streptothricin and streptomycin. Ph.D. thesis.* Rutgers, The State University of New Jersey, New Brunswick, NJ.

35. Waksman SA, Schatz A, Reilly HC. 1946. Metabolism and chemical nature of *Streptomyces griseus. J Bacteriol* **51**:753–759.

36. Waksman SA, Reilly HC, Schatz A. 1945. Strain specificity and production of antibiotic substances. V. Strain resistance of bacteria to antibiotic substances, especially to streptomycin. *Proc Natl Acad Sci U S A* **31**:157–164.

37. Reilly HC, Harris DA, Waksman SA. 1947. An actinophage for *Streptomyces griseus. J Bacteriol* **54**:451–466.

38. Hutchison D. 1949. *Production and tuberculostatic properties of streptothricin-like antibiotics. Ph.D. thesis.* Rutgers, The State University of New Jersey, New Brunswick, NJ.

39. Hutchison D, Swart EA, Waksman SA. 1949. Production, isolation and antimicrobial, notably antituberculosis, properties of streptothricin VI. *Arch Biochem* **22**:16–30.

40. Swart EA, Hutchison D, Waksman SA. 1949. Neomycin, recovery and purification. *Arch Biochem* **24**:92–103.

41. Waksman SA, Hutchison D, Katz E. 1949. Neomycin activity upon *Mycobacterium tuberculosis* and other mycobacteria. *Am Rev Tuberc* **60**:78–89.

Printed and bound by CPI Group (UK) Ltd, Croydon, CR0 4YY

27/10/2024

14580445-0001